ENGINEE

The Practical OPNET® User Guide for Computer Network Simulation

The Practical OPNET® User Guide for Computer Network Simulation

Adarshpal S. Sethi
Vasil Y. Hnatyshin

CRC Press
Taylor & Francis Group
Boca Raton London New York

CRC Press is an imprint of the
Taylor & Francis Group, an **informa** business

CRC Press
Taylor & Francis Group
6000 Broken Sound Parkway NW, Suite 300
Boca Raton, FL 33487-2742

Printed in the United States of America on acid-free paper
Version Date: 20120612

International Standard Book Number: 978-1-4398-1205-1 (Hardback)

Library of Congress Cataloging-in-Publication Data

Sethi, Adarshpal S.
     The practical OPNET user guide for computer network simulation / Adarshpal S. Sethi, Vasil Y. Hnatyshin.
          p. cm.
     Includes bibliographical references and index.
     ISBN 978-1-4398-1205-1 (hardback)
     1. Computer networks--Computer simulation. 2. OPNET. I. Hnatyshin, Vasil Y. II. Title.

TK5105.5.S4243 2012
004.601'1--dc23   2012021348

Visit the Taylor & Francis Web site at
http://www.taylorandfrancis.com

and the CRC Press Web site at
http://www.crcpress.com

To my loving wife Parveen, without whose immense patience and support this book would not have been possible.

Adarshpal S. Sethi

To my parents who were always an endless source of inspiration, encouragement, and support. To Ivy and Markiyan for their smiles and laughter that brightened my days.

Vasil Y. Hnatyshin

Contents

         10.4.1   Specifying Global QoS Profiles 301
                  10.4.1.1 Committed Access Rate Profiles..............302
                  10.4.1.2 Custom Queuing Scheduler......................303
                  10.4.1.3 RED and WRED Configuration...............305
                  10.4.1.4 FIFO Profiles 307
                  10.4.1.5 MWRR/MDRR/DWRR Profiles 307
                  10.4.1.6 Priority Queuing Profiles.........................309
                  10.4.1.7 WFQ Profiles 309
         10.4.2   Specifying Local QoS Profiles.............................309
                  10.4.2.1 Traffic Classes and Traffic Policies 310
                  10.4.2.2 WFQ/DWFQ Profiles 313
                  10.4.2.3 Priority Queue Profiles and Custom
                            Queue Profiles..314
                  10.4.2.4 MDRR Profiles and DWRR Profiles........315
                  10.4.2.5 Policer Profiles..315
                  10.4.2.6 RED/WRED Profiles..................................316
         10.4.3   Deploying Defined QoS Profiles on
                  an Interface..317
         10.4.4   Closing Remarks 319
         10.4.5   QoS-Related Statistics..320

Chapter 11 Network Layer: Routing 321

         11.1   Introduction 321
                11.1.1   Deploying Routing Protocols in a Simulated
                         Network 322
                11.1.2   Configuring Routing Protocol Attributes.................325
         11.2   Routing with RIP..326
                11.2.1   Introduction to RIP..326
                11.2.2   Local RIP Configuration Attributes.........................328
                11.2.3   RIP Interface-Specific Configuration
                         Attributes..330
                11.2.4   Configuring RIP Start Time.....................................332
                11.2.5   RIP Simulation Efficiency Mode 332
         11.3   Routing with OSPF...334
                11.3.1   Introduction to OSPF...334
                11.3.2   OSPF Attributes 335
                11.3.3   Configuring OSPF Processes 336
                11.3.4   Specifying OSPF Configuration on Router
                         Interfaces 339
                11.3.5   Configuring OSPF..340
                11.3.6   Configuring Link Costs for OSPF Routing..............340
                11.3.7   Configuring OSPF Timers...343
                11.3.8   Configuring OSPF Areas 345
                11.3.9   Configuring OSPF Area Border Routers..................346

Preface

Knowledge of how to simulate and model the performance of various systems has become an essential tool for any IT (information technology) professional. Today, investment in expensive hardware is difficult to justify without a comprehensive evaluation of its performance through simulation models. However, building a simulation model is not a trivial task. It requires a thorough understanding of simulation and modeling concepts, extensive knowledge of the properties of the modeled system, and a solid mathematical background.

OPNET® Technologies Inc. (http://www.opnet.com) is a leading company that develops and markets software for the simulation, modeling, and analysis of computer networks. IT Guru and Modeler are the most popular products for network simulation offered by OPNET. They are widely used in education for teaching basic and advanced topics in data communication and computer networks, as well as in industry and government for simulation, study, analysis, and performance prediction of various network systems. Accuracy of results, rich functionality, and simplicity of use are the main reasons why these software packages draw so many users from various areas and disciplines.

In today's technology-oriented world, almost every company relies on its private networking infrastructure in addition to the Internet. Thus, the demand for professionals who can evaluate network performance as well as identify and fix related problems has grown significantly in the past few years. OPNET software is an excellent tool for simulating real-life networks, evaluating their performance, and identifying potential problems before they arise. That is why OPNET products, IT Guru and Modeler in particular, find widespread applications in the industry. They are also heavily used in government, particularly in the Defense Department for modeling defense applications and networks.

Teachers of networking courses in universities constantly look for various new active learning tools for engaging students in classroom discussion, providing students with hands-on experiences, and getting students to become more interested in the subject area. OPNET software constitutes an excellent active learning tool that may help teachers achieve all these goals. In particular, OPNET software allows teachers to study and evaluate diverse networking systems with as simple or as complex topology as needed, illustrate various networking concepts, and show the students how the network performance changes under different conditions. It is no wonder that more than 500 universities worldwide* currently use OPNET network simulation software in teaching and research.

This book is, to our knowledge, the first text to provide a comprehensive description of OPNET IT Guru and Modeler software and how to use this software for the simulation and modeling of computer networks. The book also includes a set of laboratory projects to help learn different aspects of the software.

---

* http://www.opnet.com/university_program/participant_spotlight/universities.html.

The target audience for this book includes IT professionals and researchers who work in the area of computer communications and networks, as well as undergraduate and graduate students of courses in this area. By providing an in-depth look at the rich features of OPNET software, this book will be a valuable reference for those IT professionals and researchers who work on projects that require creating simulation models. At the same time, the book will facilitate learning OPNET for new users of the software by organizing the topics in a logical manner that corresponds to the protocol layers in a network. The laboratory projects are similarly organized and include references to the corresponding topics in the main part of the book.

Chapters 1 through 4 provide a systematic introduction to some of the basic features of OPNET, which are required to do just about any network simulation. These chapters describe the concepts of projects and scenarios, how to create a network topology, how to edit the attributes of objects and configure the topology to reflect a desired configuration, how to configure and run the simulation, and how to observe and analyze the results of the simulation.

The remaining chapters describe how to work with various protocol layers using a top-down approach, which is commonly adopted in many networking textbooks. Chapters 5 through 7 show how to work with applications in OPNET, including both standard and custom applications, and how to deploy applications and user profiles at network nodes. Chapter 8 describes the transport layer protocols: Transmission Control Protocol (TCP) and User Datagram Protocol (UDP). Chapters 9 through 11 deal with the network layer and include a description of IP addressing; advanced IP features such as multicasting, Quality of Service (QoS), and IP Version 6 (IPv6); and routing with Routing Internet Protocol (RIP) and Open Shortest Path First (OSPF). Finally, Chapter 12 describes the bottom two layers of the protocol stack, the physical and link layers, including how to work with Ethernet and wireless LANs.

Each chapter includes explanations of the OPNET features that are relevant to the chapter topic and then provides step-by-step instructions on how to work with those features during a network simulation. The logical organization of these features is intended to serve as a valuable reference tool in its own right. The only other available description of many of these features is the OPNET documentation that accompanies the software products themselves. However, the OPNET documentation is extremely large and it is not easy to find a specific topic in it. The organization of chapters in our book should facilitate the job of quickly locating instructions for performing a desired task.

The last part of the book contains a set of laboratory projects, each with a complete simulation focused on a certain topic. The topics of the laboratory projects reflect the same progression of topics, which is used to organize the main chapters of the book. Thus, the first two laboratory assignments are introductory in nature, which serve to illustrate the major steps required in any simulation. It is recommended that the reader should read Chapters 1 through 4 of the book before attempting to do these laboratory assignments (each assignment includes a list of chapters that forms the recommended reading). Laboratory assignments 3 through 6 deal with application layer modeling. The remaining assignments go down the protocol stack from the transport layer, then to the network layer, and end with the link layer.

Although there are several other laboratory manuals on the market, none of them includes a detailed description of the features of OPNET software. This book will allow the reader to gain an in-depth understanding of the complex functionality available in OPNET in addition to learning how to do various tasks in a network simulation. Moreover, most other laboratory manuals give detailed step-by-step instructions for completing each assignment, which we believe is not a good pedagogic device. In our personal experience using these manuals in the classroom, the step-by-step instructions for configuring OPNET simulations are monotonous, boring, and error-prone. It is very easy to skip or omit a step without noticing it, causing the results of the simulation to be different from those reported in the laboratory manual. Identifying and fixing these errors is very difficult even for seasoned OPNET users. Frequently, it is much easier to simply redo the whole assignment from the start than to find the error. Furthermore, the reader gets lost in the details of the step-by-step instructions, preventing the reader from noticing the overall objectives of the task. The students usually blindly follow the instructions without trying to think through each step they perform, preventing them from learning and understanding the configuration steps. In our experience, students are often able to successfully complete an assignment and yet have no clue as to what they have actually accomplished and why it was done that way.

The laboratory projects in this book deliberately do **not** contain step-by-step instructions for using the OPNET software. They instead provide the reader with the overall goals of the experiment, describe the general network topology, and give a high-level description of the system configuration required to complete the simulation. They contain references to sections in the main part of the book that contain step-by-step instructions on how to complete each subtask required in the project. By this device, a novice reader who does not know how to perform the task can easily look up the instructions on how to do it. But as the reader becomes more experienced, there will be less need to look up the instructions for each of the later laboratory assignments. We believe that this approach makes the laboratory assignments more engaging and more exciting for the reader. More importantly, the reader will more easily see the forest for the trees and will gain a better understanding of the whats and whys of the simulations rather than get lost in the details.

It should also be noted that this book only describes how to work with standard models of both IT Guru and Modeler. Modeler is in general a more complex product that allows users to design custom models that simulate new protocols or modifications of existing protocols. We do not describe those aspects of Modeler in this book.

Finally, we would like to acknowledge the reviewers of this book, Jasone Astorga and Marina Aguado, for their numerous insightful suggestions that helped to greatly improve the manuscript.

Authors

Dr. Adarshpal S. Sethi was one of the first educators to introduce OPNET software into the classroom in the late 1980s. Since then, he has used OPNET widely for teaching and research. At the University of Delaware in Newark, Delaware, he currently teaches a course on Simulation of Computer Networks that is based entirely on the use of OPNET for network simulation. Dr. Sethi received his PhD in computer science from the Indian Institute of Technology (IIT) in Kanpur, India, and he is currently a professor in the Department of Computer & Information Sciences at the University of Delaware. He has served on the faculty at IIT Kanpur, was a visiting faculty at Washington State University in Pullman, Washington, and a visiting scientist at IBM Research Laboratories in Zurich, Switzerland and at the U.S. Army Research Laboratory in Aberdeen, Maryland. Dr. Sethi is on the editorial boards of the *IEEE Transactions on Network and Service Management*, *International Journal of Network Management*, and *Electronic Commerce Research Journal*. He is also active on the program committees of numerous conferences. Dr. Sethi's research interests include architectures and protocols for network management, fault management, management of wireless networks, and network simulation and modeling. Dr. Sethi received the University of Delaware's Excellence-in-Teaching Award in 2000, an award given yearly to only 4 of over 1200 faculty at the university. He also received the OPNET Modeling Excellence Award in 1996, given by Mil3 Inc. (the former name of OPNET Technologies Inc.) for the innovative use of OPNET models in a research paper. Dr. Sethi was featured by OPNET Technologies Inc. in an educator profile on the back cover of *IEEE Communications Magazine* in June 2006.

Dr. Vasil Y. Hnatyshin has been actively using OPNET software products for more than 15 years. He has published his research using OPNET software in several journals and presented it at numerous international conferences. Dr. Hnatyshin's research interests include QoS in the Internet, Transmission Control Protocol (TCP)/IP networks, location aided routing in Mobile Ad Hoc Networks (MANET), network security, wireless communication, as well as network simulation and modeling. Dr. Hnatyshin has successfully integrated OPNET software into several courses such as Data Communications and Computer Networks and TCP/IP and the Internet Technologies, which he regularly teaches at Rowan University. Dr. Hnatyshin received his PhD from the University of Delaware in Newark, in 2003. He is currently an associate professor in the Department of Computer Science at Rowan University in Glassboro, New Jersey, where he continues his teaching and research endeavors using OPNET software products.

OPNET Trademark Information

OPNET, IT Guru, OPNET Modeler, and ACE are registered trademarks of OPNET Technologies, Inc. in the United States and/or other countries. All screenshots in this book are Copyright © 2011 OPNET Technologies, Inc. All rights reserved. Used with permission. In the screenshots of Chapter 7, examples of user profile and application behavior are for illustration only.

1 Getting Started with OPNET

1.1 OPNET IT GURU AND MODELER

OPNET Technologies Inc. is the leading developer of network simulation software and solution provider for application and network management issues. Its software products are widely used for research and development of emerging networking technologies; for performance evaluation, testing, and debugging of communication networks, protocols, and applications; and for teaching and research at numerous colleges and universities. OPNET currently provides over a dozen software products and countless specialized modules and models that can be easily customized for study and evaluation of almost any of today's network paradigms.

OPNET software has an easy-to-use graphical user interface, which can be used to build various network configurations and test their performance with simple drag-and-drop actions and a few clicks of a mouse. OPNET software contains a huge library of models that simulate most of the existing hardware devices and cutting-edge communication protocols. Such abundance of simulation models makes it possible for you to easily simulate the most complex computer networks and to configure protocols that implement the most up-to-date communication technologies. OPNET's **IT Guru** and **Modeler** are among the most popular network simulation software packages. Both these products allow you to study various computer networks using the built-in models of various communication devices, links, protocols, and commonly used networking technologies. However, unlike IT Guru, Modeler has additional functionality that allows you to create new simulation models and modify existing ones.

OPNET maintains a University Program, which provides free software licenses and discounted technical support to qualified colleges and universities all over the world. Through the OPNET University program, the IT Guru Academic Edition software package, based on IT Guru commercial Version 9.1, can be downloaded free of charge.

Additionally, the University Program provides full-featured commercial versions of IT Guru and Modeler to qualified faculty and students. To obtain these full-featured software packages, you need to submit an online application to OPNET Technologies Inc. Once the OPNET University Program approves your application, you will either receive an installation CD or you will be provided with access to download the installation files. Upon application approval, OPNET also provides a one-day maintenance license, which allows you to directly contact OPNET's Technical Support. This license can be of great help to first-time users in case there are software installation problems. Currently, IT Guru and Modeler are supported for Windows and Red Hat

Linux operating systems. For more information, visit OPNET's University Program web resources available at http://www.opnet.com/university_program.

This user guide is based on the full-featured commercial Version 16.0 of IT Guru and Modeler software packages, the newest version of OPNET software available at the time of writing this book. However, most of the features described in this user guide are also available in the previous versions of OPNET software. In this guide, we will refer to both IT Guru and Modeler software as simply OPNET and only use the specific names when it is necessary to distinguish between them.

1.1.1   Installing OPNET IT Guru and Modeler

To install OPNET, you need to verify that your computer satisfies the specified system requirements and then follow installation instructions that accompany the distribution files on the provided CD or on the web. Generally, IT Guru and Modeler installation procedures include the following steps:

1. Install the software product.
2. Install OPNET models.
3. Install product documentation.
4. Start and configure the license server.
5. Configure C/C++ compiler (Modeler only).
6. Configure the web browser to view software documentation (may be omitted).

During the software installation phase, you may be prompted to specify the type of the licensing system to be installed on your computer. Section 1.1.2 describes the licensing modes that can be configured at the time of installation.

1.1.2   OPNET License Server

All OPNET products must obtain a license prior to their execution: one license per each running instance of OPNET software. When you start OPNET, the license server is contacted. If the requested software licenses are available, then the license server will allocate the corresponding license and OPNET execution will begin. Otherwise, the request will be denied and the software will not be able to run. The OPNET license server is a process responsible for managing license requests. Typically, the license server is started automatically upon computer boot-up and runs continuously in the background.

You can configure OPNET to support one of the following licensing modes:

- *Standalone*—the current computer will operate as a license server for this computer; it will manage license requests that originate from this computer only. Such an option is useful if you want to have a certain number of licenses allocated for the exclusive use of this computer.
- *Floating: serve licenses from this computer*—the current computer will operate as a network license server; it will manage license requests

generated by this or any other computer in your network. In this case, the software licenses are shared among all computers in the network. This option is recommended for a system administrator who manages network software and will ensure that the OPNET license server is up and running whenever there are active OPNET users.

- *Floating: access licenses from remote server*—the current computer will operate as a client and will contact a license server, which resides on another computer in your network, to obtain the license upon software startup. When selecting this option, you will be required to specify the Internet Protocol (IP) address and the port number of the OPNET license server. Contact your network/application administrator to verify the license server information. This option is typically used when there is an OPNET license server already configured and running in your network.

OPNET also provides a License Management tool, which allows the network/software administrator to start/stop the OPNET license server, to add/deregister software licenses, and to carry out other license management tasks.

1.1.3 FOLDERS CREATED AT INSTALLATION

At the time of first installation, OPNET generates several new folders on your computer. One of the key folders contains the actual software components, models, documentation, and system files. This folder is named OPNET, and it is usually located in the directory where the operating system stores its applications, such as C:\Program Files in Windows. Additionally, OPNET generates several folders that contain user-specific information:

- op_admin—this folder contains such automatically generated information as backup files (subdirectory bk), temporary files (subdirectory tmp), various log files, and a software environment configuration file, which contains OPNET software preferences. We also refer to this folder as Admin.
- op_models—this is the default folder where OPNET stores user-created simulation study configuration files, usually referred to as project files. OPNET also allows saving project files in locations other than the default op_models folder.
- op_reports—this is the default folder where OPNET stores simulation study reports. OPNET allows changing this default storage location for reports by modifying the software preferences (see Section 1.2).

1.1.4 ENABLING OPTIONAL PRODUCT MODULES

Upon request, OPNET Technologies Inc. may approve licenses for optional product modules that provide additional software functionality such as terrain modeling, 3D network visualization, automation, etc. If you have obtained such licenses, a **Select**

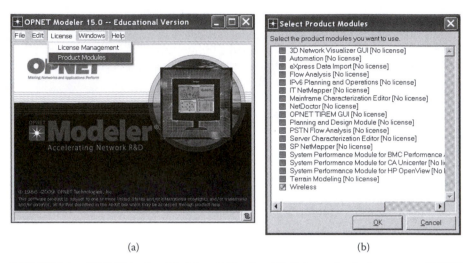

(a)   (b)

FIGURE 1.1   OPNET Product Modules: (a) **Product Modules** option in the start-up OPNET Modeler window, and (b) **Select Product Modules** window.

Product Modules window (Figure 1.1b) will be displayed when you run OPNET for the first time. This window will not be displayed after the first time. Follow these steps to enable the available product modules:

- Click on the checkboxes of available product modules to select or unselect them as desired (Figure 1.1b).
- Click the **OK** button to close the window.
- Restart OPNET so that your changes can take effect.

If you forgot to enable certain modules the first time you ran OPNET, you can still do so later by following these steps:

- Start OPNET.
- Select **Product Modules** from the **License** pull-down menu (Figure 1.1a).
- The **Select Product Modules** window (Figure 1.1b) now appears. Follow the steps described above to enable the desired modules.

1.2   MANAGING OPNET PREFERENCES

OPNET allows you to change software preferences such as system behavior, functionality, appearance (e.g., colors and sizes), supporting applications, location of model directories, etc. Each preference is represented by a variable whose value can be changed. To simplify searching for a specific configuration option, the preferences are organized into categories. A preference has a "user-friendly" name and a "technical" name called a tag. For example, the preference that defines the program used to view OPNET documentation has the user-friendly name **Path to Document Viewer Program** and the technical name **vudoc_prog**. Since the product documentation starting from Version 12.0 onwards has been changed to HTML format, you can set the value of the **vudoc_prog** preference to `Firefox` or any other desired web browser.

1.2.1 THE PREFERENCES EDITOR

OPNET preferences can be changed via the **Preferences Editor** shown in Figure 1.2. The **Preferences Editor** window consists of five sections. A search textbox titled *Search for:* is located at the top of the **Preferences Editor** window. This box allows you to search for preferences by preference names, by preference values, or by all information. You can specify the search type by selecting one of the following values from the pull-down list: `In Names`, `In Values`, or `Anywhere`. You can start the search by clicking the **Find** button.

The preferences can be arranged by *Groups* (i.e., categories for common configuration aspects of OPNET) or *Source* (i.e., the location where a particular set of preferences is defined, such as a full path to the configuration file). As shown in Figure 1.2, by default, the preferences are arranged by groups. A section in the left part of the **Preferences Editor** window contains a tree view of the preferences organized by groups or sources. You can browse this tree by clicking on the + or – sign to expand or collapse a tree branch. Clicking on a tree branch displays all the preferences that belong to the selected group or source in a *Preferences Table* located on the right side of the **Preferences Editor** window.

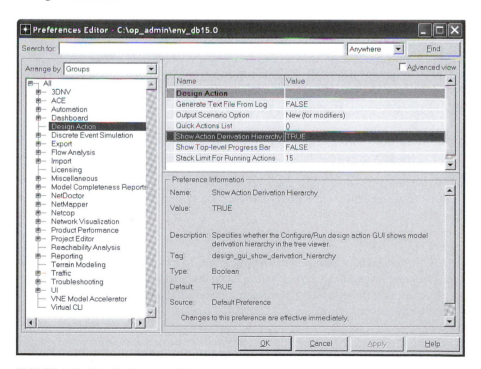

FIGURE 1.2   The **Preferences Editor**.

By default, the *Preferences Table* consists of only two columns: *Name*, which contains a user-friendly preference name, and *Value*, which contains the actual value of that preference. Selecting the *Advanced view* checkbox adds two more columns to the *Preferences Table*: *Tag*, which provides the technical name of the preference, and *Source*, which specifies where the preference is defined outside the **Preferences Editor**. The *Preference Information* pane, located right below the *Preferences Table*, displays a complete description of the selected preference.

Finally, a set of standard buttons is located at the bottom of the **Preferences Editor**:

- **OK**: Accept the changes and close the window.
- **Apply**: Apply the change and leave the window open.
- **Cancel**: Close the window without saving the changes.
- **Help**: Open a help window that provides a detailed description of the **Preferences Editor**.

1.2.2   CHANGING PREFERENCE VALUES

Execute the followings steps to view or change the values of OPNET preferences:

- Start OPNET.
- Click on the **Edit > Preferences** menu item which will open the **Preferences Editor**.
- To change the value of a preference:
  - Left-click on the *Value* field of the desired preference in the *Preferences Table*.
  - Select a value from the pull-down menu. In some cases, you may want to choose the Edit... option, which allows you to type in the value.
- Click **OK** in the **Preferences Editor** to accept the changes.

1.2.3   THE ENVIRONMENT FILE

OPNET stores the values of the preferences in a software environment file located in the Admin directory. To identify the location of this directory, follow the steps below:

- Start OPNET.
- Click on **Help > About This Application**. Click on the *Environment* tab and then expand the *System Information* category, which contains a brief description of key system parameters including the location of the Admin directory (see Figure 1.3).

The Admin directory contains the system environment file named env_db<ver>, where <ver> corresponds to the version of OPNET being used. For example, OPNET products Version 15.0 will have the environment file named env_db15.0. The

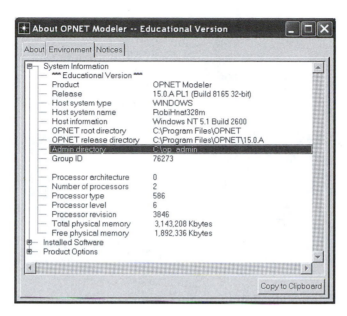

FIGURE 1.3   OPNET's environment settings.

OPNET environment file contains the names and values of most commonly used preferences stored in plain text. This file can be opened and modified with any text editor.

1.3   VIEWING DOCUMENTATION

OPNET provides detailed product documentation, which includes a description of the features available in the software and also includes a set of tutorials and links to various web resources. OPNET documentation is stored as an HTML file and can be viewed directly using a web browser that supports Javascript and HTML frames. Alternatively, the documentation can be accessed from the main OPNET window via the **Help** menu (see Figure 1.4). Refer to the installation instructions provided with OPNET for more information on configuring the web browser and viewing online help.

The OPNET documentation is organized into topic chapters and subsections. It provides several useful features (see Figure 1.5) including search, index, glossary, quick links, expanding and collapsing chapter and subsection titles, showing/hiding descriptions of procedures for using various aspects of the software, accessing product documentation chapters in PDF format, and many others.

1.4   WORKING WITH FILES AND MODEL DIRECTORIES

At various points during a simulation study, you will need to either open an existing file or create a new file and store it in a directory. Both IT Guru and Modeler work with a variety of different file types, which store the network models used in the simulation and the data collected during the simulation runs. However, IT Guru is limited to the file types (e.g., project files, generic data files, probe models,

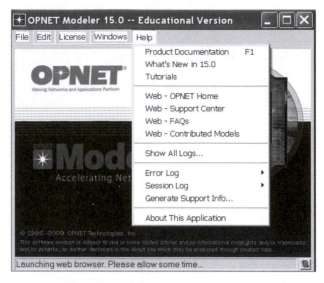

FIGURE 1.4   The OPNET **Help** menu.

FIGURE 1.5   OPNET product documentation.

etc.) that deal with creating a simulation study without modifying the underlying models, whereas Modeler has access to a broader range of file types including those that contain the implementations of the simulated devices and network technologies. Specifically, Modeler can access node models, process models, link models, external C/C++ header and implementation files, etc.

1.4.1 FILE CHOOSER MODES

Recent versions of OPNET, starting with Version 12.0, provide two different file browsing methods for choosing a file to be opened:

- *General file chooser* (Figure 1.6a)—a browser that allows searching of all the mounted storage devices on the current computer. It has the appearance of a regular OS file browser.
- *File chooser organized by model directories* (Figure 1.6b)—a browser that displays only folders known to OPNET. These are folders included in the list of OPNET model directories. Using this mode narrows the available options, thereby speeding up the process of choosing the desired file.

The file chooser organized by model directories is the default file chooser and is available in all versions of OPNET. The general file chooser mode has been made available in Versions 12.0 and later. You can switch between the modes by clicking on the icon in the lower-left hand corner of the window. In the general file chooser, this icon is a blue square with a white star in it (Figure 1.6a). In the file chooser organized by model directories, it is a square divided into four smaller multicolored squares (Figure 1.6b). Both file chooser modes allow you to browse files based on the file type, which often significantly reduces the time needed to locate a project, process model, C/C++ external code, or other file of a specific OPNET type.

1.4.2 ADDING MODEL DIRECTORIES

The default directory used by OPNET to store model files is the directory op_models (Section 1.1.3). Often you may wish to store project files on external drives or in locations other than the default model directory. In such a case, you should add to the list of OPNET model directories the folder where the files of interest reside. This allows seamless access to the OPNET files stored outside the default directory. This feature

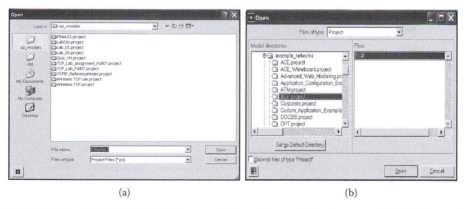

(a) (b)

FIGURE 1.6 The OPNET **File Chooser** window: (a) general file chooser and (b) file chooser organized by model directories.

is especially useful when working with the file chooser organized by model directories because, once a new folder has been added to the list of OPNET model directories, all the files located in that folder become visible via this file chooser mode.

There are several different ways to add new model directories. We list just a few below:

Approach #1: This is the simplest approach for adding a new model directory.
- Start OPNET.
- Select **File > Manage Model Files > Add Model Directory,** which opens a new window called **Browse for Folder**.
- Locate a new directory to be added and click the **OK** button.
- A window, **Confirm Model Directory,** shown in Figure 1.7, will appear. The window contains two checkboxes:
- *Include all subdirectories*: Selecting this checkbox will include not only the selected directory but all of its subdirectories as well.
- *Make this the default directory*: Selecting this checkbox will make the selected directory the default directory, which means that all new files will be saved into that directory.
- Click the **OK** button to add the selected directory.

Approach #2: This approach modifies the software preferences attribute named **Model Directories** via the **Preferences Editor**. In some older versions of OPNET, this attribute was referred to by its tag name **mod_dirs**.
- Start OPNET.
- Click on **Edit > Preferences**. This opens the **Preferences Editor** (Section 1.2.1).
- Search for attribute **Model Directories** or expand the **Miscellaneous** group* and scroll down to the **Model Directories** attribute.
- Click on the Value field of the **Model Directories** attribute. A **Model Directories** window opens up that lists all the directories currently included in this attribute.

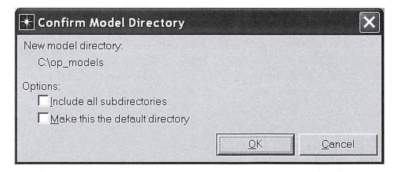

FIGURE 1.7   The **Confirm Model Directory** window.

---

* Versions prior to 12.0 do not group the attributes and thus you may need to scroll down the list to find the **mod_dirs** attribute.

- Click the **Insert** button and type the full path of the directory to be added to the list. By default, the newly inserted folder is placed at the top of the list, making it the **working** or **default directory**. If you had first selected a directory in the list before clicking **Insert**, then the new directory is inserted just before the selected directory.
- Clicking buttons **Delete**, **Move Up**, or **Move Down** allows you to delete items from the list or to manipulate their order in the list.
- Click **OK** twice to accept the changes.

Another way to add model directories is to manually edit the value of the **mod_dirs** attribute in the environment file located in the **op_admin** directory.

In each of the above approaches, after the directory is added, it is a good idea to click on **File > Manage Model Files > Refresh Model Directories**, which makes OPNET aware of the files in the newly added directory. Please note that the order in which the directories are listed in the **mod_dirs** attribute is very important as it determines the order in which OPNET software searches the directories to locate your project and model files. For example, consider the following situation. You have created three projects all named `MyFirstAssignment` and saved them in three different folders named `c:\op_models`, `c:\users\op_models`, and `c:\users\project_ files\op_models`. Let us assume that these directories are listed in the **mod_dirs** attribute in the following order:

- `c:\users\op_models`
- `c:\op_models`
- `c:\users\project_files\op_models`

Thus, when you open project `MyFirstAssignment`, the project files located in directory `c:\users\op_models` will be loaded by the OPNET software because this directory will be searched before any other directory that contains `MyFirstAssignment` project files. If you want to open the project files located in directory `c:\op_models`, then you need to place this directory in the **mod_dirs** attribute above all other directories that contain project `MyFirstAssignment` files. Alternatively, you can use the general file chooser to open the files. Generally, it is a good idea to name all your model and project files differently. Such a policy will protect you from accidentally overwriting important project and other model files or running a simulation using an incorrect model. Please refer to OPNET documentation for additional information. Specifically, you may want to examine the System Environment chapter, which contains guidelines for organizing user files and model directories.

1.5   PROJECTS AND SCENARIOS

IT Guru and Modeler allow you to conduct studies of various network technologies in a vast array of simulated environments. OPNET refers to individual simulation studies as projects. Each project is further divided into one or more scenarios. We can define a **project** as a collection of simulation studies of a particular system or technology under a range of different configuration settings. A **scenario** is a single simulation study of such a system under a specific configuration setting (namely one set of configuration parameter values).

Consider the following example. A company would like to perform a simulation study to determine how to expand its network infrastructure. For this study, the company may create a project called `NetworkExpansion`,* which is further divided into individual scenarios called `10nodes_200K`, `20nodes_200K`, `10nodes_500K`, and `20nodes_500K`. Generally, it is a good idea to choose descriptive project and scenario names. In our example, the first word of a scenario name indicates the number of nodes added to the company's infrastructure, whereas the second word indicates the capacity of the links that connect the new nodes to the rest of the company's network.

OPNET products contain various easy-to-use features for managing projects and scenarios, including the following: creating new projects and scenarios, duplicating existing scenarios, switching from one scenario to another, configuring and running simulations for a single or multiple scenarios, and comparing simulation results from different scenarios and projects. We discuss some of these features below, whereas others are described in later chapters.

1.6   WORKING WITH PROJECTS

Typically, a simulation study in OPNET starts with opening an existing project file or creating a new one.

1.6.1   Opening an Existing Project

To open an existing OPNET project, follow these steps:

- From the main OPNET window, select **File > Open** or click **Ctrl-O**.
- Choose the preferred mode of file browsing (Section 1.4.1).
- Make sure that the pull-down list called *Files of type:* has selected `Project files`.
- Double-click on the project file that you want to open (or click once to select the project file and then click the **OK** button).

You may need to browse the available directories to locate your project of interest.

1.6.2   Creating a New Project with the Startup Wizard

For creating a new project, OPNET provides a helpful feature called the **Startup Wizard**. Below are the steps to create a new project using the **Startup Wizard**:

- From the OPNET main window, select **File > New**, then select `Project` in the pull-down list, and click **OK**.

---

* Please note that earlier versions of OPNET software did not allow empty spaces in the project and scenario names.

- Enter (type in) the desired *Project Name* and *Scenario Name* (see Figure 1.8).
- Make sure that the checkbox *Use Startup Wizard when creating new scenarios* is selected. Click **OK**.

Now we describe the key components of the **Startup Wizard**. If the **Startup Wizard** has been selected, then the **Initial Topology Window** will appear. Otherwise, a default empty project will be created and you can start developing the simulation study.

- In the **Initial Topology Window**, choose Create Empty Scenario and click **Next**.
- In the **Choose Network Scale Window**, choose between World, Enterprise, Campus, Office, Logical, or Choose from maps and then click **Next**.

The windows that appear next will depend on the selection you have made. Options World, Enterprise, and Choose from maps allow defining a geographical scale based on a selected map via the **Choose Map** window, whereas options Enterprise, Campus, and Office allow specifying the dimensions of the simulated area using the width and height of a rectangular area via the **Specify Size** window.

- In the **Choose Map** window (Figure 1.9), select an item in the Border Map or choose among the available MapInfo maps, and then click the >> button to move the chosen map into the selected area. Once the geographical maps have been selected, click **Next** to proceed to the next configuration option.
- In the **Specify Size** window, specify *X Span* and *Y Span* of the network and click **Next**.

Once the dimensions of the simulated area have been specified, the **Startup Wizard** will open the **Select Technologies** and **Review** windows.

- In the **Select Technologies** window, choose the model families to be used by clicking the box under *Include?* to change it from No to Yes, and click **Next**. A model family is a collection of models that belong to a coherent set of technologies, such as *internet_toolbox*, *ethernet*, *MANET*, *Cisco*, etc. When

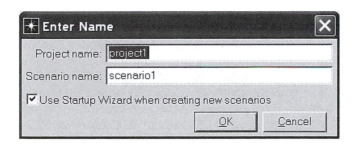

FIGURE 1.8   Creating a new project.

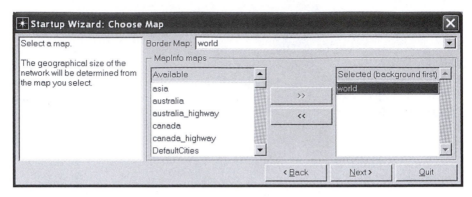

FIGURE 1.9   The **Choose Map** window.

you select one or more model families, the objects in those families will become part of the default model family, which appears when the **Object Palette** opens. Irrespective of your choice of technologies (or even if you choose no technologies at all), the entire set of objects in all technologies, not only the default model family, is always available in the **Object Palette**.
- In the **Review** window, click **Finish**.
- While using the **Startup Wizard**, you may click the **Back** button at any time to return to the previous window. You may also click **Quit** to exit the **Startup Wizard** and create a default project.

When this process is finished, the **Object Palette** and **Project Editor** windows will open. You can now proceed to choose objects from the **Object Palette** and place them in your project workspace within the **Project Editor**, creating and then configuring the network to be used in your study. Subsequent chapters of this guide describe how to use the **Object Palette** and how to create, configure, execute, and debug a simulation study, and then collect and analyze the results.

1.6.3 DELETING A PROJECT

Often a novice OPNET user may create multiple projects, some of which may need to be deleted. Perform the following steps to permanently delete all files associated with a particular project:

- From the main OPNET window, select **File > Delete Projects**. This action opens the **Delete Project** window, which contains a list of all currently accessible projects.
- To delete a particular project, click on that project of interest and then click the **OK** button in the **Confirm** window that appears. This will result in permanent removal of all project files from the hard disk.
- If needed, repeat this process to delete any additional unnecessary projects.
- Once all the desired projects have been deleted, click the **Close** button to close the **Delete Project** window.

1.7 WORKING WITH SCENARIOS

Scenarios help to organize large simulation studies into small parts that examine a specific aspect or configuration of the investigated phenomena. Scenarios are also useful when you want to repeat a simulation with a few configuration differences, such as changing the data rate of a link, the number of workstations in a LAN, the routing protocol, TCP flavor, employed technology, etc. In these cases, you can create multiple scenarios within the same project to better organize the study and to facilitate comparison of multiple configurations of the network being studied.

OPNET provides an array of features for manipulating project scenarios. In fact, the **Project Editor**, a window where the simulation study is created and configured, contains a separate pull-down menu called **Scenarios** dedicated to various options for managing project scenarios (see Figure 1.10).

The following subsections describe the most commonly used options on the **Scenarios** menu. There are additional options for generating, configuring, accessing, and comparing various reports regarding the current scenario that are not described here. For additional information on these features, refer to the OPNET product documentation's section on Publishing Network Information.

1.7.1 CREATING SCENARIOS

The first two options in the **Scenarios** menu provide alternative methods for creating scenarios:

- **New Scenario… (Ctrl+Shift+N)** creates a new empty scenario within the current project. This option follows the same procedure for creating a new scenario as that described in Section 1.6.2 for creating an empty scenario for a new project using the **Startup Wizard**.

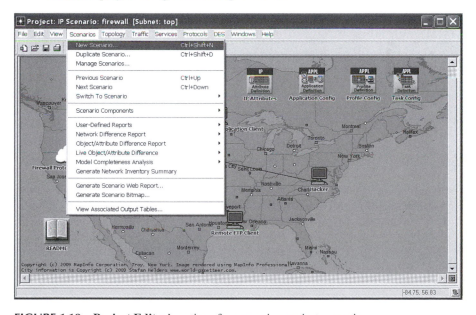

FIGURE 1.10   **Project Editor**'s options for managing project scenarios.

- **Duplicate Scenario... (Ctrl+Shift+D)** duplicates the current scenario. This option creates another scenario that is identical to the current one. This option is especially useful when studying a model that uses essentially the same setup with the exception of a few small configuration differences. In such cases, you need to duplicate an existing scenario and then modify only those few configuration parameters that are different.

1.7.2 MANAGING SCENARIOS

The next option in the **Scenarios** menu is **Manage Scenarios**, which is a comprehensive tool for managing scenarios within the current project. When performing a large and complex simulation study that consists of numerous scenarios, it is more practical to configure the simulation and start execution of all the scenarios of interest with a single click of the button instead of configuring and executing individual scenarios independently. As shown in Figure 1.11, the **Manage Scenarios** window organizes all scenarios within the project in the form of a table with the following columns:

- #—specifies a scenarios' number. Pressing Ctrl + <scenario number> switches the **Project Editor** to the scenario designated by the pressed number. Left-clicking on the number of a scenario provides a set of options for moving the selected scenario to a different location in the table.
- *Scenario Name*—specifies the name of a scenario. Left-clicking on the name of a scenario allows changing (i.e., typing in) the name of that scenario.
- *Saved*—specifies the current scenario status, which could be saved or unsaved. Left-clicking in the *Saved* cell of a scenario opens an option called <delete> which, when selected, marks the current scenario for deletion.

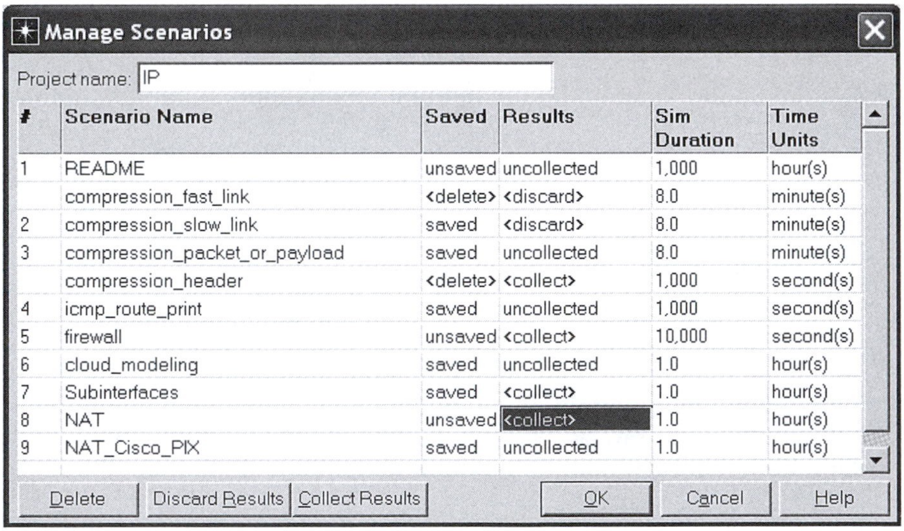

#	Scenario Name	Saved	Results	Sim Duration	Time Units
1	README	unsaved	uncollected	1,000	hour(s)
	compression_fast_link	<delete>	<discard>	8.0	minute(s)
2	compression_slow_link	saved	<discard>	8.0	minute(s)
3	compression_packet_or_payload	saved	uncollected	8.0	minute(s)
	compression_header	<delete>	<collect>	1,000	second(s)
4	icmp_route_print	saved	uncollected	1,000	second(s)
5	firewall	unsaved	<collect>	10,000	second(s)
6	cloud_modeling	saved	uncollected	1.0	hour(s)
7	Subinterfaces	saved	<collect>	1.0	hour(s)
8	NAT	unsaved	<collect>	1.0	hour(s)
9	NAT_Cisco_PIX	saved	uncollected	1.0	hour(s)

Project name: IP

Delete   Discard Results   Collect Results   OK   Cancel   Help

FIGURE 1.11   The **Manage Scenarios** options.

- *Results*—specifies the status of simulation results, which is represented by one of the following values: uncollected—there are no simulation results collected for this scenario, out of date—the simulation results for this scenario are available but are stale due to possible configuration changes, or up to date—the simulation results for this scenario are available and are fresh. Left-clicking in the *Results* cell of a scenario provides the following options:
  - <collect>—will configure the project to execute the current scenario and recollect the results. This option is only available when the value of the *Results* cell is set to uncollected or out of date.
  - <recollect>—will configure the project to execute the current scenario and recollect the results. This option is only available when the value of the *Results* cell is set to up to date.
  - <discard>—will configure the project to discard the results for the current scenario. This option is only available for scenarios whose simulation results are either up to date or out of date. This option is particularly useful when you need to save disk space since simulation results often require a lot of storage space.
- *Sim Duration*—specifies the execution length of a simulation scenario. You may left-click in a scenario's *Sim Duration* cell to change (i.e., type in) the duration of simulation (i.e., enter a new value in the cell).
- *Time Units*—specifies the units of the value specified in the corresponding *Sim Duration* cell. Left-clicking in the *Time Units* cell of a scenario allows you to choose the time units from one of the following values: second(s), minute(s), hour(s), day(s), or week(s).

If one of the scenarios is selected, then you can use the buttons **Delete, Discard Results,** or **Collect Results** to respectively mark the selected scenario for removal, have its results discarded, or have its results recollected.

There are also several additional options available in the **Manage Scenarios** tool:

- Left-clicking on the title of columns *#, Scenario Name,* and *Saved* provides options <delete all>, which marks all the project's scenarios for removal, and <keep all>, which undoes the previous action.
- Left-clicking on the title of column *Results* provides options <collect all>, which marks all the project's scenarios for execution, and <discard all>, which marks all the project's scenarios to have their simulation results discarded.
- Left-clicking on an empty cell in the columns *#* or *Scenario Name* provides a set of options for duplicating one of the existing scenarios or for creating a new one.

Finally, once you have configured the project's scenarios to your liking, you can click the **OK** button to save all the changes or click the **Cancel** button to exit the **Manage Scenarios** tool without saving the changes. When you click the **OK** button, all scenarios marked to collect or recollect the results will automatically begin executing in the order of their scenario numbers. This is a very useful feature that allows you to run multiple scenarios with a single mouse click.

1.7.3 SELECTING A SCENARIO

The next set of menu options in the **Scenarios** menu deals with selecting a scenario within the current project:

- **Previous Scenario (Ctrl+Up)** switches to the previous scenario on the list. This option does nothing if this is the first scenario in the list.
- **Next Scenario (Ctrl+Down)** switches to the next one in the list. This option does nothing if this is the last scenario in the list.
- **Switch to Scenario** allows switching to a specific scenario within the current project.

1.7.4 IMPORTING SCENARIO COMPONENTS

Another option in the **Scenarios** menu, **Scenario Components**, allows importing into or exporting from the current scenario various components including network model, probe model (e.g., statistics configured for collection), simulation results, and configuration information. By default, exported scenario components are saved in the op_models directory. However, the import operation accesses components that are located in any OPNET-visible directory. The import operation overwrites existing components of a scenario, unless the component itself is a scenario, in which case the scenario(s) is added to the current project.

To import a scenario, perform the following steps (see Figure 1.12):

- Select **Scenarios > Scenario Components > Import**.
- Choose component type Project from the pull-down list at the top of the **Import** window.

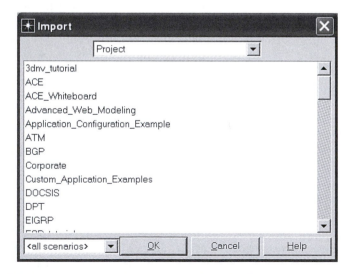

FIGURE 1.12   Importing a scenario component of type **Project**.

- Highlight the project of interest.
- From the pull-down list at the bottom left corner of the **Import** window, select the scenario of interest or choose `<all scenarios>` and click the **OK** button.

In the procedure above, if you wish to import a component type other than **Project**, then you may select a different value from the pull-down list at the top of the **Import** window. The pull-down list at the bottom left corner of the **Import** window is available only when you import a scenario component of type **Project**.

2 Creating Network Topology

2.1 INTRODUCTION

Network topology describes the interconnection of network devices or nodes and their communication channels or links arranged within a certain physical space. The literature often distinguishes between physical and logical topologies of a network. Physical topology is the actual layout of the nodes and the links that connect them, taking into consideration such factors as the physical locations of individual nodes and the actual distances spanned by the connecting communication links. On the other hand, logical or signal topology provides an abstract representation of the communication paths between the nodes (e.g., the routes taken by the data to reach various destinations) without regard for the actual physical locations and distances between individual nodes in the network. Logical topology primarily deals with protocol configuration of the network devices for the purpose of establishing channels for sending and receiving the data.

This chapter describes the features available in OPNET for creating the physical/logical topology of a simulated network. Generally, there are seven main topologies (Figure 2.1): fully-connected mesh, not fully-connected (i.e., partially connected) mesh, bus, star, ring, tree, and hybrid. In a fully connected mesh topology, every node has a direct link to every other node. In a partially connected mesh, network nodes have direct links to some nodes, but not necessarily all the other nodes. Bus topology has a single backbone link or bus with the nodes tapping into that link via drop lines. Star topology has a central device connected to all the other nodes in the network. There are no direct links between noncentral nodes. Ring topology has nodes organized in the form of a circle where each node is only connected to two nodes on either of its sides. In tree topology, any two nodes are connected via a single path. Finally, hybrid topology has the nodes connected to one another using some combination of any of the above topologies.

One could distinguish between nodes that have multiple links (or interfaces) and nodes that have a single link. Hubs, switches, routers/gateways, and other similar network devices are responsible for connecting other nodes and thus have multiple links. End nodes, such as workstations and servers, are usually connected to other devices or networks via a single link, although sometimes, the end nodes may be multihomed (i.e., connected to more than one network) and thus may have multiple links. For example, to improve reliability and/or to provide resilience to link failure, a company may have multiple redundant links connecting their servers to the network(s).

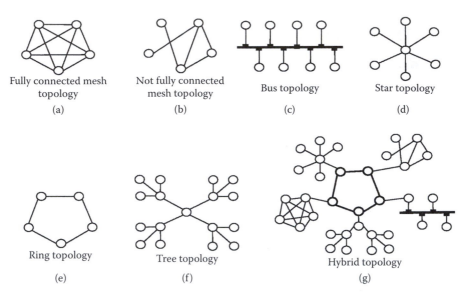

FIGURE 2.1   Network topologies. (a) Fully connected mesh topology, (b) not fully connected mesh topology, (c) bus topology, (d) star topology, (e) ring topology, (f) tree topology, and (g) hybrid topology.

OPNET provides a large library of models for network devices, communication links, LANs, and network clouds. The models are differentiated based on their type (e.g., link, hub, switch, router, and server) and supported communication interfaces. Examples of available models include Ethernet server, point-to-point protocol (PPP) workstation, IP router with Ethernet and PPP interfaces, 100BASE-T LAN, Ethernet/ Fiber Distributed Data Interface (FDDI) switch, 100BASE-T Ethernet link, 56 kbps PPP link, and many others. This chapter describes various features available in OPNET for creating and editing network topology. Specifically, the sections that follow in the remainder of this chapter provide instructions for creating network topology using **Object Palette** and **Rapid Configuration** tool, describe commonly used node and link models, show how to work with subnets, and give an overview of the annotation tool.

2.2   OBJECT PALETTE TREE UTILITY FOR CREATING NETWORK TOPOLOGY

The **Object Palette** utility provides access to all OPNET models. Through this utility, you can create desired network topologies, organize frequently used models for easier access, and create custom device models. Figure 2.2 shows a view of the **Object Palette Tree** with the default object palette set to *internet_toolbox*. The **Object Palette Tree** window opens automatically when you create a new project using the **Startup Wizard** (Section 1.6.2). You can also open the **Object Palette** from within the **Project Editor** by either clicking on the *Open Object Palette* icon in the toolbar or selecting the **Topology > Object Palette** option from the pull-down menu.

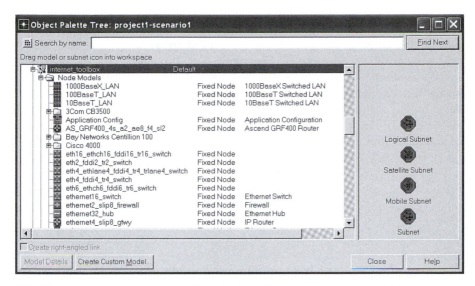

FIGURE 2.2   Default view of the **Object Palette Tree**.

You can either browse the **Object Palette Tree** or use the search facility to identify the node and link models needed for your simulation study. Once a model is found, you can drag and drop it into the **Project Editor**'s workspace (or simply the project workspace). The **Object Palette Tree** organizes available models into several palettes or categories:

- *Node Models* palette contains available node models of the communicating devices such as hubs, switches, routers, gateways, workstations, and servers.
- *Link Models* palette contains models of the links such as 1000BASE-T Ethernet link, T1 duplex link, and 16 Mbps Token Ring.
- *Path Models* palette contains models for specifying network paths that support such technologies as High Assurance Internet Protocol Encryptor (HAIPE), Multiprotocol Label Switching (MPLS), and Public Switched Telephone Network (PSTN).
- *Demand Models* palette contains models for specifying traffic flows and connections such as IP voice traffic flow, PSTN voice traffic flow, and IP security.
- *Wireless Domain Models* palette contains models for representing such wireless domains as sparse grid, full grid, and mobility.
- *Shared Objects* palette contains a collection of node, link, path, demand, and domain models grouped according to common properties. For example, *3Com* collection contains the models of devices produced by the 3Com Corporation, *applications* palette contains models needed for specifying and deploying applications, and *internet_toolbox* group contains node, link, and utility models commonly used for modeling the Internet.

2.2.1 MODEL NAMING CONVENTIONS

Each of the categories of models in the **Object Palette Tree** is further divided into subgroups organized by name, machine type, object ID, vendor, interface type, link transmission mode, and other parameters. These categories often overlap, which means that the same model can be retrieved from multiple palettes. Each model name usually includes such information as node or link type (e.g., server, gateway, and LAN), available interfaces (e.g., Asynchronous Transfer Mode [ATM], Serial Line Internet Protocol [SLIP], FDDI, Token Ring, and Ethernet), and the number of interfaces of each type.

Figure 2.3 shows a partial list of the models available in the *internet_toolbox* palette, the default palette or model family selected by the **Startup Wizard** during project creation. Let us examine the highlighted node model named *CS_4000_3s_e6_fr2_sl2_tr2* by deconstructing its name. The first part of the name indicates that

FIGURE 2.3   Some of the models in the *internet_toolbox* subcategory.

this is a model of the CISCO C4000 router, which is a fixed node (i.e., not a mobile or satellite node). The model name of this device is deconstructed as follows:

- *CS*—Cisco (3C means 3Com, AS means Ascend, etc.)
- *4000*—model of the Cisco router
- *3s*—three slots
- *e6*—six Ethernet interfaces
- *fr2*—two Frame Relay (FR) interfaces
- *sl2*—two SLIP or simply IP interfaces
- *tr2*—two Token Ring interfaces

Similarly, *eth6_ethch6_fddi6_tr6_switch* is a switch that has six Ethernet, six EtherChannel, six FDDI, and six Token Ring interfaces; *ethernet4_slip8_gtwy* is an IP gateway router that has four Ethernet interfaces and eight IP interfaces; *ethernet_wkstn* is a client node that has a single Ethernet interface; and *ppp_server* is a server that has a point-to-point link supporting a single SLIP interface.

The naming conventions of the link models are very similar to those of the nodes. The model's name provides adequate information to determine the supported communication protocol and link capacity, while the model description indicates if the link provides duplex or simplex communication. For example, as Figure 2.3 shows, *1000BaseX* is a model of the duplex Ethernet 1000BASE-X link; *FR_T1* is a duplex FR link with T1 (or DS1) capacity; and *PPP_28K* is a model of the point-to-point duplex 28 kbps link that supports a SLIP connection.

Bus topologies require slightly different link models. Specifically, to create a bus topology, you need bus and bus tap models, which are available in the **Object Palette Tree** under *Link Models...Bus Models* and *Link Models...Bus Tap Models* categories respectively.

OPNET also provides some models of complete networks. For example, *10BaseT_LAN* models a LAN in which the nodes are connected via 10BASE-T Ethernet links, and *atm32_cloud* represents an ATM network with 32 ATM interfaces, while *ip32_cloud* is an IP network with 32 SLIP interfaces. The *ip32_cloud* model is commonly used to simulate the Internet. Finally, utility nodes such as *Application Config, IP VPN Config, IP Attribute Config, Profile Config*, and others do not represent any specific real-life devices but are provided by OPNET to simplify setup and configuration of various applications and networking technologies.

2.2.2 MODELS IN THE *INTERNET_TOOLBOX* PALETTE

The models available in the *internet_toolbox* palette are often sufficient to work with the most common networking protocols and technologies. The following list provides a brief description of the common uses for some of these models:

- *ppp_wkstn* and *ethernet_wkstn* are used to model client workstations, the end nodes that run various applications over SLIP and Ethernet links respectively.

- *ppp_server* and *ethernet_server* are used to model servers, which are also end nodes, but they provide services for various client applications over SLIP and Ethernet links respectively. Server and workstation objects allow modeling of the client–server paradigm.
- *1000BaseX_LAN*, *100BaseT_LAN*, and *10BaseT_LAN* models are used to simulate Ethernet LANs that run over 1000BASE-X, 100BASE-T, and 10BASE-T Ethernet connections respectively. By default, each of the above network models simulates the operation of a LAN with 10 end nodes.
- *ethernet32_hub* models an Ethernet hub device, a layer 1 repeater that connects up to 32 Ethernet end nodes to form a single Ethernet network segment.
- *ethernet16_switch* models an Ethernet switch, a layer 2 device with 16 interfaces.
- *ethernet4_slip8_gtwy* models a gateway router, a layer 3 device with four Ethernet and eight SLIP interfaces. Ethernet interfaces are usually used to connect to the local subnetworks (e.g., LAN objects, hubs, or switches), while IP interfaces are usually used to model connections to IP networks and to such node devices as routers, servers, and workstations.
- *ethernet2_slip8_firewall* also models a gateway router but with additional firewall features. This model has two Ethernet and eight SLIP interfaces.
- *ip32_cloud* models an IP cloud and is commonly used to represent the connectivity of the Internet when the exact topology is of no interest. This model has 32 serial IP interfaces.
- *1000BaseX*, *100BaseT*, and *10BaseT* are models of link objects that connect devices with Ethernet interfaces.
- *PPP_28K*, *PPP_33K*, *PPP_56K*, *PPP_DS1*, and *PPP_DS3* are models of duplex point-to-point links with various data rates that can connect two nodes that run IP.
- *Application Config* and *Profile Config* models are used for configuring applications and user profiles in the simulated network. We provide more details about these models in Chapters 5 and 7 respectively.

2.3   WORKING WITH THE OBJECT PALETTE TREE

This section provides instructions for opening the **Object Palette Tree**, finding the device and link models needed for the simulation study, and creating network topology.

2.3.1   OPENING THE OBJECT PALETTE

To open the **Object Palette Tree** from within the **Project Editor** (i.e., a project scenario is open), either of the following two approaches can be used:

Approach #1: Click on the *Open Object Palette* icon that looks like ▩.*
Approach #2: Select **Topology > Open Object Palette** from the pull-down menu.

---

* The **Project Editor**'s toolbar that contains the icons for commonly used operations is configurable. Thus, it is possible that the *Open Object Palette* icon is not configured to be visible, in which case Approach #1 cannot be used.

2.3.2   SEARCHING FOR MODELS IN THE OBJECT PALETTE

The **Object Palette Tree** provides a search facility to help find models based on their names. For example, to find models of switch devices, type the word `switch` in the *Search by name:* textbox and click the **Find Next** button. The search will start from the current position in the tree and will continue down toward the bottom. If the model is found, then the search utility will highlight it. Click the **Find Next** button to locate the next model in the tree that contains the word `switch` in its name. If the model is not found, then the window will remain unchanged and no model will be highlighted.

Not all model names contain complete words that describe the model properties. For example, many link models do not contain the word *link* in their name, while devices with Ethernet interfaces may contain only the characters *eth* in their name instead of the complete word *ethernet*. For this reason, you should choose the search criteria with care or browse the **Object Palette Tree** to find the model of interest.

When browsing through available models in the **Object Palette**, often the object's name and the brief model description are not sufficient to determine if the model is appropriate for the current simulation study. To obtain additional details about the model of interest, highlight the model by clicking on it and then click on the **Model Details** button; alternatively, right-click on the model and select the **View Model Details** option from the pop-up menu.

2.3.3   CREATING CUSTOM MODELS

Sometimes the models available in the **Object Palette** do not contain devices with the necessary number or combination of required interfaces. For this purpose, OPNET allows you to create your own custom node models. Use the following steps to create custom device models:

- Open the **Object Palette** window (see Section 2.3.1).
- Click on the **Create Custom Model** button to open the **Create Custom Device** window shown in Figure 2.4.
- Select the type of device you would like to create (e.g., router, bridge, switch, or cloud). In some cases, you may have to make a subselection, for example, a switch may be ATM, Ethernet, or FDDI.
- Specify the values of the new device parameters (also called attributes). These parameters usually define such information as the number and the type of interfaces in the new device. Certain devices may provide an option for specifying other information.
- If necessary, you can change the icon for representing the new device. Clicking on the icon for the device opens an **Icon Palette** window that allows you to select a different image for the new device icon.
- By default, the new device is saved into the *my_model_list* palette, but you may select a different palette to contain the new model by selecting one from the pull-down list called *Destination palette*.

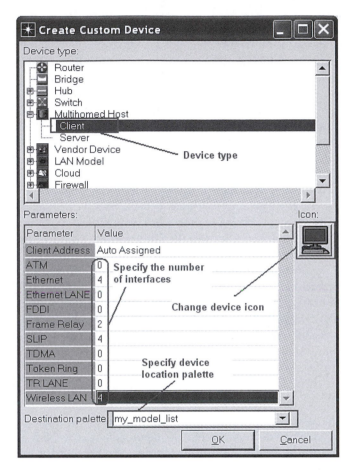

FIGURE 2.4   Create Custom Device window.

- Once all attributes, including the destination palette, of the new device are specified, click the **OK** button to create the device.
- Finally, type the name of the new device in the provided prompt and click **OK**. OPNET will notify you that the new device has been successfully added and will open the **Object Palette Tree** in the folder where the new device resides. The new device files are stored in the default directory on the local disk (usually op_models).

2.4   CREATING NETWORK TOPOLOGY

Once the necessary node and the link devices are identified and prepared, you are ready to create the network topology for your simulation study using the **Object Palette**. Sections 2.4.1 through 2.4.4 provide instructions for adding and deleting nodes and links in a network topology, and for performing other editing operations on these objects. All of these subsections assume that the **Object Palette** is open (see Section 2.3.1).

2.4.1   CREATING NETWORK TOPOLOGY: ADDING NODES

- Browse or scroll the **Object Palette Tree** and select the node model that you would like to use in your simulation study by clicking on it. An icon of the selected model appears in the right-hand part of the **Object Palette Tree** window.
- Clicking on the **Model Details** button opens a new window that contains a detailed description of the selected model, which could be useful in determining if the object accurately represents the device that you want to simulate.
- To create a new node in a network topology, drag the node model's icon into the project workspace and drop it in a desired location. Alternatively, you can click on the icon of interest and then click again within the project workspace in the location where you would like to place the object.
- After you have placed a node with any one of the above techniques, you can repeatedly click in the project workspace to add multiple instances of the same object.
- Right-click anywhere in the project workspace to end the operation.
- The above steps can be repeated on as many objects as necessary to place nodes of various object types in the network topology.

OPNET also provides models of utility objects that do not represent any real-life networking devices. These models are provided to facilitate the process of defining and configuring various networking technologies and applications. The task of adding utility objects to the project workspace is also performed with drag and drop, the same way as for regular nodes. However, unlike regular nodes, utility objects cannot be connected via link, demand, or path models because utility objects do not model any real networking devices and are provided only for configuration purposes.

2.4.2   CREATING NETWORK TOPOLOGY: ADDING LINKS

Use the following steps to add links between nodes in a network topology:

- Select the link model of interest in the **Object Palette Tree** by clicking on its icon.
- Click on the first node in the project workspace that you wish to connect with this link. A line representing the link will appear and will follow the mouse wherever you take it. Alternatively, you can drag and drop the link model onto the first node to achieve the same effect.
- Drag the line to the second node to be connected by the link and click on it. This creates a link of the desired type joining the two selected nodes.
- You can continue connecting other nodes with the same link type using the same technique, that is, click on the first node to be joined, then drag the line to the second node and click on it.
- You can end this operation by right-clicking anywhere in the project workspace.

- Sometimes you may want to draw a link along a curved path instead of a straight line. To achieve this effect, click on the first node to be connected. Next, instead of clicking on the second node, click on a location within the project workspace where you would like the link's path to go. While defining the link's path, you may click within the project workspace as many times as needed. Finally, click on the second node to complete the link's path.

2.4.3 Creating Network Topology: Deleting Nodes or Links

In addition to adding objects to a network topology, you may sometimes want to remove undesired objects from the created network. If the deleted object is a node that has links attached to it, then the links will be removed as well. There are two ways to delete an object:

Approach #1: Select the object of interest by clicking on it and then press the DELETE key on your keyboard.

Approach #2: Select the object of interest by clicking on it and then, from the pull-down menu of the **Project Editor**, select **Edit > Delete**.

2.4.4 Creating Network Topology: Other Editing Operations

The **Project Editor** allows selecting one or more objects and doing various operations on them. To select objects, use any of the following techniques:

Approach #1: Select an object in the workspace by clicking on it. Then select further objects by either shift-clicking or control-clicking on them (i.e., click on the next object while holding the **Shift** or **Control** key on your keyboard).

Approach #2: Left-click anywhere in the workspace and then drag the cursor to define a rectangle, releasing the mouse at the end. All objects (nodes and links) within the rectangle will be selected.

After the objects are selected, you can move them to a new location in the workspace by simply dragging them with the mouse. The **Edit** menu in the **Project Editor** also supports such operations on selected objects as **Cut** (Ctrl+X), **Copy** (Ctrl+C), **Paste** (Ctrl+V), **Undo** (Ctrl+Z), and **Redo** (Ctrl+Y).

2.5 THE RAPID CONFIGURATION TOOL

Rapid Configuration is a useful tool for creating network topologies with a large number of nodes. The **Rapid Configuration** tool is accessible from within the **Project Editor**.

2.5.1 Creating Network Topology with the Rapid Configuration Tool

Use the following steps to create a network topology with the **Rapid Configuration** tool:

- From the pull-down menu, select **Topology** > **Rapid Configuration...**
- When the **Rapid Configuration** window appears, select the desired topology configuration from the pull-down list and click the **Next** button to configure the selected topology. The topology configuration list includes the following options: Bus, Full Mesh, Randomized Mesh, Ring, Star, Tree, or Unconnected Net.
- The **Rapid Configuration** window also provides an option to specify a seed value for the random number generation. This feature is very important since some topologies may require randomness for node placement or other properties. To specify a seed, click the **Seed** button. In the **Seed Selection** window, type in the desired seed value, or click **Generate** to automatically provide a random value for the seed. Click **OK** to complete the selection.
- After you select the network topology and click **Next**, you will see a new window that allows specification of such parameters as node and link models (e.g., the star topology asks you to specify models for the center and periphery nodes, and the tree topology asks for intermediate and leaf node models), number of nodes, node positioning, and other parameters.
- If the node and/or link pull-down lists do not contain the models needed to create the corresponding network topology, then click on the **Select Models** button and choose the necessary technology family from the provided model list.
- Once all the necessary values are set, click **OK** to create the desired network topology and place it in the project workspace.

2.5.2 EXAMPLE: CREATING ETHERNET LAN WITH THE RAPID CONFIGURATION TOOL

As an example of using the **Rapid Configuration** tool, here are the steps for creating an Ethernet LAN using a star topology with 10 Ethernet workstation periphery nodes connected to an Ethernet switch:

- In the **Project Editor**, select **Topology** > **Rapid Configuration** from the pull-down menu.
- Choose Star as your configuration topology and click **Next**.
- In the new window that appears, set the following values:
  - *Center node model*: ethernet16_switch
  - *Periphery node model*: ethernet_wkstn
  - *Number*: 10—this is the number of periphery nodes. This value should not exceed 16 since the switch that we used has only 16 Ethernet interfaces.
  - *Link model*: 10BaseT (this will give a data rate of 10 Mbps)
  - *Center X*: 12 (you can use the default value that is already provided)
  - *Center Y*: 12 (you can use the default value that is already provided)
  - *Radius*: 40 (you can use the default value that is already provided)
- Click **OK**.

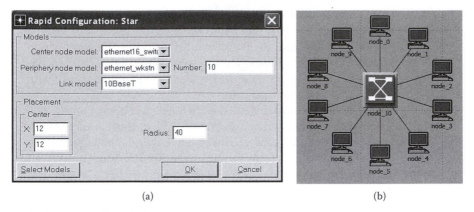

(a)   (b)

FIGURE 2.5   (a) **Rapid Configuration** tool for creating a star topology. (b) Network with star topology created with this tool.

Figure 2.5 shows the **Rapid Configuration** settings and the resulting star topology. If you make a mistake and connect the nodes using the wrong link type or connect more nodes to a device than the number of interfaces available at that device, then OPNET may or may not provide any warning about an invalid network topology. See Section 2.6.2 for some common mistakes to avoid and how to verify the validity of the created network topology.

2.6   CONFIGURING LINK OBJECTS

Link objects connect node devices in a network. Links are defined through the data link layer (i.e., layer 2) technology that validates compatibility of the nodes attached via the link. Most of the time, you will select a link model that is appropriate for creating the desired topology, for example, you may choose the link model *PPP_56K* to join two nodes with a point-to-point link with transmission rate 56 kbps. But in some situations, after you add a link object in a network topology, you may have to configure it according to your needs.

Link models contain various basic attributes such as transmission rate or propagation speed. Some of these parameters are configurable and others are not. This depends on the type of the link model used in the simulation and on the parameter itself. Link models that contain modifier *adv* or *int* (i.e., advanced or intermediate model derivation level) usually allow you to change such commonly used parameters as the transmission rate of the link. However, link models such as *1000Base_X* and *100Base_T* do not allow setting transmission rate regardless of the modifier used in the model's name.

2.6.1   Changing Basic Link Properties

Basic properties of point-to-point links that can be configured include transmission rate and propagation delay. Use the following steps to change any of these to desired values:

- Right-click on the link object and select the **Edit Attributes** option.
- In the **Edit Attributes** window, select the *Advanced* checkbox in the lower right corner.
- The propagation delay is specified by the attribute named **delay**, which is normally set to the value `Distance Based`. This means that the propagation delay is automatically configured based on the distance traversed by the link. If you would like to set the delay to a specific value independent of the distance or the positions of the nodes, then click on the value field of the attribute, select `Edit...`, and type in the desired value.
- The transmission rate is specified by the **data rate** attribute. This attribute is not visible in all link models. If the current link model does not contain the **data rate** attribute, then
  - Change the **model** attribute (Section 3.4.5) of the current link to another model with the same name but which contains an *int* or *adv* modifier. This is done by clicking on the value field of this attribute and selecting `Edit...` from the pull-down menu. Browse the list of models that appears and select the new desired model. Make sure that the new link model uses the same link protocol (i.e., layer 2 technology).
- Set the **data rate** attribute value by typing in the desired link data rate in bits per second. Alternatively, you may select one of the preconfigured data rate values, if available.
- Click the **OK** button to save the changes and close the **Edit Attributes** dialog box.

2.6.2 VERIFYING LINK CONNECTIVITY

Simply placing nodes into a project workspace and connecting them with links is not sufficient to create a working model of a network topology. There are a variety of constraints on how nodes can be connected with links, and the simulation will not proceed if these constraints have not been followed. In general, the link type must be compatible with the node types that it joins, and both nodes must have available interfaces of the link's type. For example, you cannot connect an Ethernet workstation with an Ethernet server using an FDDI or a Token Ring link; instead, you need to use an Ethernet link. OPNET will provide a warning message if you attempt to join two node models that have no available interfaces that match the type of the connecting link. As Figure 2.6 illustrates, the warning message provides two choices:

- *Add link and ports*—the node model will be updated to include an interface (e.g., port) that allows a connection using the selected link.
- *Add link without assigning ports*—the node model will not be updated, and as a result, the added link will become nonfunctional.

There are other situations in which OPNET may not give a warning message about improperly connected links, which may produce an incorrect network topology (e.g., if you try to change the **model** attribute of an already connected node or link object). For this reason, once the network topology has been created, it is a good idea to

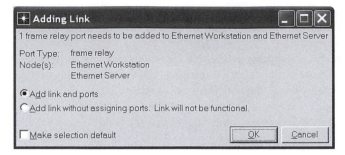

FIGURE 2.6   Warning message while adding an incompatible link.

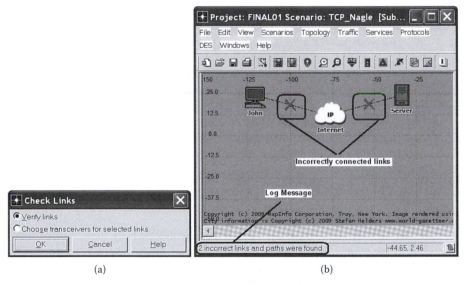

(a)   (b)

FIGURE 2.7   (a) **Check Links** window opened by the **Verifying Link Connectivity** utility. (b) A network with two incorrectly configured links found by this utility.

verify that all the nodes in the network are connected properly. OPNET provides a simple utility for performing this task, called **Verifying Link Connectivity**. Here are the steps to use this utility:

- From the pull-down menu, select **Topology > Verify Links** or, alternatively, press Ctrl+L. This will open the **Check Links** window (see Figure 2.7a).
- In this window, choose the *Verify links* option and click **OK**. You can also click the **Help** button to find additional information regarding the **Check Links** window.

If all the links are connected correctly, the message "All links and paths are connected properly" will appear at the bottom of the **Project Editor** window. Otherwise, if say, two incorrectly connected links were identified, then these links will be marked with red crosses and the message at the bottom of the **Project Editor** window will read "2 incorrect links and paths were found." Figure 2.7b illustrates such a situation.

The usual suspects for incorrect link connections are as follows:

- Connecting the link to a node that does not support the link's communication protocol. For example, directly connecting an Ethernet server (e.g., *ethernet_server* model) to an IP router (e.g., *ethernet4_slip8_gtwy* model) using a point-to-point link (e.g., *PPP_DS1* model) will cause the link to become nonfunctional since the Ethernet server does not support the point-to-point SLIP protocol required by the link model.
- Attaching more nodes to a device than the number of communication interfaces available on that device. For example, connecting 20 Ethernet servers to an Ethernet switch with 16 interfaces (e.g., *ethernet16_switch* model) will cause four connecting links to become nonfunctional since the switch supports only 16 interfaces and the additional four servers have nothing to connect to.
- Connecting a link to the wrong port. Some devices such as routers (e.g., *ethernet4_slip8_gtwy* model) support multiple interface types. Thus, even though the router may support IP interfaces, connecting a point-to-point link to the Ethernet port will cause the link to become nonfunctional. Generally, if available, OPNET automatically will select the correct interface for the connection being created. However, incorrect link to port assignment may still occur in certain situations, for instance, when using the **Rapid Configuration** tool (Section 2.5).

2.7 FAILING AND RECOVERING NETWORK ELEMENTS

OPNET provides a utility for failing and recovering network elements during the course of a simulation. Often it is important to determine how the system behaves when one or more links go down, or if a key node such as a server or a router fails. Clearly, one can simply remove the elements to be failed, run the simulation and collect the results, and then add the removed elements back. However, removing and then adding the objects back into the network topology can be quite a cumbersome task. Furthermore, removing and then adding network elements does not allow studying a situation where the link or node objects fail and then recover in the middle of the simulation, for example, when you want to know how the network and its protocols behave immediately before and after the element failure or recovery. To address such situations, OPNET provides a separate utility node for failing and recovering elements in the network. There are two approaches to failing/recovering objects in the network:

Approach #1: Fail/Recover object(s) for the entire simulation:
- Select the objects to be failed/recovered by clicking on them (shift/control-clicking if multiple objects are to be selected).
- From the pull-down menu, select **Topology > Fail Selected Objects** or **Recover Selected Objects** to fail or recover the selected objects. Alternatively, click the ▣ icon to fail and the ▣ icon to recover the selected objects.

Approach #2: Fail/Recover object(s) at specific times during the simulation:

- Open the **Object Palette.**
- Expand the *Shared Object Palettes* group, then *utilities*, then *Node Models*, where you will find the *Failure Recovery* node. You can also directly browse for this object.
- Add the *Failure Recovery* node to the project workspace.
- Right-click on the node and select the **Edit Attributes** option.
- You may want to change the name of the node to better reflect its purpose. For example, you can name this node `Failure and Recovery Node`.
- To fail/recover links, expand its **Link Failure/Recovery Specification** attribute.
- To fail/recover nodes, expand its **Node Failure/Recovery Specification** attribute.
- For each failure or recovery event, add a row in the corresponding attribute. For example, to fail the node named `node_0` at time 150 seconds and then recover it at time 200 seconds, you need to set the number of rows in the **Node Failure/Recovery Specification** attribute to 2.
- Expand each of the rows and set the following attribute values: **name** of the node or link to be failed or recovered, **time** of failure/recovery, and **status**, which is either `Fail` or `Recover`.
- When failing/recovering nodes, you may also want to set the **Node Failure Mode** attribute that has the following values:
  - `Node and attached Links`—fails/recovers both the node and the links attached to it.
  - `Node Only`—fails/recovers only the node, while links remain unaffected.
  - `Attached Links Only`—fails/recovers only the links attached to the node.
- Click **OK** to save the changes.

Figure 2.8 shows a sample setting of the *Failure Recovery* node with circular boxes around the attributes whose values have been changed. Using the above configuration, `node_0` will fail at time 150 seconds and then recover at time 200 seconds. Notice that the **Node Failure Mode** attribute is set to its default value of `Node Only`.

2.8   SUBNETS

OPNET products contain a very useful feature for grouping elements of the network topology into separate subnetworks. OPNET provides a special subnet object for holding other objects such as nodes, links, and other subnet objects. With the help of the subnet object, you can arrange nodes in the network in a hierarchical fashion. If subnet A contains subnet B, then subnet A is called the parent subnet of subnet B, while subnet B is a child subnet of A. Multiple levels of hierarchy can be specified through one or more subnet objects, each representing a physical or logical part of

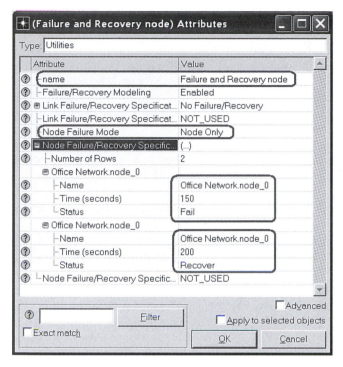

FIGURE 2.8   Sample attribute settings of the *Failure Recovery* node.

the larger network. The highest level of the subnetwork hierarchy is referred to as the top level or global subnet. Subnetworking provides a powerful tool for organizing complex networks in a fashion that facilitates quick and easy access to and viewing of different parts of the network.

When dealing with large networks, organizing topology into subnetworks usually is the first order of business. First, you need to add one subnet object for each subnetwork you will have in your topology. Next, you should populate each of the created subnets with such network elements as nodes, links, and perhaps other subnets. Utility objects are responsible for configuration of the whole simulation study and thus should be added only once, usually in the top level of the network or in one of the top's child subnets.

2.8.1   Adding a Subnet Object

Approach #1: From **Project Editor**
- From the pull-down menu of the **Project Editor**, select **Topology > Subnets**.
- From the submenu, choose one of the following:
  - **Create Fixed Subnet**—creates a physical subnetwork in which the nodes are not allowed to move during the simulation. Fixed subnets represent static networks and are usually connected to other nodes or subnets via one or more physical links.

- **Create Logical Subnet**—creates a logical or virtual subnetwork that disregards such physical topology characteristics as geographical node positions. Logical subnets are also usually connected to other nodes or subnets via one or more links.
- **Create Mobile Subnet**—creates a mobile subnetwork in which the nodes are allowed to move during simulation. Mobile subnets and the nodes placed within them are not allowed to be connected to other nodes or subnets via physical links.
- **Create Satellite Subnet**—creates a satellite subnetwork. Like mobile subnets, satellite subnets and the nodes placed within them are not allowed to be connected to other nodes or subnets via physical links.

- Click on the location within the project workspace where you would like to place the subnet object. Some versions of OPNET may ask you to specify the name of the newly added object.
- Add as many subnet objects as you desire and right-click anywhere in the project workspace to end the procedure.

Approach #2: From **Object Palette**

- Open the **Object Palette** (see Section 2.3.1).
- Drag and drop anywhere in the project workspace one of the subnet icons that appears to the right in the **Object Palette** window. Note that the icon named *Subnet* represents a fixed subnet.
- You can add more than one subnet of the selected type by clicking again in the project workspace.
- Right-click anywhere in the project workspace to end adding the selected subnet.

2.8.2   MOVING AROUND THE NETWORK HIERARCHY

- Entering a subnet (going down into the hierarchy):
  Approach #1: Double-click on the subnet icon.
  Approach #2: Right-click on the subnet icon and select **Enter Subnet** from the pop-up subnet icon menu.
- Traveling to the parent subnet (moving up the hierarchy):
  Approach #1: Right-click anywhere in the project workspace and select **Go to Parent Subnet** from the pop-up menu.
  Approach #2: Click on the *Go to Parent Subnet* icon on the toolbar that looks like ⊙.
  Approach #3: From the main pull-down menu, select **View > Subnets > Go to Parent Subnet**.
  Approach #4: Press Ctrl+0 (number zero).

For example, consider the network topology shown in Figure 2.9. Figure 2.9a shows the top-level network topology in the scenario cloud_modeling of the project called IP as indicated by the window's title. The topology shown in Figure 2.9a consists of five subnetworks called client_site1, ...,

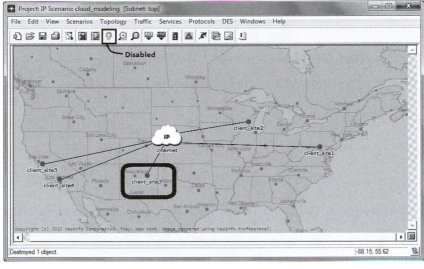

(a)

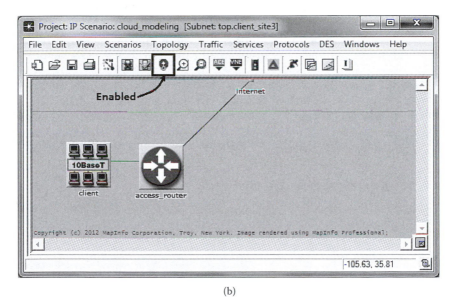

(b)

FIGURE 2.9 Example of a network topology with subnets: (a) the top-level network and (b) the client_site3 subnet.

client_site5, respectively, all connected to the node that models the Internet. Notice that in Figure 2.9a, the *Go to Parent Subnet* icon is disabled because there is no subnetwork above the top level. Figure 2.9b illustrates the client_site3 subnet of the same scenario as also pointed out by that window's title. Notice that in the right image, the *Go to Parent Subnet* icon is enabled. Clicking that icon in the project workspace of the `client_site3` subnet will move you back to the parent subnet, which is the top-level subnet shown in Figure 2.9a.

2.8.3 CREATING A NETWORK TOPOLOGY WITH SUBNETWORKS

- At the starting network level, add node, link, and utility objects as needed.
- If the current network level contains subnetworks within its structure, then add a subnet object for each subnetwork.
- Build the network topology for each of the subnet objects:
  - Enter the subnet.
  - Create the network topology within the subnet by dragging and dropping objects from the **Object Palette** (Section 2.3) or by using the **Rapid Configuration** tool (Section 2.5).
- To connect the objects in a parent subnet with the object within the child subnet:
  - Go to the parent subnet.
  - Add a link to connect the desired node in the parent subnet with the child subnet.
  - Connecting a link to a subnet opens the **Select Node** window, which lists all the nodes contained in the child subnet.
  - From the pull-down list in the **Select Node** window, choose the node to which you would like to connect the link and click the **OK** button.

To connect a node within a subnet with another node in the parent subnet, you need to move to the parent subnet first and only then can you add the desired link. Similarly, to connect nodes in two child subnets, you need to move to the common top-level parent subnet first and only then you can add a link to connect the nodes in the corresponding child subnets.

2.8.4 MOVING OBJECTS BETWEEN SUBNETS

To move already existing objects into a different subnet located at the same hierarchical level:

- Add a subnet icon (if not already present).
- Select the objects to be placed into the new subnet.
- Right-click on the subnet icon and select **Move Selected Nodes into Subnet** from the pop-up menu.

To move objects from the child subnet into the parent subnet:

- Select the objects to be moved.
- From the main menu, select **Topology > Subnets > Move Selected Nodes into Parent Subnet**.

To move already existing objects around the subnet hierarchy:

- Select the objects to be moved.

- From the main menu, select **Topology > Subnets > Move Selected Nodes into Specific Subnet**.
- From the **Choose Subnet** window, select the subnetwork to which you want to move the nodes and click **OK**.

You should be careful when moving objects around the subnet hierarchy because the **Project Editor** may place the objects one on top of another or into locations that are not readily visible. To view the moved objects, you may need to zoom in or zoom out within the project workspace. Additionally, left-clicking on the icon in the bottom right corner of the **Project Editor**'s window brings up a small pane that displays the whole project workspace and locations of all the nodes placed within it. Another left-click on the same icon makes this pane go away. There are other advanced subnet features available in OPNET, but they are beyond the scope of this book. More information on these advanced features is available in the OPNET documentation.

2.9   CREATING TOPOLOGY ANNOTATION

OPNET products include an annotation tool that helps to better organize the network topology and provides various visual and textual hints. The annotation tool allows you to label the objects in the network; draw lines, circles, and rectangles using different colors; and fill drawn objects with colors of your choice.

As shown in Figure 2.10, the **Annotation Palette** consists of the following five objects, each of which can be placed into the project workspace:

- *Box and ellipse objects* allow you to draw rectangular and elliptical shapes in the project workspace. These shapes can be designed to have different color, size and placement, fill color, rotation, and other characteristics.
- *Line objects* are used to draw lines with various attributes such as color, position, arrowhead at the start or the end of the line, and line style, for example, solid or dashed.
- *Text objects* allow you to add textual descriptions of the items in the project workspace. Their attributes include font size, text color, position, and other characteristics.
- *Icon objects* allow you to place any of OPNET's icons into the **Project Editor**.

While most operations for working with the **Annotation Palette** are self-explanatory, Sections 2.9.1 through 2.9.3 provide a brief description of the key

FIGURE 2.10   Annotation Palette.

Annotation Palette operations. Figure L2.1 in Laboratory Assignment #2 shows an example of a network topology with annotation objects.

2.9.1 Adding Annotation Palette Object to the Project Workspace

- Open **Annotation Palette** by selecting **Topology > Open Annotation Palette** from the pull-down menu in the **Project Editor** window.
- Click on the annotation object to be added:
  - For box and ellipse objects, drag the object's outline in the project workspace.
  - For line objects, left-click in the project workspace where you would like your line to start. Next, left-click to end the line at a point and start another line segment at the same point, double-left-click to end the line, or right-click to cancel drawing the line. You can add as many line segments as you need by left-clicking in the workspace multiple times before ending the operation with a double-left-click.
  - Starting with Version 14.0, when adding a text object, OPNET automatically opens the editor for typing the text. Once the text is typed in, close the window by selecting **File > Commit** from the editor's pull-down menu (or by clicking on the button in the top right corner to close the window). Next, drag the outline of the text object to the location within the project workspace where the text object is to be added. Older versions of OPNET may first ask you to specify the object outline and only then will allow you to type in the text.
  - For icon objects, simply left-click in the project workspace where you would like to place the icon.
- Repeat the operation multiple times if you desire to add more objects of the same type.
- Right-click to end the operation.

2.9.2 Modifying Attributes of Annotation Palette Objects

To modify attributes of any **Annotation Palette** object in the project workspace, right-click on the object of interest and select **Edit Attributes** from the pop-up menu.

- Frequently used attributes for *box* and *ellipse* objects are as follows:
  - **Color**—specifies the line color of the object.
  - **Fill**—specifies the fill color of the object.
  - **Width/Height**—specifies the dimensions of the object.
  - **Rotation**—allows you to rotate the object by the specified number of degrees.
- Frequently used attributes for *line* objects are as follows:
  - **Color**—specifies the color of the line.
  - **Drawing style**—allows you to smooth the line segments by setting the value of this attribute to `spline`.

- **Line style**—draws the object as a solid or as a dashed line.
  - **Head arrow/tail arrow**—adds an arrow at the start or at the end of the line.
- Frequently used attributes for *text* objects include the following:
  - **Color**—specifies the color of the text.
  - **Font**—allows you to specify the size of the text.
  - **Background color**—sets the fill color of the box within which the text is placed.
  - **Rotation**—allows you to rotate the object by the specified number of degrees.
- Frequently used attributes for *icon* objects are as follows:
  - **Icon name**—specifies the icon displayed in the project workspace. You cannot type in the name directly, but if you click on the value of this attribute, a window is opened, which allows you to select the icon of your choice.

2.9.3   SHOWING/HIDING ANNOTATION PALETTE OBJECTS IN THE PROJECT WORKSPACE

To show/hide all annotation objects from the **Project Editor**'s pull-down menu, select **View > Annotations > Show in Subnet**. This toggles the annotation objects, alternately hiding and showing them.

2.10   REMOVING NODE CLUTTER

OPNET provides several features that allow removing clutter and overlapping in network topologies that contain a large number of icons. The following list describes each feature and provides instructions for using it:

- **Automatic Icon Scaling** adjusts the size of the icons automatically when icons are placed too close to one another and their images start to overlap. To enable/disable this feature, from the **Project Editor**'s pull-down menu, select **View > Layout > Automatic Icon Scaling**. Automatic icon scaling can also be set through the **Preferences Editor** (see Section 1.2) by changing the value of the preference named **Disable Icon**.
- **Autosizing** automatically adjusts the size of the icons in the workspace. This feature can adversely influence the simulation performance and thus is automatically turned off when the number of icons in the scenario exceeds a certain threshold value. This threshold can be changed through the **Preferences Editor** by modifying the value of the preference named **Element Count Threshold for Aggressive Icon Autosizing**. The default value of this preference is 500.
- **Automatic Label Placement** is another feature that reduces the clutter of the created network topology by automatically placing the object labels so that they do not cover one another. To enable this feature, select **View >**

Layout > Automatic Label Placement. This feature can also be controlled through the **Preferences Editor** (Section 1.2) by changing the values of the preferences with the following tags:

- **title_autoplacing.try_small_font**—uses smaller label font to reduce overlapping.
- **title_autoplacing.directions**—places the label at a different location (e.g., top, bottom, left, or right) in an attempt to reduce overlapping.
- **title_autoplacing.disable**—controls if automatic label placement is enabled.

- **Interactive Icon Scaling** allows you to manually adjust the size of the icons in the project workspace. If you do not select any object in the workspace, then all icons in the current scenario will be affected; otherwise, the icon scaling is applied only to selected objects. To use this feature, select **View > Layout > Scale Node Icons Interactively**.

3 Configuring Network Topology

3.1 INTRODUCTION

Building a network topology as described in Chapter 2 is only the first and usually one of the simplest steps in developing a simulation study. The subsequent steps include such tasks as configuring individual links and network devices, setting up and deploying protocols and applications, and defining the profiles of users in the network. In OPNET, every network element is represented as an object. Characteristics or properties of the network elements are specified through object **attributes**. Configuration of network devices is performed by modifying the attribute values of the corresponding objects in the simulation model. The accuracy of a simulation study depends not only on how precisely the placement of various nodes and their connecting links matches the real-life network but also on correct configuration of the devices in the created network topology. The closer the OPNET-created representation of the simulated network is to the real network, the more accurate the simulation results will be. OPNET provides device and link models to support most of the common networking protocols. However, you do not need to specify the values of all attributes of all the protocols and technologies used in your simulation study, as each protocol is usually preconfigured with the most frequently used default values. Most of the time, you only need to change some of those default values.

For example, if you want to develop a simulation study that examines the influence of TCP's Maximum Segment Size (MSS) value on the performance of a File Transfer Protocol (FTP) application, then you would need to create a topology that represents the network to be studied, configure and deploy the FTP application, and only change the default values of the MSS parameter in the TCP protocol configuration on the selected nodes in the network. On the other hand, if you wish to compare the performance of various routing protocols, then the simulation study may require several scenarios, each of which utilizes a different routing protocol. In this case, you may configure the simulation to have one scenario deploying Routing Information Protocol (RIP), a second scenario with the Intermediate System to Intermediate System (IS-IS) protocol, a third scenario with Open Shortest Path First (OSPF), and so on. In this situation, although the routing protocol changes from one scenario to another, you might not need to change the device models used in the simulation. Such situations are quite common since device models usually support various protocols that can be used interchangeably (e.g., routing protocols: RIP and OSPF; transport protocols: TCP and User Datagram Protocol [UDP]). As a result, you can often change the protocol used in a scenario simply by modifying the device configuration.

OPNET software is very complex and it frequently allows different alternate procedures for accomplishing the same task. Specifying protocols and configuring their attributes can often be performed using different sets of instructions. This chapter primarily concentrates on instructions for configuring network elements by changing their attribute values. Specifically, this chapter describes the different types of object attributes, explains how to set attribute values in the various network elements in a variety of ways, and offers examples of some commonly used device attributes. Protocol-specific alternate procedures for configuration are described in later chapters that deal with each protocol layer.

3.2   OBJECT ATTRIBUTES

The simplest way to change the property settings of an object is to modify the values of its attributes. This section describes the types of attributes that an object may have and how to access these attributes in any particular object.

3.2.1   TYPES OF ATTRIBUTES

Generally, there are three types of attributes that any object may possess:

- **Basic** or **noncompound attributes** represent a single property of the corresponding object. A noncompound attribute has no subattributes and has a single value assigned to it. Examples of noncompound attributes are buffer size represented as an integer value, IP address represented as a string, application start time represented as a double precision floating point number, and so on.
- **Compound attributes** group subattributes based on some common characteristic. Compound attributes contain one or more subattributes, which themselves could be either basic or compound. A compound attribute may also have a value assigned to it. The value of a compound attribute is a collection of the values of all its subattributes. A compound attribute value usually has a name associated with it. For example, the attribute **TCP Parameters** consists of several subattributes such as receive buffer size, MSS, maximum acknowledgment (ACK) delay, availability of Fast Retransmit feature, availability of Fast Recovery feature, and so on. A set of values assigned to all these subattributes represents the value of the compound attribute **TCP Parameters**. To simplify configuration of the TCP protocol, OPNET provides preconfigured values for the **TCP Parameters** attribute. Some of these values have names such as Default, Tahoe, Reno, New Reno, SACK, NT (3.5/4.0), and so on, which correspond to a TCP version/flavor whose selection automatically determines the subattribute value settings. Alternatively, you can choose to specify the subattribute values individually.
- **Grouping attributes** or simply **attribute groups** are similar to compound attributes, except that grouping attributes cannot have values associated with them. The only responsibility of the **attribute groups** is to organize into a single category those attributes that have a common purpose or that

belong to the same protocol. For example, the attribute group **IP Routing Protocols** consists of several compound subattributes, each of which represents configuration of one of the IP routing protocols such as Border Gateway Protocol (BGP), Interior Gateway Routing Protocol (IGRP), Enhanced Interior Gateway Routing Protocol (EIGRP), IS-IS, and so on. The attribute group **IP Routing Protocols** cannot have values assigned to it.

3.2.2 THE OBJECT POP-UP MENU

Any object placed in the project palette including networking devices, links, demands, domain objects, utility nodes, various annotation items, and so on has an **Object Pop-up Menu** associated with it. Right-clicking on the object opens this menu. It contains a set of options or actions that are frequently performed on this object, including an option called **Edit Attributes,** which allows you to modify the object's attribute values.

The list of available actions in the **Object Pop-up Menu** varies depending on the object's type. For example, one of the operations for link objects is redefining the link's path, but this operation is not applicable to device or utility nodes. The following list provides a description of the most commonly used operations available in the **Object Pop-up Menu** (see Figure 3.1):

- **Edit Attributes**—this is one of the key operations that open the object's **Edit Attributes** dialog box (default view), which allows you to view and modify the object's primary attributes (i.e., properties). The default view in the **Edit Attributes** dialog box only displays the primary or frequently used attributes. However, the dialog box contains a checkbox called *Advanced,* which allows you to switch between advanced and default views of the object's attributes.
- **Edit Attributes (Advanced)**—opens the object's **Edit Attributes** dialog box with the advanced view enabled. The advanced view contains both primary and advanced attributes. Advanced attributes are not available in the default view of the **Edit Attributes** dialog box. Unselecting the *Advanced* checkbox in the advanced view will switch you back to the default view.
- **Set Name**—this operation allows you to change the object's name, making the simulation study more user-friendly. The object's name can also be changed by changing the value of the **Name** attribute in the **Edit Attributes** dialog box.
- **View <Object> Description**—opens a window that contains a general description of the object's model. The object could be a node, link, demand, and so on.
- **Edit <Object> Model**—this option is only available in OPNET Modeler. It opens the node view of the object and allows you to edit the node's model.
- **Select**—selects or highlights the current object. The same effect can be achieved by simply left-clicking on the object.
- **Select Similar <Objects>**—selects all objects of the same type (namely, those that have the same simulation model). This option is useful for changing the attribute values in multiple objects simultaneously (see Section 3.4.2).

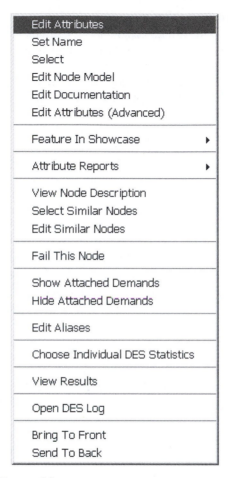

Edit Attributes
Set Name
Select
Edit Node Model
Edit Documentation
Edit Attributes (Advanced)
Feature In Showcase ▶
Attribute Reports ▶
View Node Description
Select Similar Nodes
Edit Similar Nodes
Fail This Node
Show Attached Demands
Hide Attached Demands
Edit Aliases
Choose Individual DES Statistics
View Results
Open DES Log
Bring To Front
Send To Back

FIGURE 3.1   Object Pop-up Menu.

- **Fail This <Object>**—marks the current object as failed so that it becomes unavailable during the simulation. Objects that could be marked as failed include nodes, links, and demands. OPNET does not allow failing of subnet objects.
- **Recover This <Object>**—if the current object was previously marked as failed, then this option recovers the object by making it operational during the simulation. Objects that could be recovered include nodes, links, and demands. OPNET does not allow recovering subnet objects.

If you right-click in a project palette location that contains multiple objects, then OPNET provides a different pop-up menu that lists all the objects placed in that location. You can then select the object to be modified. Figure 3.2 illustrates such a situation, where right-clicking on a spot that contains four demand objects brings up a menu listing these demands. The figure then shows the first demand in the list being selected for editing.

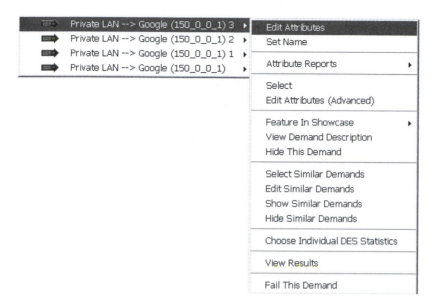

FIGURE 3.2   Selecting an object from the **Multiple Object Pop-up Menu**.

3.3   THE EDIT ATTRIBUTES DIALOG BOX

The **Edit Attributes** dialog box, accessed from the **Object Pop-up Menu**, lists attributes that describe OPNET model properties including configuration parameters for various networking protocols and technologies available within the object. The dialog box is organized in the form of a table with three columns:

- *Attribute Description*: The leftmost column of the dialog box is either empty or contains an icon, which is a circle with a question mark inside it. Clicking on this icon opens up a window with a brief description of the attribute. If the column does not contain such an icon for an attribute, then there is no description available for that attribute. Grouping attributes usually do not have any attribute description associated with them.
- *Attribute Name*: The central column contains attribute names. Each attribute may be any of the three types described in Section 3.2.1, namely, basic, compound, or grouping attributes. Attribute groups and compound attributes are shown in a tree view, in which leaf nodes are the basic attributes with values assigned to them, while the non-leaf nodes are compound or grouping attributes. Compound and grouping attributes can be expanded to view their subattributes or collapsed to have a more compact view of the object's attributes (see Section 3.3.2).
- *Attribute Value*: The rightmost column displays the attribute value. Usually, most objects have all of their attributes preconfigured with certain default values. However, you may change the default attribute values to more accurately reflect the system you wish to model.

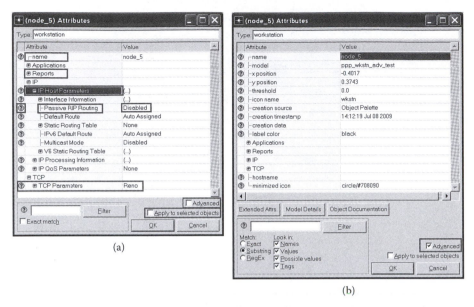

FIGURE 3.3   **Edit Attributes** dialog box: (a) default view and (b) advanced view.

Figure 3.3a illustrates an example view of the **Edit Attributes** dialog box. As the figure shows, at the top of the dialog box, there is a noncompound attribute **name** that defines this object's name. The attribute **Reports** is a grouping attribute and thus has no value assigned to it. The compound attribute **TCP Parameters**, on the contrary, has a preset value Reno assigned to it. This value indicates that the node is configured to support the Reno flavor of TCP.

Figure 3.3a also shows two checkboxes highlighted with a rectangle near the lower right corner. Selecting checkbox *Advanced* provides additional filtering options and makes all of the available attributes visible (Figure 3.3b). Selecting checkbox *Apply to selected objects* applies attribute changes to all the objects that have been selected. This is a convenient method to change the value of a particular attribute in multiple objects, but it requires that the attribute be set to the same value in all the objects. The objects must be preselected before opening the **Edit Attributes** dialog box. If for some reason the attribute value of a selected object cannot be changed, then OPNET will not change the value, but it will not provide any warning messages. It is up to you to verify that the changes have been applied to all the selected objects. Also, if you change any attribute that is part of a compound attribute, then the entire compound attribute value is replaced within all the selected objects, which may lead to undesired results.

3.3.1   ACCESSING ATTRIBUTE DESCRIPTION

Use the following steps to view a detailed description of an attribute:

- Right-click on the object of interest and select the **Edit Attributes** option to open the **Edit Attributes** dialog box.

- Browse through the attributes available in the dialog box to find the attribute of interest.
- Basic and compound attributes have an icon (in the form of a circle with a question mark inside it) located in the leftmost column of the **Edit Attributes** dialog box. Click on the icon to open a window with a description of the selected attribute.

3.3.2 WORKING WITH COMPOUND AND GROUPING ATTRIBUTES

Compound and grouping attributes that are not expanded in the **Edit Attributes** dialog box are marked by an icon with the plus sign inside it. To expand such an attribute, use one of the following methods:

Approach #1: Click on the plus sign or the small arrow to the left of the attribute's name. This causes the attribute to expand and changes the icon so that it now contains a minus sign.

Approach #2: Left-click on the value field of a compound attribute to view and/ or select one of the available configuration options. Selecting option Edit... will expand the current attribute by opening a pop-up window in which you can configure all of the subattributes of the current compound attribute. In some compound attributes, left-clicking on the value field will directly open the pop-up window with the subattribute values. This approach is not applicable to grouping attributes since they do not have a value.

Approach #3: Right-click on a compound or grouping attribute's value field and select the option **Expand Row**. The same can be done by right-clicking on the attribute's name field as well. Generally, right-clicking on a compound attribute provides the following options (for a grouping attribute, only the **Expand Row** option is available):

- **Expand Row**—expands the compound attribute.
- **Promote Attribute To Higher Level**—allows you to configure this attribute at the simulation level. Refer to Section 3.5 for more details.
- **View Attribute**—opens a read-only pop-up window with the attribute's configuration.

For example, as Figure 3.3 shows, the grouping attribute **IP**, when expanded, contains three additional compound subattributes: **IP Host Parameters**, **IP Processing Information**, and **IP QoS Parameters**. The compound attribute **IP Host Parameters** is also expanded and it contains several compound and basic subattributes. **Passive RIP Routing** is a subattribute of **IP Host Parameters**. This is a basic attribute with its value set to Disabled.

3.3.3 ATTRIBUTES WITH MULTIPLE INSTANCES

Certain objects may contain multiple instances of the same attribute. For example, a server may support multiple application services, each of which may need to be configured separately. In such a situation, an attribute in question may have a child

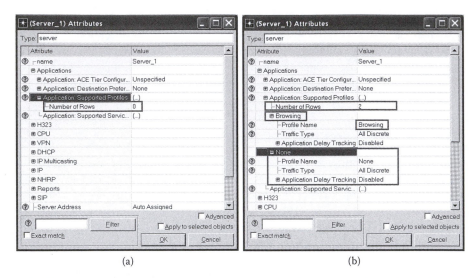

(a)   (b)

FIGURE 3.4   Attribute with multiple instances: (a) initially when **Number of Rows** is 0 and (b) when **Number of Rows** is changed to 2.

(i.e., subattribute) called **Number of Rows**. The value of the attribute **Number of Rows** corresponds to the number of instances of the parent attribute to be created. By default, the value of **Number of Rows** for most attributes is 0. If that value is changed, then a new set of attributes will be added at the same level as the **Number of Rows** attribute. Figure 3.4 shows an attribute that requires multiple instances.

As Figure 3.4a shows, initially, the attribute **Number of Rows** is set to 0. Two new compound attributes appear once the value of the attribute is changed to 2 (Figure 3.4b). By default, the name of the added attributes is **None**, and it is identical to the value of the **Profile Name** attribute. After the value of the **Profile Name** attribute is changed to Browsing, the name of the added compound attribute also becomes **Browsing**. When dealing with multiple instances of the same attribute, it is common for a newly added attribute instance to change its name to the value of one of its subattributes.

3.3.4   FILTERING ATTRIBUTES

The **Edit Attributes** dialog box also provides a feature for filtering attributes based on selection criteria. The filtering feature differs when using the regular and advanced versions of the **Edit Attributes** dialog box. With the regular **Edit Attributes**, you can only specify filtering criteria as a substring that will be matched against the attribute's name, value, or a tag. Only attributes with a partial or complete match of the search criteria will remain visible in the **Edit Attributes** dialog box.

The advanced **Edit Attributes** dialog box filtering feature allows you to specify how the matching criteria are defined (e.g., exact string match, partial string match, or regular expression) and which attribute properties should be searched for a match (e.g., the attribute's name, value, possible values, tag, or any combination of the above fields). Recall from Chapter 1 that a tag is the "technical" keyword associated with

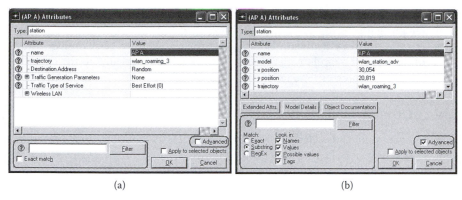

(a)   (b)

FIGURE 3.5   Filtering option in the **Edit Attributes** dialog box: (a) regular and (b) advanced.

the model's attribute. OPNET documentation recommends using protocol names, interface names, or attribute values as search criteria for the most efficient filtering of the attributes. Figure 3.5 shows the appearance of the filtering feature in regular and advanced **Edit Attributes** dialog boxes.

3.3.5   FINDING ATTRIBUTES USING REGULAR EDIT ATTRIBUTES FILTERING FEATURE

- Right-click on the object of interest and select the **Edit Attributes** option to open the **Edit Attributes** dialog box.
- Type in the search criteria in the textbox next to the **Filter** button.
- Click the **Filter** button to apply the filtering criteria.
- To remove the filtering selection, clear the textbox and click on the **Filter** button again.
- For additional information regarding the filtering feature, click on the icon in the form of a circle with a question mark inside it, located to the left of the filter's search criteria textbox.

3.3.6   FINDING ATTRIBUTES USING ADVANCED EDIT ATTRIBUTES FILTERING FEATURE

- Right-click on the object of interest and select the **Edit Attributes (Advanced)** option to open the advanced view of the **Edit Attributes** dialog box.
- If the **Edit Attributes** dialog box is in the regular view, then click on the *Advanced* checkbox to switch to the advanced view.
- If needed, change the matching criteria by clicking on one of the following options of the *Match* radio-button: exact, substring, or RegEx.
- If needed, change the type of the attribute properties to be searched, by selecting or unselecting the checkboxes under the title *Look in:* (e.g., Names, Values, Possible Values, or Tags).
- Type in the search criteria in the textbox next to the **Filter** button.
- Click on the **Filter** button to apply the filtering criteria.

- To remove the filtering selection, clear the textbox and click the **Filter** button again.
- For additional information regarding the filtering feature, click on the icon in the form of a circle with a question mark inside it, located to the left of the filter's search criteria textbox.

3.4   CONFIGURING OBJECT PROPERTIES

The most common method for configuring an object is to change its attribute values. This section describes how to change attribute values for a single object as well as convenient methods for doing the same for multiple objects simultaneously.

3.4.1   CHANGING ATTRIBUTE VALUES OF A SINGLE OBJECT

- Right-click on the object of interest.
- From the object pop-up menu, select **Edit Attributes** or **Edit Attributes (Advanced)**.
- You may check the *Advanced* checkbox to make all of the object's attributes visible.
- Browse the **Edit Attributes** dialog box (Section 3.3) to locate the attribute of interest.
- You may want to expand compound attributes (Section 3.3.2) to reach the subattributes of interest.
- To change the value of a basic attribute, left-click on the attribute's value field and perform one of the following:
  - Type in the new value.
  - Select a value from the provided pull-down menu.
  - Select the Edit... option from the pull-down menu and type in a new value.
- Once all the attribute values have been changed, click **OK** to close the window and save the changes.

3.4.2   CHANGING ATTRIBUTE VALUES OF MULTIPLE OBJECTS

Often, it is necessary to change the same attribute value in multiple objects. Consider a situation where you wish to simulate a network topology with 50 workstations running the Reno flavor of the TCP protocol. Clearly, configuring 50 workstations individually would take too long. Fortunately, OPNET provides several methods for configuring multiple objects simultaneously.

- Select the objects to be modified. All these objects do not have to use the same model, but they should at least have one common attribute. To select objects in the network topology, perform one of the following actions:
  Approach #1: Left-click on the project workspace and then drag the mouse over the area where the objects of interest reside.
  Approach #2: While holding the CTRL key, left-click on the objects that you would like to select.

Approach #3: Right-click on one of the objects to be selected and then choose the option **Select Similar Nodes** from the **Object Pop-up Menu** (Section 3.2.2). This operation will select all objects in the current scenario that have the same model. It should be noted that even objects that are located on a different level of the network hierarchy (e.g., subnet), but which have the same model name will be selected as a result of this operation.

Approach #4: Choose **Edit > Select All in Subnet** in the top-level pull-down menu. This will select all objects in the current subnet.

Approach #5: Choose **Edit > Select Objects...** in the top-level pull-down menu. This will open a **Define Selection** window that allows you to find and select nodes in the whole scenario, including all subnets, based on certain criteria such as object type or availability of an attribute.

- Having selected all the desired objects, now right-click on any of the selected nodes and choose the **Edit Attributes** option.
- Check the *Apply Changes to Selected Objects* checkbox so that changes performed on this node will propagate to all the nodes that were selected in the previous step. This is an important step that is often forgotten when you are in a hurry. Forgetting to check this checkbox will cause the changes to occur only in the one node that was right-clicked in the previous step and not in all the other preselected objects.
- Perform the necessary changes of the object's properties, following the steps given in Section 3.4.1.
- Click the **OK** button when finished. The changes performed on this node will be applied to all the selected nodes. OPNET may prompt you to confirm if you want to perform the changes on all the selected objects.

There are a few additional points of caution about editing attributes of multiple objects. If you have selected multiple objects and you change the value of any sub-attribute of a compound attribute, then the entire compound attribute will be copied to all the selected objects. This may lead to undesired effects. Also, if OPNET determines that the requested changes cannot be applied to one or more of the selected objects, then no changes will be made; however, you will not be notified about such a failure.

3.4.3   EDITING SELECTED OBJECTS

OPNET provides another option for changing attribute values of objects by editing selected objects in the network:

- First select all the desired objects whose attributes are to be modified by using any of the five approaches described in the first step of Section 3.4.2. These objects do not have to be of the same model.
- Right-click in the project workspace and, from the pop-up menu, select the **Edit Selected Objects** option. The **Objects Attributes** dialog box window that appears contains a table with one row per selected object. The columns in the table contain all the attributes available in all the selected objects. Notice that the attributes with the same name appear in the same column.

The current values of the attributes for each object are also shown. If a particular object does not have a certain attribute, then its value is not shown and an empty (blue) box appears instead.

- You can change the values of the attributes for any of the objects in this table to any desired values.
- After all the necessary attributes have been updated, click **OK** to close the window and save the changes.

3.4.4 EDITING SIMILAR NODES OR LINKS

The last method we describe can be used to change attributes of similar objects only. This method is only available for node and link objects.

- Right-click on one of the objects of interest (it must be either a node or a link) and, from the object pop-up menu, choose **Edit Similar Nodes** or **Edit Similar Links**, as the case may be.
- This action also opens the **Objects Attributes** dialog box window, but it will contain only similar objects, that is, objects that have the same model. All objects with the same model in the current scenario will be displayed.
- Change the values of the attributes as desired and click **OK** to save the changes.

3.4.5 THE MODEL ATTRIBUTE OF AN OBJECT

The most important object property is defined through an advanced attribute called **model**, which defines the network element that is being modeled in the current object. OPNET allows the value of this advanced attribute to be changed. However, one must be extremely careful when changing the **model** attribute value because such a change may lead to inconsistencies in the created network system. For example, consider a situation in which a properly configured network topology, *Email_client* node, has its **model** attribute changed from Ethernet workstation (*ethernet_wkstn*) to point-to-point protocol workstation (*ppp_wkstn*). Such a modification will result in an incorrect system configuration because the *ethernet_wkstn* node model requires an Ethernet link, while *ppp_wkstn* only supports a SLIP interface.

On the other hand, in certain situations, changing an object's model can be extremely helpful. If you recall, OPNET objects often have various levels of derivation. Models with *int* modifier or no modifier at all may have certain attributes hidden, while models with *adv* modifier have all of the model attributes available to the user. Thus, changing an object's model from the one with *int* modifier or no modifier at all to the same model name but with *adv* modifier may provide access to additional attributes and may allow the user to configure the simulated system more accurately.

The following steps are used to change an object's **model** attribute:

- Right-click on the object of interest and select the **Edit Attributes (Advanced)** option to open the advanced view of the **Edit Attributes** dialog box.
- If the **Edit Attributes** dialog box is in regular view, then click on the *Advanced* checkbox to switch to advanced view.

- Click on the value field of the attribute called **model** (second attribute from the top of the table).
- OPNET may provide a warning message cautioning you that changing of the model attribute may render some attributes invalid or inconsistent. Click the **Yes** button if you are set on changing the **model** attribute.
- A scroll-down list of available models will appear. Scroll down the list and choose a new model for the current object.
- Select the `Edit...` option if the model of interest is not available in the provided list.
    - Choosing the `Edit...` option opens a window that is very similar to the **Object Palette** window (Section 2.3) and contains all the models available in OPNET.
    - Browse through the list of the models and choose a new model for the current object.
    - Click the **OK** button to close the window.
- Click the **OK** button to accept the changes and close the **Edit Attributes** dialog box.
- It is strongly recommended that you verify the network configuration for consistency after changing the **model** attribute (see Section 2.6.2 on how to verify links).

3.5   PROMOTING OBJECT ATTRIBUTES

OPNET software relies on a model hierarchy that contains several levels of abstraction. Each level of the hierarchy usually has a special editor associated with it. The lowest level of the OPNET model hierarchy uses *process models* to specify various protocols and network technologies. Process models are implemented using a variation of the C++ programming language called *Proto-C* along with extended finite state machine transition diagrams. The **Process Editor** provides functionality to develop and compile process models and the external code files that are often associated with them. Thus, the lowest level of the OPNET model hierarchy, which we call the *process model level*, allows you to model individual network protocols and technologies with the **Process Editor**.

The next level in the OPNET model hierarchy, which we call the *node level*, defines individual network devices using the **Node Editor**. At the node level, a network device is modeled as a collection of one or more modules joined via packet streams or statistical wires, which provide the means for intermodule communication. Generally, every module has a process model associated with it, which defines the operation of that module. Thus, the **Node Editor** provides access to the node level of the OPNET model hierarchy where you can model individual network devices.

The **Project Editor** belongs to the next level of the OPNET model hierarchy, which we call the *network level*. At this level, you can specify the topology of the simulated system and configure the various components of that system. The **Project Editor** allows you to organize node models, which have been created at the previous level with the **Node Editor**, into a network topology, and then change attributes of the objects in the created topology to accurately represent the network being modeled in the simulation study.

The final level of the OPNET model hierarchy deals with the configuration of the simulation study or the whole simulated system. This level of hierarchy, which we call the *simulation level*, does not have a separate editor associated with it. Usually, the **Project Editor** provides the means for specifying such simulation configuration parameters as the duration of simulation, statistics to be collected during the simulation run, seed for the random number generator, and other attributes that influence the whole simulation. In addition to the **Process Editor**, **Node Editor**, and **Project Editor**, OPNET contains other editing utilities such as **Link Editor**, **Path Editor**, **Demand Editor**, **Packet Format Editor**, **ICI Editor**, **Distribution (PDF) Editor**, and **Probe Editor**. However, these editors are less frequently used and are not described in this book. Figure 3.6 illustrates the OPNET model hierarchy.

The availability of these editors is directly dependent on which OPNET product you are using. Specifically, OPNET Modeler incorporates all of the above-mentioned editors, while IT Guru provides access only to **Project Editor**, **Probe Editor**, **Distribution (PDF) Editor**, and a few others. IT Guru does not provide facilities for changing process models and creating custom network devices using the **Process Editor** and **Node Editor** respectively. Nevertheless, most object attributes are defined and used within the process models at the lowest level of the model hierarchy. The values of these attributes can be directly set at the process model and node levels. However, having the values of the attributes modifiable only at the lowest levels of the model hierarchy, or having model attributes preset for every node model, would significantly limit the flexibility of the OPNET products. For this reason, OPNET provides a mechanism that allows you to set the attribute values at the network and simulation levels using the **Project Editor**. The process of configuring model attributes so that their values can be specified at a higher level of the model hierarchy is called *promotion*.

The attribute values can be promoted to the node, to the network, and even to the simulation level. The attributes can also be promoted from a child to its parent subnetwork. Attributes promoted to the simulation level are considered to be attributes of the whole simulation scenario and thus are set at the simulation run-time.

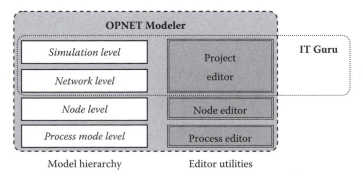

Model hierarchy   Editor utilities

FIGURE 3.6 OPNET model hierarchy.

Promoting attributes to the simulation level is a very convenient feature because it allows for:

- Changing commonly used parameters in a single place instead of clicking on various objects in the network topology.
- Setting up automated simulation runs with different attribute values.
- Iterating through a range of possible attribute values instead of duplicating scenarios and changing values of certain attributes.
- Aggregating promoted attributes over multiple objects, which is similar to setting attribute values of multiple objects simultaneously.

OPNET does not allow promoting attributes that do not have a value associated with them, and therefore, grouping attributes cannot be promoted. Additionally, advanced attributes and the attribute **name** cannot be promoted. Figure 3.7 illustrates attribute promotion.

Promoted attribute values can be set at the simulation level or at the parent subnet level.* Generally, as an attribute is promoted to different levels of the model hierarchy, its name changes according to the following naming conventions:

```
network_type.subnet01.subnet02. ... subnet0n.node_name.attribute_name
```

For example, consider a simulated *Campus Network* that contains subnet *sub_top* with the child subnet *sub01*. Also assume that node *Admin01* with attribute **TCP Parameters** is located within subnetwork *sub01*. Table 3.1 shows how the name of the attribute **TCP Parameters** changes as it is promoted to different levels of the model hierarchy.

At the simulation level, promoted attributes can be configured with one or more values, or even have their values specified in the form of a range. In addition, OPNET provides a wildcard feature, which allows you to set the values of promoted attributes that are common to multiple objects in the network. The rest of this section provides instructions for promoting and unpromoting attributes, configuring values of promoted attributes at various levels of the model hierarchy, and using the wildcard notation.

3.5.1 PROMOTING AN OBJECT ATTRIBUTE

- Right-click on the object of interest and choose the **Edit Attributes** option.
- Right-click on the value field of the attribute of interest and select the option **Promote Attribute To Higher Level** (Figure 3.7a). The value of the attribute will change to `promoted` (Figure 3.7b).
- You can promote as many attributes within the object as needed.
- Click the **OK** button when finished.

---

* In OPNET Versions 13.5 and 15.0, promoted attributes are not visible at the parent subnet level until they have been accessed or viewed through simulation-wide object attribute feature.

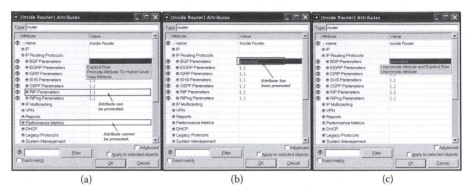

FIGURE 3.7   Attribute promotion: (a) how to promote an attribute to the next higher level, (b) an attribute that has been promoted, and (c) how to unpromote an attribute.

TABLE 3.1

Example of Attribute Naming Convention

Levels of Model Hierarchy	Attribute's Name
Admin01	*TCP Parameters*
sub01	*Admin01.TCP Parameters*
sub_top	*sub01.Admin01.TCP Parameters*
simulation level	*Campus Network.sub_top.sub01.Admin01.TCP Parameters*

3.5.2 UNPROMOTING AN OBJECT ATTRIBUTE

- Right-click on the object of interest and choose the **Edit Attributes** option.
- There are several ways to unpromote an attribute value:
  Approach #1: Right-click on the value field of the promoted attribute of interest and select **Unpromote Attribute** or **Unpromote Attribute and Expand Row** (only available for compound attributes) (Figure 3.7c).
  Approach #2: Left-click on the value field of the promoted attribute, which will allow you to set a value for the attribute, automatically causing it to be unpromoted.
- You can unpromote as many attributes within the object as needed.
- Click the **OK** button when finished.

3.5.3 CONFIGURING PROMOTED OBJECT ATTRIBUTES AT THE SIMULATION LEVEL

- Click the *Run* icon ![run] or press Ctrl+Shift+R to open the **Configure/Run DES** window.
- When the **Configure/Run DES** window appears, you may need to switch to detailed view by clicking the **Detailed...** button or pressing Alt+D.
- When the detailed view **Configure/Run DES** window appears, expand the **Inputs** control page that appears in the tree view in the left-hand pane

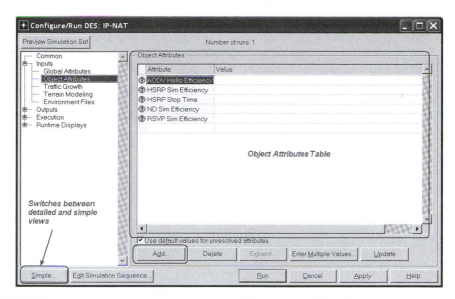

FIGURE 3.8 *Object Attributes Table* of the **Configure/Run DES** window.

and then click on the **Object Attributes** option. The *Object Attributes Table* is now displayed in the right-hand pane. Figure 3.8 illustrates this step. In some older versions of OPNET, commonly used attributes such as **AODV Hello Efficiency** may not appear in the *Object Attributes Table* as shown in Figure 3.8. Clicking the **Update** button updates the table to display only those attributes that are available and have been added to the *Object Attributes Table*.

- Add promoted attributes:
  - Click the **Add** button to open the **Add Attributes** window, which allows you to add promoted attributes into the *Object Attributes Table*.
  - The **Add Attributes** window lists all the attributes that have been promoted to the simulation level.
  - Click on the field in the *Add?* column to mark the attribute as to be added into the *Object Attributes Table*.
  - Click the **OK** button to add the marked attributes to the *Object Attributes Table*.

3.5.4 SPECIFYING VALUES FOR PROMOTED ATTRIBUTES AT THE SIMULATION LEVEL

Once the promoted attributes have been placed in the *Object Attributes Table*, you can specify values for them.

- Specifying a single value for a promoted attribute in the *Object Attributes Table* is no different than specifying an attribute value through the **Edit**

Attributes dialog box: left-clicking in the *Value* field provides a list of available preset values, while double-left-clicking, two consecutive left-clicks or choosing the Edit... option, allows you to specify a custom value. When editing a compound attribute, you may be provided with a separate window for specifying subattribute values.

- To specify multiple values for a single attribute:
  - Click on the attribute name to select it.
  - Click the button **Enter Multiple Values,** which will bring up another window for setting values of the selected attribute.
  - Multiple values can be set either explicitly by specifying one value per row or, for numeric values, by specifying the initial value in the *Value* column, the maximum value in the *Limit* column, and the incrementing step in the *Step* column.
  - Repeat these actions to set multiple values in other attributes.
  - Click the **OK** button to accept the chosen attribute values and close the window for setting multiple attribute values.

Figure 3.9 illustrates how the attribute **TCP Parameters[0].Receive Buffer** is configured to have multiple values. In the case shown in Figure 3.9, a total of 17 values have been specified for this attribute. The first three values are provided one value per row, while the other 14 values have been specified in the form of a range starting from 70,000 to 200,000 (inclusive) with the incremental step of 10,000. Also notice the button **Details**, which is circled in Figure 3.9. Clicking the **Details** button opens up a window that contains a description of the attribute being configured.

- Once the values of all promoted attributes have been set, click the **Apply** button to save the changes or click the **Run** button to execute the simulation. The **Configure/Run DES** window remains open if you click the **Apply** button. To close the **Configure/Run DES** window, click the **Cancel** button or click the *Close* icon in the top right corner of the window.

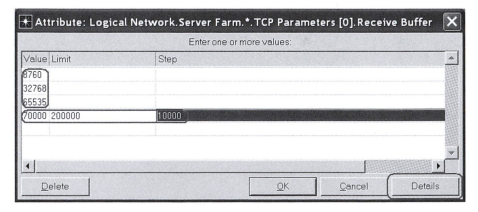

FIGURE 3.9 Setting multiple attribute values.

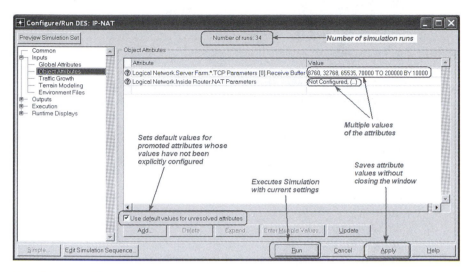

FIGURE 3.10   Final configuration of promoted attributes.

Figure 3.10 shows the final configuration of the promoted attributes. The top of the window shows the number of simulation runs required to collect data for the given attribute configuration. This value is 34 since there are 17 different values of the attribute **TCP Parameters[0].Receive Buffer** and two values for the attribute **NAT Parameters** (the values are separated by commas; the first value is Not Configured and the second value is represented by (...)). Thus, the total number of simulation runs corresponds to the number of possible combinations ($17 \times 2 = 34$) for choosing different sets of attribute values. Also notice a checkbox called *Use default values for unresolved attributes* located in the bottom section of the window just above the **Add...** button. Selecting this checkbox will force all the promoted attributes whose values have not yet been set to have default values assigned to them. This option is very useful because it ensures that all the attributes have set values. Running the simulation without assigned values for some of the attributes may lead to unpredictable and possibly incorrect results.

Note: In older versions of OPNET software, it is advisable to select the checkbox called *Save vector and environment files for each run in set*. Selecting this checkbox ensures that the results are saved for every simulation run. Starting from OPNET Version 13.5, the results of all simulation runs are automatically saved and this checkbox is not present in the **Configure/Run DES** window.

3.5.5   CONFIGURING PROMOTED ATTRIBUTES AT THE SUBNET LEVEL

Promoting attributes of objects within a subnet and then setting their values when you are in the parent subnet is a highly useful feature when you need to reconfigure attributes of objects in the subnet multiple times. With the help of this feature, instead of drilling down into the subnet and then changing attribute values one object at a time, you can promote the attributes and then change them in one place as needed. The following steps explain this procedure:

- Promote desired attributes in the object(s) within the current subnet.
- Move to the parent subnet (right-click in the project workspace and select **Go To Parent Subnet**).
- Right-click on the subnet icon and select the **Edit Attributes** option.
- The **Edit Attributes** dialog window that appears will contain a list of all promoted attributes that belong to objects located within the selected subnet.
- Edit the attribute values and press the **OK** button to save the changes and close the window.

3.5.6 USING THE WILDCARD OPTION FOR ASSIGNING VALUES TO MULTIPLE PROMOTED ATTRIBUTES

Consider an example where the simulated *Logical Network* contains a subnet *Server Farm* with three nodes in it named *Admin01*, *Admin02*, and *Admin03*. All of these nodes have promoted the attribute **TCP Parameters** to the simulation level. As a result, the following three attributes are available for configuration in the *Object Attributes Table* of the **Configure/Run DES** window:

- **Logical Network.Server Farm.Admin01.TCP Parameters**
- **Logical Network.Server Farm.Admin02.TCP Parameters**
- **Logical Network.Server Farm.Admin03.TCP Parameters**

The names of these promoted attributes are the same with the exception of the node name to which each attribute belongs. To simplify configuration of these attributes, you can employ the wildcard character * to represent the different node names within the common part of the attribute names:

- **Logical Network.Server Farm.*.TCP Parameters**

Thus, instead of having one attribute per object listed in the *Object Attributes Table* of the **Configure/Run DES** window, you can aggregate the promoted attributes that have the same name but belong to different objects. The result of such an aggregation is a single listing of an attribute using the wildcard character. This wildcard aggregation, however, applies to *promoted* attributes only. Attributes with the same name that have not been promoted are not influenced by the wildcard aggregation. Assigning a value to the wildcard aggregated attribute causes all of the promoted attributes with that name to have the same value assigned to them. You may also choose to narrow the scope of the aggregated attribute by replacing the wildcard in the attribute's name with a specific object or subnet name.

The following steps describe how to use the wildcard character for configuring the promoted attributes at the simulation level:

- Open the **Configure/Run DES** window and select the **Object Attributes** option.
- In the *Object Attributes Table*, click on the **Add** button to add the promoted attributes into the table.
- In the **Add Attribute** window, click on the attribute of interest.
- If the selected attribute is specified with the help of the wildcard character, then the button **Expand** will become visible.
  - Click the **Expand** button to open the **Expand Wildcard** window.
  - Click on the wildcard symbol and, from the pull-down menu that appears, select a specific object for which you would like the attribute to be configured. This will narrow the scope of the current attribute by adding the additional specific attribute to the list displayed in the **Add Attribute** window.
  - Click the **OK** button to save the changes and return back to the **Add Attribute** window.
- If the selected attribute is specified without the use of a wildcard, then the button **Wildcard...** will become available.
  - Click the **Wildcard...** button to open the **Find Wildcard** window.
  - Click on the part of the attribute's name to which you would like to apply the wildcard and, from the pull-down menu, select the wildcard character. This action will expand the scope of the current attribute to include all promoted attributes that satisfy the new wildcard attribute name.

(a)   (b)

FIGURE 3.11   (a) The **Expand Wildcard** window and (b) the **Find Wildcard** window.

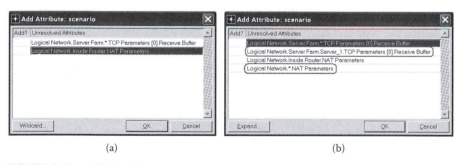

(a)   (b)

FIGURE 3.12   Add Attribute window: (a) before performing wildcard operations, and (b) after the wildcard symbol is replaced with the *Server_1* node in the **Receive Buffer** attribute, and *Inside Router* is replaced with the wildcard in the **NAT Parameters** attribute.

- If there is only one promoted attribute with the specified name, then this action has no effect, otherwise a new attribute will be added to the list in the **Add Attribute** window.
- Click **OK** to save the changes and return back to the **Add Attribute** window.

Figure 3.11 shows the **Expand Wildcard** and **Find Wildcard** windows, while Figure 3.12 shows how the **Add Attribute** window looks before and after the wild-card symbol is replaced with the *Server_1* node in the **Receive Buffer** attribute, and *Inside Router* is replaced with the wildcard in the **NAT Parameters** attribute.

4 Configuring and Running a Simulation

Developing a simulation study usually involves creating a model of the system to be simulated, selecting simulation statistics, configuring simulation attributes, running the simulation, and examining the collected results. Chapters 1–3 of this book have described how to develop basic models of simulated systems using OPNET software. In this chapter, you will now learn how to run the simulation, including how to configure the simulation prior to running it, and view and analyze the results after it has finished running.

Before a simulation can be executed, you must select statistics that you want to observe during the simulation run. Section 4.1 introduces simulation statistics in OPNET, and Section 4.2 describes the steps for selecting and configuring statistics for a simulation study. After the statistics have been selected, there are additional simulation attributes that must be configured. Examples are the amount of time for which the simulation should be run, or the initial seed to be used for random numbers to be generated during the simulation. Section 4.3 explains how these attributes can be configured. Section 4.3 also describes the next step, that is, how to execute the simulation. After the simulation has finished running, you will be interested in observing the results. Sections 4.4 through 4.7 describe how to display and analyze the collected simulation results. In some situations, the simulation execution may be unsuccessful because of the errors in constructing the model or in configuring the simulation. Section 4.8 describes how simple debugging may be done using the discrete event simulation (DES) Log, which records certain important messages generated during the simulation run.

4.1 SIMULATION STATISTICS IN OPNET

A simulation **statistic** is a collection of one or more values that describe a certain aspect of the process behavior during simulation. Simulation statistics help to find answers to the questions that motivated the simulation study. They provide insight into what may have occurred in the modeled system during the simulation run.

In OPNET, simulation statistics are usually stored on the local disk as a results file. The OPNET simulation process collects the statistic value and the time when the value was recorded and stores these as a pair. Each value describes the process state upon the occurrence of a certain event or at a specific instance in time. OPNET refers to such statistics as **vector** data. The results file for vector data is then a list or vector of value and time pairs. It is also possible for the simulation to collect statistics that consist of a single number, which usually is the sample mean, last value, time average, variance, and minimum or maximum value of the collected statistic of interest.

OPNET refers to such statistics as **scalar** data. Generally, the content of the results collected (which may be vector or scalar data) depends on how the simulation is configured to collect a particular statistic.

For example, consider a statistic that records the amount of traffic sent by the IP module in a certain node. Depending on the statistic's configuration, the values stored in the results file may be any of the following:

a. A single value, which corresponds to the total amount of traffic sent by IP
b. A collection of values containing the amount of IP traffic sent at specific time instances or upon specific event occurrences (e.g., upon a packet being scheduled for departure from IP)
c. A collection of values, each of which corresponds to the amount of IP traffic sent, averaged over a certain period of time (e.g., every second)

The statistics available for collection in OPNET are divided into two categories:

- Object Statistics, which describes the process behavior within a specific object of the simulated system. Depending on whether the object is a Node, a Link, or a Demand, such a statistic is called either a **Node Statistic**, a **Link Statistic**, or a **Demand Statistic**.
- Global Statistics, which describes the behavior of the particular protocol in the whole simulated system. A global statistic is shared by all the objects in the simulation, which are of that specific type, and all these objects contribute to the total value of the statistic.

An example of a Node Statistic (or an object statistic for a node) is **Client Ftp.Traffic Sent (bytes/sec)**. This statistic measures the rate of traffic sent in bytes/second by the client FTP applications of a specific node. If the statistic is collected at multiple nodes, then the traffic will be measured and collected for each node individually. An example of a global statistic is **Ftp.Traffic Sent (bytes/sec)**. This statistic measures the total rate of traffic sent in bytes/second by FTP applications at all nodes in the simulated network, all added together into a single quantity.

4.1.1 STATISTIC COLLECTION MODES

A statistic's **collection mode** specifies the manner in which the statistic's values will be collected. Knowing the collection mode of a statistic is vital for understanding the meaning of the reported values for that statistic, and it may critically affect the answers to the questions that the simulation study is trying to resolve. A statistic may be collected as either a vector data or a scalar data. When collected as **vector data**, there are three main statistic collection (or capture) modes in OPNET:

- **all values**—all values of the statistic are recorded for every relevant event occurrence.

- **sample**—collects only a sample of all the statistic values. The manner of sampling may be by count (e.g., every fifth data point is recorded) or by time interval (e.g., data is recorded every 6 seconds).
- **bucket**—all values are observed over a bucket specified by a time interval or a count (e.g., 10 seconds or 50 values), and a single result value is generated for each bucket (e.g., sum or average of all values in the bucket). For many commonly used statistics, this is the default collection mode.

When collected in **all values** mode, statistic values are recorded for every event that influences the statistic of interest. Therefore, this mode provides the most accurate description of what happened in the simulation study. However, the all values mode results in a huge amount of data collected, which is often difficult to interpret, and therefore, these results frequently are not useful in understanding the events that occurred during the simulation.

For example, the departure of an IP packet on a node's outgoing interface is the event that directly influences the **IP.Traffic Sent (packets/sec)** statistic. The results file will contain the vector data with a list of time-value pairs, where each pair records the time of the event and the statistic value for each event observed. If the statistic is recorded in all values mode, its values are collected for every packet departure from the IP layer. Since the statistic value is the number of packets sent, this value is 1 for every event. The result is a long list of time-value pairs in which the time is the departure time of a packet and the value is 1 for every pair. Not only does this cause a large amount of data to be collected, but the data is difficult to interpret and understand. For instance, if the data is transmitted at an average rate of 1000 packets/second, then a 10-minute simulation will result in about 600,000 data points (time-value pairs) for the **IP.Traffic Sent (packets/sec)** statistic collected in all values mode.

The second collection mode, **sample** mode, collects one sample value either by time interval, for example, every 10 seconds, or by count, for example, every 10th event. This technique considerably reduces the amount of data collected. However, sampling must be designed with care to make sure it is representative of the property to be monitored. If, for example, link utilization is to be monitored, and if packet flow over the link does not have a deterministic pattern (such as one packet transmission every 5 seconds), then sampling by time interval will work well. On the other hand, sampling of link utilization on every fifth packet transmission will not provide representative values as utilization will be 1 for each sample. Sampling by count is better suited for discrete quantities such as queue length, where you could obtain a sample of the queue length for say every sixth arrival or departure from the queue.

The third collection mode, **bucket** mode, is the most commonly used method for collecting simulation statistics. In bucket mode, similar to sample mode, the statistic values are collected over periods defined either as time intervals or by count. Only one value, called a **bucket**, is recorded for each period. However, bucket mode takes all observations of the statistic in a bucket and processes them to form a single value, which is then recorded. The processing may produce the sum of all the values, or their average, or it may use some other function that is applied on all the values in the bucket. The size of the output vector is proportional to the number of buckets: one value per bucket, just as it is in sample mode. Also, similar to sample mode, bucket mode must be used carefully to avoid reaching incorrect conclusions.

When a statistic is collected as **scalar data**, it results in a single value for the entire simulation. A scalar statistic value is usually obtained from all the observed statistic values by applying sum, average, probability, minimum, maximum, or other statistical functions. Examples of statistics collected as scalar data include the total number of packets sent by a node, average queue length, minimum segment delay, and so on. By itself, a scalar value provides a very narrow view of the process behavior. For this reason, when collecting scalar values, the same simulation is usually executed multiple times with one of the system's configuration parameter values being varied. The results of these multiple simulation runs are then collected and recorded in a form of value pairs: *<value of configuration parameter that was changed>* and *<the resulting scalar value of the statistic>*. For example, consider a simulation study that attempts to examine the influence of the transmission rate on the maximum queue occupancy. In such a study, the simulation is executed several times with maximum queue occupancy collected as a scalar statistic while the transmission rate value is varied. The results vector will contain a collection of value pairs: one pair for each simulation run. Each pair will contain the recorded values of maximum queue occupancy and the corresponding transmission rate configured in that simulation run. The results of such a study could be plotted as a graph: maximum queue occupancy versus transmission rate.

4.1.2 DECIDING WHICH STATISTICS TO COLLECT

OPNET provides a wide variety of simulation statistics and the ability to configure their collection modes. However, OPNET does not automatically select the statistics to be collected during a simulation. By default, a simulation collects no statistics at all. The simulation still executes even if no statistics were selected for collection. Such a simulation is not very useful as it provides no information about what has occurred in the modeled system. It is the responsibility of the user to determine which statistics are pertinent to the study and should be collected during the simulation run.

Therefore, when you design a simulation study, you must decide which statistics should be chosen for collection during the simulation run. Clearly, it is not a good idea to configure the simulation to collect all the available statistics. Such a configuration will significantly slow down the execution of the simulation and will consume a significant amount of computing resources. Instead, the simulation should be configured to collect only those statistics that are necessary for the simulation study.

But which statistics should be collected? The first and most obvious choice is to select statistics that are directly related to the configuration of the simulation model. For example, if the simulation has been configured to run an FTP application, then it is prudent to select FTP statistics. Similarly, if the simulation study has explicitly configured certain protocols and technologies, then statistics that describe the behavior of these and other closely related protocols and technologies are good candidates for collection. Usually, individual statistics are combined based on certain common characteristics such as the protocol. However, even if a protocol was explicitly configured for a simulation study, it may not be a good idea to collect all the available statistics of the particular protocol.

The next choice is equally obvious and deals with statistics that help to answer the simulation study questions. For example, if the goal of the simulation is to examine the influence of an FTP application on the performance of the network layer, then the appropriate network layer statistics should be collected even if the simulation employs only default configuration values of the network layer protocols.

Finally, it is often a good idea also to select statistics that are not directly related to the simulation study but that may provide an insight into the operation of the simulated network. Such statistics can aid in debugging and validating the simulation model. For example, it is often useful to collect Link Statistics (e.g., throughput and utilization) to help determine how the traffic travels though the network links and how much traffic is being generated by individual nodes. Traffic patterns in the network may help identify nodes that are configured incorrectly. For example, if a node is configured to generate traffic, yet the throughput on its adjacent links is close to zero, then it is likely that the simulation model is configured incorrectly. Additionally, link utilization may explain reported end-to-end response times (e.g., traffic delay increases exponentially as link utilization approaches 100%). On the other hand, you should not collect statistics that are not used in the simulation study. Therefore, there is no reason to collect e-mail or UDP statistics if neither of these protocols is employed in the studied network.

Overall, determining statistics to be collected in a simulation is an iterative process. When the simulation is still being developed and may require debugging, you will be wise to collect more statistics than necessary so that you may better understand the behavior of the simulated system. Once most of the configuration issues have been fixed and the simulation exhibits stable behavior, some of the statistics may no longer be needed and may be removed from the configuration to speed up the simulation execution. The final simulation may collect only those statistics that help illustrate the main points of the study.

4.2   SELECTING SIMULATION STATISTICS

4.2.1   Choose Results Window

The OPNET GUI provides a **Choose Results** window for selecting statistics to be collected during simulation. There are three ways to open the **Choose Results** window:

- Right-click on any object (Node or Link) and select **Choose Individual DES Statistics** from the pop-up menu.
- Right-click anywhere in the workspace and then select **Choose Individual DES Statistics** from the pop-up menu.
- Go to the **DES** menu in the **Project Editor** and select **Choose Individual Statistics…**

The appearance of the **Choose Results** window and the type of statistics available for selection differ slightly depending on which of these three methods was used to open it. Recall from Section 4.1 that statistics may be either Global or Object Statistics, where Global Statistics collect information about the behavior of a particular process in the simulated system as a whole, while Object Statistics collect information regarding the process behavior in a single object. Object Statistics are further classified as Node, Link, or Demand Statistics depending on whether the object of interest is a node, a link, or a demand object respectively.

When the **Choose Results** window is opened by right-clicking on an object, only statistics for that object may be selected. Therefore, the node statistic **IP.Traffic Dropped (packets/sec)** selected by right-clicking on a particular node in the network will report the number of packets dropped per second in the IP layer of that node. When the **Choose Results** window is opened by right-clicking in the project workspace or through the **DES** menu, then both Global and Object Statistics may be selected, however, the Object Statistics will be collected individually for all objects in the network topology to which that statistic pertains. For example, if the global statistic **IP.Traffic Dropped (packets/sec)** is selected, then it will report the total number of packets dropped per second in all the nodes of the simulated network. On the other hand, if the Node Statistic **IP.Traffic Dropped (packets/sec)** is selected by right-clicking in the project workspace, then this statistic is collected for all nodes in the network, but the values are available individually for each node instead of being totaled up as in the case of the global statistic. For this reason, Object Statistics collected in this manner are also sometimes called **Scenario-wide Statistics**.

Irrespective of how the **Choose Results** window is opened, further instructions for choosing specific statistics are the same in all cases. Figure 4.1a illustrates the **Choose Results** window with Scenario-wide Statistics. As Figure 4.1a shows, the **Choose Results** window contains Global, Node, and Link Statistics categories, which are further divided into subcategories that group together statistics based

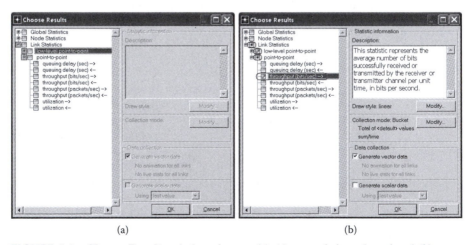

(a)   (b)

FIGURE 4.1   Choose Results window shown with (a) no statistics selected and (b) some statistics selected.

on common protocol, technology, application, and so on. It is also possible for the **Choose Results** window to contain other statistic categories such as Demand Statistics or Module Statistics. These statistic categories become available for collection only when the simulation is configured to include certain features. For example, you can collect Demand Statistics only when the simulated network model contains demand objects. Instructions for processing statistic categories are the same regardless of the number and type of the categories displayed in the window. To simplify our discussion, we describe only the Global, Node, and Link Statistics categories, which are the most commonly used in simulation studies. Global and Node Statistics often have very similar subcategories, usually aggregated based on the protocol type (e.g., IP, TCP, BGP, and e-mail). Link Statistics contain only two subcategories: *low-level point-to-point* and *point-to-point*, which collect information about physical link data transmission.

Another way to describe the statistic organization in the **Choose Results** window is as a three-level tree. The topmost level contains Global, Node, and Link general categories, which are further divided into subcategories based on the technology or networking protocol. The final tree level contains the actual statistics. As shown in Figure 4.1a, the entries in the top two levels contain icons in the form of a plus or minus sign, which allows expanding or collapsing the category view (i.e., clicking on the plus sign expands the category view and changes the icon to contain a minus sign).

Next to the icon with the plus or minus sign, each level contains a selection checkbox, which indicates whether the statistics in the specified category or an individual statistic have been chosen for collection. There are three possible settings of this checkbox:

- *Clear*—indicates that all statistics within the category or an individual statistic have not been selected for collection during the simulation. In the window shown in Figure 4.1b, no statistics have been selected in the Global and Node categories and many individual statistics in the Link category (e.g., **utilization –>**) are not selected as well.
- *Marked with a green checkmark*—indicates that either an individual statistic or all statistics within the category have been selected for collection during the simulation. In Figure 4.1b, the Link Statistic **throughput (bits/ sec) –>** and all statistics in the *low-level point-to-point* category will be collected during the simulation.
- *Marked with a green dot*—indicates that some, but not all, statistics in the given category or subcategory are selected for collection. The green dot designation is not applicable to individual statistics. As shown in Figure 4.1b, only some statistics in the *point-to-point* and Link Statistics categories have been chosen for collection.

Clicking on the checkbox of an individual statistic toggles the selection status of that statistic. Clicking on the checkbox of a second-level category selects or unselects *all* statistics within that category. OPNET does not allow selection of all statistics in the topmost-level categories (Global, Node, or Link) with a single click on the corresponding checkbox. To select all statistics within one of the top-level categories, you must individually select all the corresponding subcategories.

4.2.2 SELECTING SIMULATION STATISTICS FOR A SINGLE SPECIFIC NETWORK OBJECT

- Right-click on the object of interest located within the **Project Editor**.
- From the pop-up menu, select **Choose Individual DES Statistics**, which will open a **Choose Results** window with the statistics available for the selected object.
- Browse the statistics tree within the **Choose Results** window by clicking on the plus or minus signs to expand or collapse statistic categories until the statistic(s) of interest is/are located. Expanded statistics will have the plus sign changed to a minus sign.
- Click on the checkbox of an individual statistic to select it. Repeat the process until all statistics of interest have been selected. To select all statistics in a subcategory, click on the corresponding checkbox.
- Click **OK** to close the window and save the changes. The chosen statistic(s) will be collected only for the selected object.

4.2.3 SELECTING SIMULATION STATISTICS FOR THE WHOLE SCENARIO

- Right-click anywhere within the workspace of the **Project Editor**. From the pop-up menu, select **Choose Individual DES Statistics**, which opens a **Choose Results** window. Alternatively, go to the **DES** menu of the **Project Editor** and select **Choose Individual Statistics...**
- Browse the statistics tree within the **Choose Results** window and select the desired statistic(s). These steps are the same as those in Section 4.2.2, except that you will now be browsing the categories Node Statistics and/or Link Statistics.
- Click **OK** to close the window and save the changes.

The above procedure will cause the selected statistic(s) to be collected individually in all applicable network objects.

4.2.4 SELECTING GLOBAL SIMULATION STATISTICS

- Right-click anywhere within the workspace of the **Project Editor**. From the pop-up menu, select **Choose Individual DES Statistics**, which opens a **Choose Results** window. Alternatively, go to the **DES** menu of the **Project Editor** and select **Choose Individual Statistics...**
- Browse the statistics tree within the **Choose Results** window and select the desired statistic(s). These steps are the same as those described in Sections 4.2.2 and 4.2.3, except that you will now be browsing the category Global Statistics.
- Click **OK** to close the window and save the changes.

4.2.5 STATISTIC INFORMATION AND DATA COLLECTION PANES

The right-hand portion of the **Choose Results** window contains statistic configuration parameters and some additional information about the selected statistic. As shown in Figure 4.2, this portion of the window contains two panes labeled *Statistic information* and *Data collection*.

As Figure 4.2 illustrates, the *Statistic information* pane includes a brief description of the selected statistic and also displays the draw style and collection mode for the statistic along with options to change both these settings. The **statistic draw style** controls how the statistic will be plotted in a graph when the results are displayed. The **Modify** button located to the right of the *Draw style* textbox allows you to change the statistic's draw style. The **statistic collection mode** is displayed below the draw style section and controls how the statistic values are collected. There is another **Modify** button located to the right of the *Collection mode* textbox, which allows you to change the statistic's collection mode.

Finally, as shown in Figure 4.2, the **Choose Results** window contains a *Data collection* pane immediately below the *Statistic information* pane. This pane has two checkboxes: *Generate vector data* and *Generate scalar data*. You can select either one of the checkboxes or both. If the *Generate scalar data* option is selected, then it is also necessary to specify how the scalar value will be computed. In this case, you may select from the following functions for computing the scalar value: `last value`, `sample mean`, `time average`, `variance`, `min value`, or `max value`.

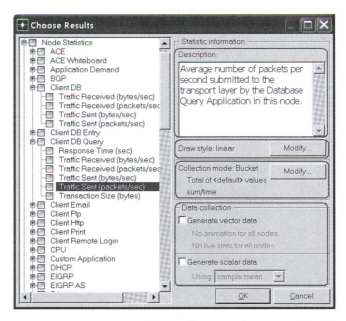

FIGURE 4.2 *Statistic information* and *Data collection* option panes of the **Choose Results** window.

4.2.6 STATISTIC DRAW STYLES

Clicking on the **Modify** button next to the *Draw style* textbox opens up a window, as shown in Figure 4.3, which allows you to select a statistic draw style. Click on the desired draw style listed in the window and click on the **Close** button to save the changes. The selected draw style will be used to display the data collected for this statistic in a graphical form.

The available draw styles are as follows:

- Linear—the data points are joined by straight line segments when displayed in a graph.
- Discrete—the data points are displayed as discrete dots with no lines joining them.
- Sample-hold—extends each point into a horizontal line, which essentially maintains (or holds) the sample value until the next sample is displayed.
- Bar—an extension of the sample-hold style in which the horizontal line drawn for each sample is filled in all the way down to the horizontal axis.
- Square-wave—draws horizontal lines for each point, just as in the sample-hold style, but then joins them with vertical lines to the next sample point. In effect, this is a bar style that is not filled in.
- Bar chart—a conventional bar graph, where a vertical line is drawn from each data point to the horizontal axis.

4.2.7 STATISTIC COLLECTION MODES

The option for configuring the **Statistic Collection Mode** is located below the draw style section. Clicking on the **Modify** button located to the right of the *Collection mode* textbox opens a **Statistic Collection Mode** window that allows you to change the statistic's collection mode. By default, the **Statistic Collection Mode** window opens in the regular view, which only allows modification of some basic collection mode properties. Clicking the *Advanced* checkbox switches the window into an advanced view, which allows modification of additional properties. Figures 4.4a and 4.4b illustrate the regular and advanced views, respectively, of the **Statistic Collection Mode** window.

The **Capture mode** option is available only in the advanced view, and it allows you to configure how the data is to be collected. The following capture modes may be selected:

FIGURE 4.3   Statistic *Draw Style* options.

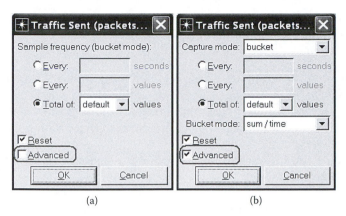

FIGURE 4.4   **Statistic Collection Mode** window: (a) regular view and (b) advanced view.

- *All values*—collect all data points.
- *Sample*—records only sample data points from the set of all available data. The number of samples is specified through either a time interval (e.g., collect data value every 2 seconds) or a sample count (e.g., collect every 10th data value). Data points that are not sampled are ignored.
- *Bucket*—the default collection mode that groups data points that occur within a period called a bucket and then applies a statistical function to each group of values. The resulting output vector contains one value for each bucket. As with the sample mode, the period may be specified as a time interval or a sample count.
- *Glitch removal*—if multiple statistic values are recorded at the same time instance, then this collection mode removes all duplicate data points retaining only the last value recorded.

The three radio buttons that follow are only applicable for bucket and sample collection modes. These radio buttons are available in both the regular and the advanced views of the window. In the advanced view, if the all values or glitch removal modes are selected, these buttons are grayed out, indicating that they are not applicable to those modes. Also, in that situation, you cannot switch back to the regular view. These radio buttons specify how frequently the values are to be aggregated or sampled for the bucket and sample modes respectively:

- *Every N seconds*—a value will be recorded after every **N** seconds.
- *Every N values*—only every **N**th value will be recorded.
- *Total of N values*—the output vector will contain a total of **N** value pairs, which means that a value will be recorded every (*duration of simulation*/**N**) time units. In this case, **N** is called *values per statistic*. Its default value is 100, which can be changed by editing the preference **num_collect_values** (see Section 1.2.1) or by changing the corresponding attribute value in the **Config/Run DES** window (see Section 4.3.1).

The pull-down menu to the right of the *bucket mode* textbox (available in the advanced view only) is applicable to the bucket collection mode only. It specifies the bucket collection mode function, which could be one of the following:

- `max value`—records the largest value collected within the bucket period.
- `min value`—records the smallest value collected within the bucket period.
- `sum`—records the sum of the values collected within the bucket period.
- `count`—records the total number of values collected within the bucket period.
- `sample mean`—records the average of the values collected within the bucket period.
- `time average`—records the time average of the values collected within the bucket period. The time average is weighted according to the time between the current value and the next value.
- `sum/time`—records the sum of all values collected within the bucket period divided by the length of the bucket period.
- `summary`—records the following four values for each bucket period: sample mean, max value, min value, and standard deviation. The resulting four sets of values are plotted in separate graphs.

The *Reset* checkbox (available in both regular and advanced views) specifies if the bucket value from the previous period is reset to zero before computation of the next bucket value. Selecting the *Reset* checkbox (which is the default) ensures that the bucket values only depend on the data values collected during that specific bucket period. Unselecting the *Reset* checkbox will incorporate the previous bucket value in computing the value during the current bucket period. This could be helpful in certain situations. For example, if bucket mode is used with the *max* function, then selecting the *Reset* checkbox will produce the maximum value within each bucket, but unselecting the *Reset* checkbox will produce the maximum value generated from the start of the simulation up to the point of each bucket.

4.2.8 MODIFYING STATISTIC COLLECTION PROPERTIES

- In the **Choose Results** window, select the individual statistic of interest.
- In the *Statistic information* pane:
  - Click on the **Modify** button located to the right of the *Draw style* label to change the statistic display style in the results graph.
  - Click on the **Modify** button located to the right of the *Collection mode* label to change how the statistic values are to be collected.

- In the *Data collection* pane:
  - Select the *Generate vector data* checkbox if you want to have statistics collected in a form of a data vector.
  - Select the *Generate scalar data* checkbox if you want to have statistics reported as a single value. Leaving both *Generate vector data* and *Generate scalar data* checkboxes unchecked will result in no data collected for the current statistic.

- Repeat the same process on other individual statistics.
- Click **OK** to close the window and save the changes.

The above instructions do not apply to the statistic categories. In general, all selected statistics will retain their default configuration unless they are explicitly modified using the preceding instructions. For most simulation studies, the default configuration of standard statistics is sufficient and requires no changes. However, if you introduce a new statistic or if the simulation study requires a different representation of the collected results, then you may need to follow the above instructions and change the statistic configuration according to your needs.

4.3   CONFIGURING AND RUNNING A SIMULATION

This section describes simulation attributes, which we call a **simulation set**, and steps for configuring and running a simulation. A simulation set is a set of attributes that describe the configuration of a simulation run (e.g., duration of simulation) and model attributes promoted to the simulation level. The simulation set consists of the following:

- *Global model attributes* describe aspects of the simulated system and not of a specific object within the system. For example, global simulation attributes may include which IP routing protocol is used in the whole simulated network, Dynamic Host Configuration Protocol (DHCP) logging (i.e., recording DHCP activity), IP routing table import option (i.e., the file name that contains the IP routing table imported into the model), and so on.
- *Simulation-wide model attributes* describe aspects of a simulation model that are common to multiple objects. Usually these are the attributes that were promoted to the simulation level (Section 3.5).
- *Simulation-run attributes* describe configuration of the simulation run for a given simulation system. Simulation-run attributes include configuration parameters such as duration of simulation, random number generation seed, the number of values per statistic, traffic growth specification, animation, use of the OPNET debugger (ODB), and others.

Configuring the simulation set and executing the simulation are tasks usually accomplished through the **Configure/Run DES** window or the **Simulation Sequence Editor**. However, simulation scenarios can also be run through the **Scenarios > Manage Scenarios** pull-down menu option in the **Project Editor** as described in Section 1.7.2. To open the **Simulation Sequence Editor**, select the **DES > Configure/Run Discrete Event Simulation (Advanced)** option from the pull-down menu. The **Configure/Run DES** window can be opened by performing any of the following actions:

- Click the *running man* icon (🏃),
- Press Ctrl+R, or
- Select **DES > Configure/Run Discrete Event Simulation...** option from the pull-down menu.

The main difference between these simulation configuration facilities is that **Configure/Run DES** allows configuration of one simulation set per simulation model while the **Simulation Sequence Editor** is a more advanced option, which may be used to create multiple simulation sets for the same simulation model. In most cases, a single simulation set is sufficient for conducting a simple study. It is worth noting that the number of simulation sets does not necessarily equal the number of simulation runs. A simulation set is a collection of values for a given simulation, while the number of simulation runs is the number of times the simulation will execute for the given simulation configuration. Some simulation attributes can be configured with more than one value, which will result in multiple executions of the same simulation set.

4.3.1   CONFIGURE/RUN DES WINDOW: SIMPLE MODE

The most common way to configure a simulation is with the **Configure/Run DES** window. This window contains two view modes: simple and detailed. The simple mode setting allows configuration of only a few key simulation parameters while the detailed mode provides a complete set of configuration controls, including an option for specifying multiple attribute values. As Figures 4.5 and 4.6 show, both modes contain a button that allows switching from the detailed to the simple view modes and vice versa. In the simple mode, the button **Detailed...** at the bottom of the **Configure/Run DES** window (see Figure 4.5) switches to the detailed view mode. Similarly, in the detailed mode, the button **Simple...** in the bottom left corner of the window (see Figure 4.6) switches the window into the simple view mode.

As shown in Figure 4.5, the simple mode allows specifying two commonly used simulation set attributes. These are as follows:

- *Duration*—how long the simulation will run or the amount of time the modeled system will be operational within the simulated environment. Usually the duration of simulation is different from the actual time it takes to complete the simulation run. For example, consider a network model with the duration of simulation attribute value set to 10 hours. If the network model is fairly simple, then the simulation run will complete in a few seconds. On the other hand, if the modeled system contains a large number of nodes that run complex protocols, then the simulation run will take much longer to complete because a complex model will typically result in a larger number of generated events, which take longer to process.

- *Values per statistic*—the maximum number of data points reported for each selected statistic. As stated in Section 4.2.7, this attribute is used to compute the bucket length for the statistics collected in the bucket and sample modes. This attribute value is ignored if all simulation statistics are configured to be collected in all values or glitch removal collection modes.

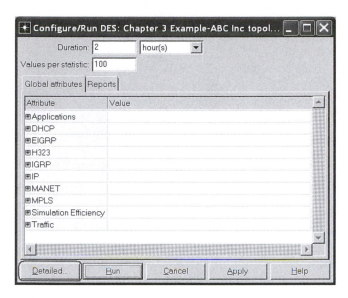

FIGURE 4.5   Simple view of the **Configure/Run DES** window.

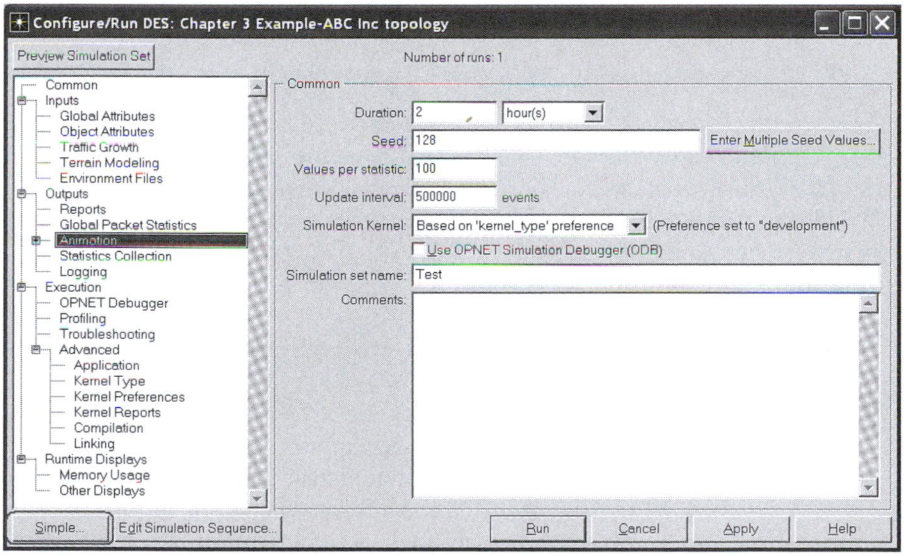

FIGURE 4.6   Detailed view of the **Configure/Run DES** window.

The simple view of the **Configure/Run DES** window contains two tabs through which two additional groups of attributes can be specified. These groups are as follows:

• *Global attributes*—global attributes for the various protocols used in the simulation are accessible here in a tree view. Each protocol name can be expanded to view the available attributes. There is also a subgroup called

Simulation Efficiency, which contains attributes to control the efficiency of a simulation for various routing protocols (see Sections 11.2.5 and 11.3.11). Finally, the subgroup *Traffic* contains attributes related to background traffic and traffic flows (see Section 6.6).

- *Reports*—this group allows you to select various statistics reports for viewing in a web browser. Each statistics report is a set of statistics collected during the simulation but prepared appropriately so that it can be viewed in a browser. We do not discuss reports further in this book.

The buttons at the bottom of the **Configure/Run DES** window have the following meanings:

- **Detailed…**—switches to the detailed view of the **Configure/Run DES** window.
- **Run**—saves the simulation set values, closes the window, and executes the simulation.
- **Cancel**—closes the window without saving the changes.
- **Apply**—saves the simulation set values without closing the window.
- **Help**—provides a description of the options in the **Configure/Run DES** window.

4.3.2 CONFIGURE/RUN DES WINDOW: DETAILED MODE

Figure 4.6 illustrates an example of the **Configure/Run DES** window in detailed mode. The detailed view mode window consists of two distinct sections. The left part of the window contains configuration categories organized in the form of a tree, while the right portion of the window contains a set of controls for setting the values of a selected configuration category.

The following configuration categories are available:

- *Common* category contains the most frequently used simulation set attributes.
- *Inputs* category contains several subcategories that can be used to specify configuration items such as global attributes, promoted object attributes, traffic growth, terrain modeling, and environment files. The global attributes are the same as those available under this tab in the simple mode (see Section 4.3.1). The promoted object attributes contains all attributes that have been promoted to the simulation level (see Section 3.5); values for these attributes can be provided here. The traffic growth subcategory allows you to specify how fast/slow the amount of traffic increases over time in the given simulation model. This subcategory is useful to determine bounds for the service level agreements (SLAs). Terrain modeling requires additional terrain modeling licenses and thus it is not discussed here. An environment file is a plain text file that contains simulation set configuration options and the values of promoted attributes. To simplify the task of configuring multiple simulation sets, it is possible to create several environment files and

then use the *Environmental Files* subcategory to specify the environmental file(s) to be used in the current simulation run.

- *Outputs* category also consists of several subcategories that allow configuring reports to be created upon simulation completion, the time period for collecting simulation statistics, simulation log data collection, animation, and so on. Most of these configuration attributes are self-explanatory.
- *Execution* category deals with configuration of various OPNET facilities for simulation execution. For example, this category allows configuring the ODB and profiler, specifying troubleshooting information, setting details of the simulation kernel (i.e., parallel vs. sequential, 32 vs. 64 bit address space, etc.), providing compilation and linking options, and others.
- *Runtime Displays* category contains attributes for configuring simulation progress reports that are displayed in the **Simulation Progress** window during simulation execution.

By default, the **Configure/Run DES** window opens in the detailed view mode with the *Common* category selected for configuration. The *Common* category is one of the most frequently used simulation configuration views and it allows specification of commonly used configuration attributes:

- *Duration*—duration of simulation (described in Section 4.3.1).
- *Seed*—the seed value for the random number generator. The button located to the right of the seed textbox allows you to specify multiple seed values. This feature is useful when you would like to obtain more statistically accurate results (i.e., more precise confidence interval) by rerunning the same simulation with a different set of seed values.
- *Values per statistic*—the maximum number of data points reported for each selected statistic (described in Section 4.3.1).
- *Update interval*—determines the frequency with which simulation progress reports are displayed in the simulation console. As the simulation executes, OPNET displays a short progress report in the simulation console window. The value of *Update interval* specifies the number of simulation events between consecutive progress reports. A typical size of an OPNET DES is a few million events. Therefore, the value of *Update interval* is usually set to a few hundred thousand events, which may result in 10 or more updates of the simulation console over the life of the simulation.
- *Simulation Kernel*—specifies the type of simulation kernel to be used to execute the simulation. There are two kernel types: development and optimized. The **development kernel** allows close monitoring and debugging of the simulation, while the **optimized kernel** optimizes the code for fast execution. The development kernel is usually used for development and debugging of the simulated model, while the optimized kernel is generally used for executing large simulations multiple times after they have been validated. The ODB is not available when executing a simulation with the optimized kernel. The default value of the **Simulation Kernel** configuration attribute is based on the value of the **kernel_type** preference (Section 1.2).

By default, the **kernel_type** preference has the value `development`. Explicitly setting the **Simulation Kernel** attribute value to `development` or `optimized` overrides the **kernel_type** preference.

- *Use OPNET Simulation Debugger (ODB)*—is a checkbox, which when selected, executes the simulation using the ODB. This option is disabled when the **Simulation Kernel** attribute value is set to `optimized`.
- *Simulation set name*—is an alphanumeric name of the current simulation set. The value of this attribute has no influence on the simulation execution. However, choosing a good and descriptive simulation set name could be beneficial when using the **Simulation Sequence Editor** to create multiple simulation sets.
- *Comments*—is a textbox that you can use to specify comments or reminders regarding the current simulation set. This attribute also has no effect on the simulation run.

In our experience, most simulations only require setting the values of simulation set attributes such as the duration of simulation, seed, values per statistic, certain global and promoted object attributes, and perhaps selected reports. The rest of the simulation attributes usually remain unchanged and set to their default values.

The detailed view mode of the **Configure/Run DES** window contains a few additional buttons, not available in the simple mode, which we describe in the following list:

- **Simple...**—switches the **Configure/Run DES** window into the simple mode.
- **Preview simulation set**—opens a text window that displays a command line invocation with associated program arguments for execution of the current simulation set. This window also displays the content of the environment file and the list of non-default values of global and promoted object attributes. This information could serve as an example for writing scripts to run the simulation outside the OPNET GUI.
- **Edit Simulation Sequence**—opens the **Simulation Sequence Editor**.

4.3.3 SIMULATION SEQUENCE EDITOR

The **Simulation Sequence Editor** is an OPNET utility that can be used to specify and manipulate (e.g., copy, paste, cut, delete, select, or execute) multiple simulation sets. Each simulation set is identified by its name and the number of simulation runs to be executed using this set. Right-clicking on the name of the simulation set and then selecting the **Edit Attributes** option opens the **Configure/Run DES** window with a detailed view of the simulation set. As shown in Figure 4.7, the **Simulation Sequence Editor** has three simulation-specific shortcut icons:

- The *blue arrow* icon (➡), which creates another simulation set.
- The *running man* icon (🏃), which executes all simulation sets that have their checkboxes selected.
- The *log book* icon (📖), which opens the simulation log.

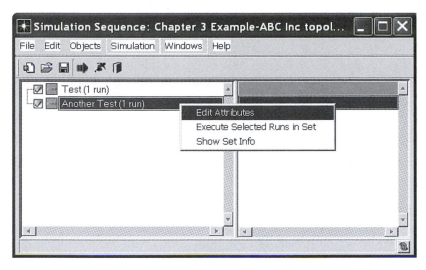

FIGURE 4.7   The **Simulation Sequence Editor**.

All these options are also available through the pull-down menu of the **Simulation Sequence Editor**.

As an alternative to the **Simulation Sequence Editor**, you can also execute simulation sets through the **Manage Scenarios** option available in the **Project Editor** (see Section 1.7.2). We review instructions for executing simulations through the **Manage Scenarios** option in Section 4.3.5. Finally, OPNET also allows simulation execution from the command prompt. However, we do not discuss this feature because it is outside the scope of this book.

4.3.4   CONFIGURING AND EXECUTING A SINGLE SIMULATION SCENARIO

- From within the **Project Editor**, click on the *running man* icon (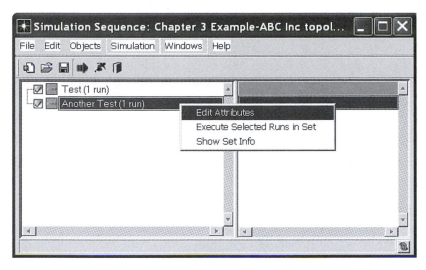), press Ctrl+Shift+R or, from the pull-down menu, select **DES** > **Configure/Run Discrete Event Simulation…**
- When the **Configure/Run DES** window appears, you may need to switch to the detailed view by clicking the **Detailed…** button or by pressing Alt+D.
- In the **Configure/Run DES** window, modify, if needed, the simulation set attributes.
- Click the **Run** button to execute the simulation.

4.3.5   CONFIGURING AND EXECUTING MULTIPLE SIMULATION SCENARIOS THROUGH MANAGE SCENARIOS

- In the **Project Editor**, from the pull-down menu, select **Scenarios** > **Manage Scenarios**.

- In the **Manage Scenarios** window that appears, set the *Results* column tag of the scenarios of interest to <collect> or <recollect>.
- After you have selected all the scenarios for which you want to (re)-collect results, click the **OK** button. This will cause all the selected scenarios to run in sequence. Executing or re-executing a scenario will overwrite any previously collected results.

4.3.6 SETTING VALUES FOR PROMOTED ATTRIBUTES

Sometimes, attributes from lower levels of the model hierarchy may be promoted to the simulation level, as described in Section 3.5. If such promoted attributes are present, then values must be provided for them before the simulation can be run. OPNET allows you to specify multiple values for a single promoted object attribute. If you do that, then the simulation will be executed multiple times, once for each value of the promoted attribute. If many promoted attributes are present, the number of simulation runs equals the product of the number of values for each attribute in the simulation set. If each attribute has only one value, then the simulation will run only one time. However, if we have k attributes, each of which has N_i values, then the total number of runs is defined by the following formula:

$$\text{Number of simulation runs} = N_1 \times N_2 \times \ldots \times N_k = \prod_{i=1}^{k} N_i$$

Only those attributes that have been promoted to the top subnet can be configured with multiple values. Refer to Section 3.5 for additional details regarding attribute promotion.

Use the following steps to set values for promoted attributes

- From within the **Project Editor**, click on the *running man* icon (🏃), press Ctrl+Shift+R or, from the pull-down menu, select **DES > Configure/Run Discrete Event Simulation...**
- When the **Configure/Run DES** window appears, you may need to switch to the detailed view mode by clicking the **Detailed...** button or by pressing Alt+D.
- In the detailed view of the **Configure/Run DES** window, expand the *Inputs* category that appears in the left part of the window and then click on the *Object Attributes* subcategory.
- To add promoted attributes:
  - Click the **Add** button to open the **Add Attributes** window, which allows you to add promoted attributes into the *Object Attributes Table*.
  - The **Add Attributes** window lists all the attributes that have been promoted to the top subnet level (i.e., simulation level).
  - Click on the field in the *Add?* column to mark the attribute to be added into the *Object Attributes Table*.

- Repeat these steps to add more than one attribute.
- Click the **OK** button to add the marked attributes to the *Object Attributes Table*.

- The values of the promoted attributes can be configured after the attributes have been placed inside the *Object Attributes Table*. Specifying a single value for a promoted attribute is no different than specifying an attribute value through the **Edit Attributes** dialog box: left-clicking in the *Value* field provides a list of available preset values, while double-left-clicking, two consecutive left clicks, or choosing the Edit... option allows you to specify a custom value. When editing a compound attribute, you may be provided with a separate window for specifying sub-attribute values.
- To specify multiple values for a single attribute:
  - Click on the attribute name to select it.
  - Click the button **Enter Multiple Values**, which will bring up another window for setting values of the selected attribute.
  - Multiple values can be set either by explicitly specifying one value per row or, for numeric values, by specifying an initial value in the *Value* column, the maximum value in the *Limit* column, and the incrementing step in the *Step* column.
  - Repeat these actions to set multiple values for other attributes.
  - Click the **OK** button to accept the chosen attribute values and to close the window.

- Once the values of all the promoted attributes have been set, click the **Run** button to execute the simulation. If at least one attribute of the simulation set is configured with multiple values, then the text label at the top of the **Configure/Run DES** window will list the *Number of runs* to be greater than one.

4.3.7 SIMULATION EXECUTION

Once you click on the **Run** button, the simulation will begin execution and a new window, called the **Simulation Execution** window, will appear. If the simulation was started from the **Configure/Run DES** window in simple view mode, then the **Simulation Execution** window will also appear in a simple mode as shown in Figure 4.8. The simple mode of the **Simulation Execution** window consists of three panes:

- The *simulation progress* pane contains a progress bar, which displays the ratio between the number of simulation seconds elapsed and the total number of simulation seconds. This pane also shows the elapsed simulation time, total number of simulation events generated so far, average and current simulation speed in units of events/seconds, elapsed real time since the start of simulation, estimated real time until simulation completion, and the total number of DES Log entries generated so far. The *simulation progress* pane also contains an **Update Progress Info** button, which when clicked on immediately updates the simulation progress with the most up-to-date data.

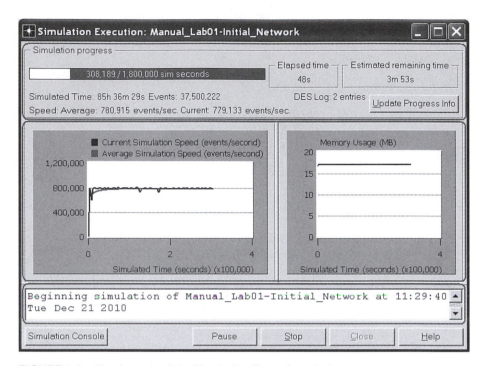

FIGURE 4.8   Simple mode of the **Simulation Execution** window.

- The *simulation speed and memory usage* pane contains two graphs: one that displays the actual and the average simulation speed and another one that shows the memory consumption during the current simulation run.
- The *messages* pane displays messages generated by the simulation except for those that describe automatic simulation progress.

There are five buttons at the bottom of the **Simulation Execution** window:

- **Simulation Console** button opens a separate simulation console window that contains all messages generated by the simulation including automatic progress updates and simulation initialization messages.
- **Pause** button suspends simulation execution. When simulation is suspended, the **Pause** button is changed to the **Resume** button, which when pressed resumes simulation execution.
- **Stop** button terminates simulation execution, resulting in only partial simulation results being collected (i.e., results are collected only until simulation was stopped).
- **Close** button closes the **Simulation Execution** window. It is disabled during simulation execution and becomes available only after the simulation has been completed or stopped.
- **Help** button opens a help window that provides a brief description of the features available in the **Simulation Execution** window.

The **Simulation Execution** window in detailed mode, shown in Figure 4.9, contains a single pane with several tabs under the *simulation progress* pane, which is identical in both detailed and simple view modes:

- *Simulation speed* tab contains a graph that shows the actual and the average simulation speed in units of events/second.
- *Live Stats* tab allows you to display certain statistics as they are being collected during the simulation. To display a statistic in the *Live Stats* tab, you need to (1) configure the simulation to collect this statistic and (2) configure the statistic to be generated as a live statistic by clicking on the *Generate live statistic* checkbox in the *Data Collection* pane of the **Choose Results** window (Section 4.2.5). Notice that only Global and individual Object Statistics (i.e., the **Choose Results** window is opened by right-clicking on a single object) can be configured to be generated as live statistics (i.e., OPNET cannot generate live statistics for multiple nodes).
- *Memory Usage* tab displays memory consumption by the current simulation.
- *Messages* tab displays the messages generated by the simulation except for those that describe automatic simulation progress.
- *Memory Sources* and *Memory Stats* display detailed information about memory blocks managed by OPNET. These tabs are not available by default. You need to configure the simulation set attribute **Runtime Displays… Memory Usage** to have these tabs become available in the **Simulation Progress** window.
- *Dependencies* tab displays the list of all the files opened by the simulation. This tab is also not available by default, but you can configure your simulation to display it in the **Simulation Progress** window by selecting the

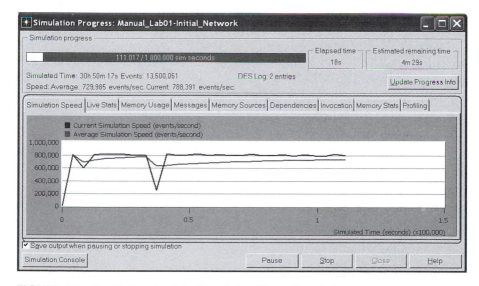

FIGURE 4.9   Detailed mode of the **Simulation Execution** window.

Generate list of component file dependencies checkbox under the **Runtime Displays...Other Displays** simulation set attribute.

- *Invocation* tab displays the command prompt invocation and all associated parameters for executing the current simulation.
- *Profiling* tab displays the simulation's profiling information. This tab is also not available by default, but you can configure your simulation to display it in the **Simulation Progress** window by configuring the **Execution... Profiling** simulation set attribute.

The remaining **Simulation Progress** window controls are identical in both detailed and simple view modes. Once the simulation completes, the simulation progress pane will display the message `Simulation Completed` and will have the **Close** button enabled.

4.4   RESULTS BROWSER

OPNET provides several utilities for displaying, examining, and comparing the results collected during simulation. The **Results Browser** is an OPNET utility for selecting and displaying collected simulation results in a graphical form. To open the **Results Browser**, select the **DES > Results > View Results...** option from the **Project Editor**'s pull-down menu or simply click the *View Results* icon (🖻).

As Figure 4.10 illustrates, the **Results Browser** window contains three tabs: *DES Graphs*, *DES Parametric Studies*, and *DES Run Tables* (the title of the tab also contains a number in parentheses that refers to the simulation run number). The *DES Graphs* tab provides facilities for creating various views of the collected statistics and displaying them in the form of standard graphs. The *DES Parametric Studies* tab allows comparing statistics with one another. Additionally, if the simulation has attributes configured with multiple values and as a result, it has been executed multiple times, then the *DES Parametric Studies* tab contains options for examining how a particular statistic varies with changes of the attribute values. Finally, the *DES Run Tables* tab contains output table reports and provides an option for generating web reports based on these output tables.

This section primarily concentrates on the result reporting features available through the *DES Graphs* tab. The *DES Graphs* tab consists of four distinct panes as shown in Figure 4.10:

- *Source* pane allows you to specify the project and/or the scenario whose results will be used to create the graphs,
- *Results* pane is where you select which statistics are to be displayed,
- *Preview* pane, which provides a preview of the selected statistics, and
- *Presentation* pane, which configures how the selected statistics are to be displayed in the graph.

Sections 4.4.1–4.4.3 describe each of these panes in more detail.

4.4.1   SOURCE PANE OF THE RESULTS BROWSER

The *Source* pane in the *DES Graphs* tab of the **Results Browser** consists of two pull-down lists called *Results for:* and *Show results:*, as well as the source tree of

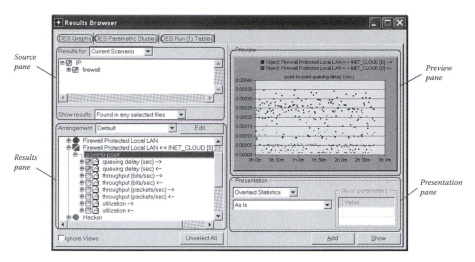

FIGURE 4.10   Results Browser window.

available projects and scenarios. Through the *Results for:* list, located at the top left corner of the window, you can specify whether the simulation results will be retrieved from `Current Scenario`, `Current Project`, or `All Projects`. Depending on your selection, the source tree will display the simulation results for the current scenario only, for all scenarios within the current project, or for all scenarios within all available projects. By default, only the current scenario is selected in the source tree.

The source tree of available projects and scenarios is organized into three levels as shown in Figure 4.11:

- List of project names (if the `Current Scenario` or `Current Project` option was selected from the *Results for:* pull-down list, then only the current project name will appear),
- List of scenario names for each project and
- List of simulation runs for each scenario.

The source tree branches can be expanded or collapsed by clicking on the icon with the plus or minus sign. The checkbox located next to the project, scenario, or the simulation run name allows you to select or unselect the source of the simulation results. Clicking on the project's checkbox alternately selects and unselects all scenarios within that project. Similarly, clicking on the scenario's checkbox selects/unselects all simulation runs within that scenario.

The *Show results:* pull-down list is located at the bottom of the *Source* pane. It is used to specify which statistics will be available in the *Statistics Tree* when multiple statistic sources have been selected. This list has two options:

- Display only those statistics that are common to all selected simulation runs
- Display statistics available in any selected simulation runs

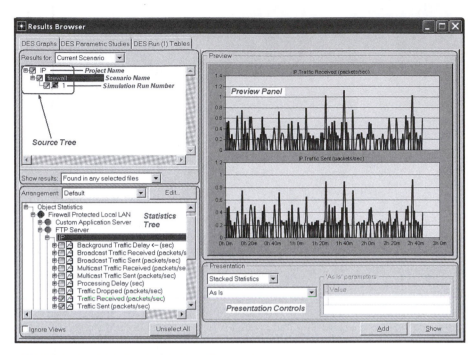

FIGURE 4.11   *DES Graphs* tab of the **Results Browser** window.

Since the results of each simulation run are saved into a separate file, the first option is called `Common to all selected files` and the second option is called `Found in any selected file`.

4.4.2   RESULTS PANE OF THE RESULTS BROWSER

The *Results* pane of the **Results Browser** allows you to select the statistics to be displayed. This pane contains the following items:

* *Arrangement:* a pull-down list that specifies how the statistics are organized within the **Results Tree**.
* **Edit...** button next to the *Arrangement* list that allows custom configuration of statistics arrangement.
* The **Statistics Tree**, which lists all statistics for the selected simulation runs.
* *Ignore Views* checkbox specifies whether or not statistics collected for hidden objects are displayed in the **Statistics Tree**. OPNET provides an option, through the **View** pull-down menu in the **Project Editor**, which allows objects such as links, demands, and domains to be made invisible. Selecting the *Ignore Views* checkbox will include results of all objects including the hidden objects in the network. Leaving this checkbox unselected will list the statistics only for those objects that are visible in the network.

- **Unselect All** button allows you to unselect all the statistics that have been selected previously. This option is very useful when you have selected several statistics that are located in different branches of the **Statistics Tree** and that are not visible simultaneously. For example, displaying a second graph requires first unselecting the statistics that have been displayed in the first graph and then selecting the new ones. Instead of browsing through the **Statistics Tree** searching for statistics that have been used in the first graph, clicking the **Unselect All** button clears all selections in the **Statistics Tree** and allows you to directly proceed to choosing statistics for the second graph.

Browsing and selecting statistics in the **Statistics Tree** is performed the same way as in the **Source Tree:**

- Click the icon with the plus sign to expand a branch of the tree
- Click the icon with the minus sign to collapse the branch
- Click the checkbox to select/unselect the item (i.e., statistic to be displayed)

Let us consider an example shown in Figure 4.12, where the scenario `firewall` from the project `IP` has **TCP Receive Buffer** attribute of the *Remote HTTP client* node configured with two values of `8760` and `32768` bytes. Such a configuration will execute the simulation twice (i.e., once for each value of the **TCP Receive Buffer** attribute) and will store the simulation results in two separate files. OPNET differentiates between such simulation runs by creating a unique name for each run, which consists of *<Project Name>-<Scenario Name>-DES-<run number>*. Therefore, in the above situation, the simulation runs will be named *IP-firewall-DES-1* and *IP-firewall-DES-2* respectively.

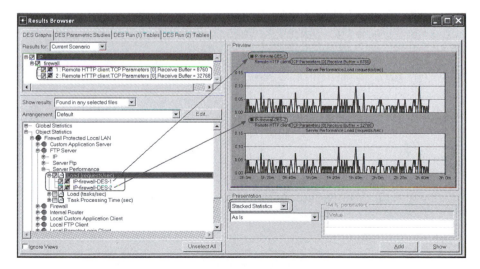

FIGURE 4.12   Simulation results for multiple runs of the same scenario.

If you select a statistic such as **FTP Server.Server Performance. Load (requests/ sec)** to be displayed, then OPNET automatically selects the results for all simulation runs within that scenario.

4.4.3   PREVIEW AND PRESENTATION PANES OF THE RESULTS BROWSER

The *Preview* pane is the third pane of the **Results Browser** window. This pane automatically displays the statistics that have been selected in the **Statistics Tree**. How the statistics are displayed in this pane depends on the selections made in the *Presentation* pane.

The *Presentation* pane is the fourth and last pane in the **Results Browser** window. This pane contains two pull-down lists. The first pull-down list presents two options that control the display of graphs when multiple statistics have been selected:

- Stacked Statistics places each selected statistic into a separate graph. For multiple statistics, these graphs are stacked one on the top of the other as shown in Figures 4.11 and 4.12.
- Overlaid Statistics places each selected statistic into the same graph as shown in Figure 4.10. When this option is chosen, each statistic in the graph is plotted with a different color.

The second pull-down list in the *Presentation* pane provides a set of predefined filters that may be selected to display the simulation results. Figure 4.13 provides a complete list of available filters while the following list provides a brief description of the most commonly used filters. Consider the following notation: V_i^k is the *i*th value of the *k*th selected statistic (the values are counted starting from 0), T_i^k is the time when V_i^k was collected, and S is the total number of selected statistics, then

- *As Is* filter displays the statistic results as they have been collected during the simulation.
- *adder* filter displays a graph that contains the sum of values of one or more statistics. The *i*th value displayed by the *adder* filter is computed as follows:

$$adder(i) = \sum_{k=1}^{S} V_i^k$$

- *average* filter displays the running mean of the values for a selected statistic. The *i*th value displayed by the *average* filter of the *k*th selected statistic is computed as follows:

$$average(i) = \frac{\sum_{j=0}^{i} V_j^k}{(i+1)}$$

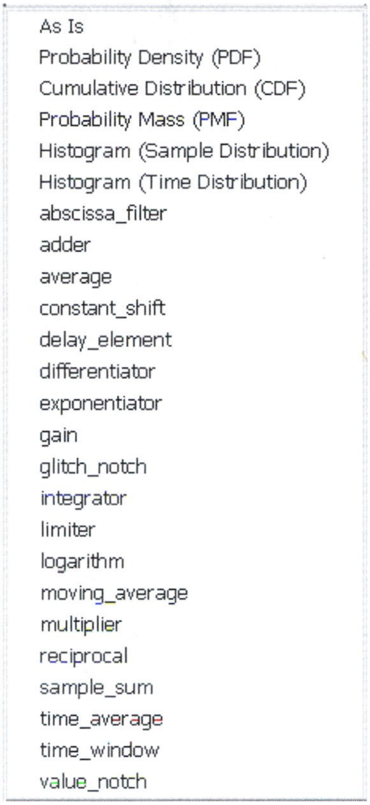

```
As Is
Probability Density (PDF)
Cumulative Distribution (CDF)
Probability Mass (PMF)
Histogram (Sample Distribution)
Histogram (Time Distribution)
abscissa_filter
adder
average
constant_shift
delay_element
differentiator
exponentiator
gain
glitch_notch
integrator
limiter
logarithm
moving_average
multiplier
reciprocal
sample_sum
time_average
time_window
value_notch
```

FIGURE 4.13   List of predefined filters.

- *time_average* filter displays the running continuous average of statistic values for a selected statistic. In effect, each value is weighted by the amount of time that the statistic had that value. The *i*th value displayed by the *time_average* filter of the *k*th selected statistic is computed as follows:

$$time\_average(i) = \frac{\sum_{j=0}^{i-1}\left(V_j^k \times (T_{j+1}^k - T_j^k)\right)}{\sum_{j=0}^{i}(T_{j+1}^k - T_j^k)}$$

- *sample_sum* filter displays the running total of statistic values. The *i*th value displayed by the *sample_sum* filter of the *k*th selected statistic is computed as follows:

$$sample\_sum(i) = \sum_{j=0}^{i} V_j^k$$

Refer to the OPNET documentation's Predefined Filters chapter for additional information regarding all predefined filter models for displaying statistic results.

4.4.4 ANALYSIS PANELS

Once the necessary statistics and their presentation styles have been selected in the **Results Browser**, clicking the **Show** button in the bottom right of the **Results Browser** window creates a new window called an **analysis panel** that contains the graphs of the selected statistics. Figure 4.14 shows an analysis panel with the simulation graphs created based on the selections shown in Figure 4.12.

The **Add** button located to the left of the **Show** button at the bottom of the **Results Browser** window allows adding selected statistics to an already created analysis panel. All you have to do is select the statistics of interest, click the **Add** button, and then click in the analysis panel. However, selected statistics are added differently depending on where in the analysis panel you click. The analysis panel consists of two parts: the graph area (shown in white color) where the actual simulation results are plotted as a graph, and the panel area (shown in gray color), which forms the boundary of the graph area on all four sides. If you click in the panel area (gray color) after clicking the **Add** button, then the selected statistics are added as if you had chosen the *Stacked statistics* presentation option, that is, a separate graph is plotted for these statistics within the same panel. On the other hand, if you click in the graph area (white color), then the selected statistics are added to the existing graph as if you had chosen the *Overlaid Statistics* presentation option. Figure 4.15 shows how the analysis panel appears after adding new statistics using both of these approaches to the analysis panel shown in Figure 4.14.

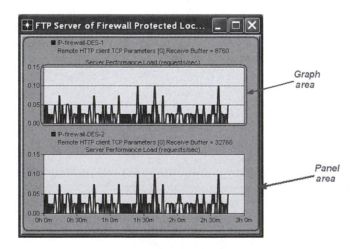

FIGURE 4.14   Analysis panel for the statistics selected in Figure 4.12.

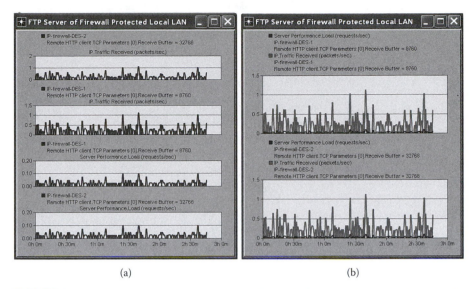

FIGURE 4.15   Adding new statistics to the existing analysis panel of Figure 4.14: (a) clicking in the panel area uses the *Stacked Statistics* presentation option and (b) clicking in the graph area uses the *Overlaid Statistics* presentation option.

4.5   VIEWING SIMULATION RESULTS WITH THE RESULTS BROWSER

Sections 4.5.1 and 4.5.2 provide step-by-step instructions for viewing simulation results with the **Results Browser**. OPNET also provides several other features including comparing simulation results collected in different scenarios and projects, finding top results (i.e., finding which objects in the simulation produce largest or smallest statistic values), and viewing simulation results with the **Time Controller**. Instructions for performing these operations are provided in Sections 4.5.3 through 4.5.7.

4.5.1   VIEWING SIMULATION RESULTS FOR THE CURRENT SCENARIO

- Open the **Results Browser** window:
  Approach #1: Right-click anywhere within the project workspace of the **Project Editor** and, from the pop-up menu, select **View Results**.
  Approach #2: From the pull-down menu in the **Project Editor**, select **DES > Results > View Results...**
  Approach #3: In the **Project Editor**, click the *View Results* icon ( ).
- In the **Results Browser** window, select the statistics to be displayed and their presentation.
- Click the **Show** button (it will not close the **Results Browser** window) to display the analysis panel with the graphs of the selected statistics.
- Click the button **Unselect All** to clear up the statistic selection (optional).

- Repeat these steps as many times as necessary.
- When finished, close the **Results Browser** window.

4.5.2   VIEWING SIMULATION RESULTS FOR A SPECIFIC OBJECT IN THE NETWORK

- Select the object of interest and then right-click on it.
- From the pop-up menu, select **View Results**.
- The **Results Browser** window that appears will include only statistics collected for the object of interest.
- Select the statistics to be displayed and their presentation.
- Click the **Show** button to display a panel with a graph of the selected statistics.
- Click the button **Unselect All** to clear up the statistic selection (optional).
- Repeat these steps as many times as necessary.
- When finished, close the **Results Browser** window.

4.5.3   VIEWING SIMULATION RESULTS FOR SCENARIOS IN THIS AND OTHER PROJECTS

- Open the **Results Browser** window (see Section 4.5.1).
- From the pull-down list called *Results for:*, select `Current Project` if you want to examine statistics collected in scenarios of the current project or select `All Projects` if you would like to examine statistics collected in all scenarios of all available projects.
- The **Source Tree** displays the list of projects and scenarios that are available for examination. Select scenarios for which you would like to examine the results.
- Select the statistics to be displayed and their presentation.
- Click the **Show** button to display a panel with a graph of the selected statistics.
- Click the button **Unselect All** to clear up the statistic selection (optional).
- Repeat these steps as many times as necessary.
- When finished, close the **Results Browser** window.

4.5.4   COMPARING SIMULATION RESULTS

- Open the **Results Browser** window with the *Results for:* pull-down list set to either `Current Project` or `All Projects`:

    Approach #1: Open the **Results Browser** window (see Section 4.5.1) and set the *Results for:* pull-down list to either `Current Project` or `All Projects`.

    Approach #2: From the pull-down menu in the **Project Editor**, select **DES > Results > Compare Results…**

- The **Source Tree** displays the list of projects and scenarios that are available for examination. Select scenarios for which you would like to compare results.
- Select the statistics to be displayed and their presentation.
- Click the **Show** button to display a panel with a graph of the selected statistics.
- Click the button **Unselect All** to clear up the statistic selection (optional).
- Repeat these steps as many times as necessary.
- When finished, close the **Results Browser** window.

4.5.5   ADDING NEW STATISTICS TO EXISTING GRAPHS

- Open the **Results Browser** window and show a panel with selected statistics (see Section 4.5.1).
- Select other statistic(s) to be displayed and their presentation.
- Click the **Add** button to add the selected statistic(s) to an existing analysis panel.
- Click on the graph (in either the graph or the panel areas) to which you would like to add the selected statistic(s).
- Repeat previous steps as many times as necessary.
- When finished, close the **Results Browser** window.

4.5.6   FINDING TOP RESULTS

- Open the **Select Statistics for Top Results** window (Figure 4.16a):
  Approach #1: Right-click anywhere within the project workspace of the **Project Editor** and, from the pop-up menu, select **Find Top Results**.
  Approach #2: From the pull-down menu in the **Project Editor**, select **DES > Results > Find Top Statistics...**

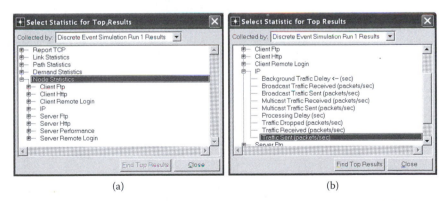

(a)   (b)

FIGURE 4.16   Select Statistics for Top Results window. (a) The category *Node Statistics* has been expanded and then (b) the statistic **Traffic Sent (packets/sec)** in the *IP* category is selected.

- In the **Select Statistics for Top Results** window, choose the statistic for which you would like to find the top results (Figure 4.16b).
- You may also want to specify which simulation run you would like to examine by selecting one of the provided options from the *Collected by:* pull-down list shown in Figure 4.16a.
- When you have selected the statistic of interest, the **Find Top Results** button will become available. Click the **Find Top Results** button to open the **Top Objects** window.

The **Top Objects** window, shown in Figure 4.17, lists selected statistic values for all applicable objects in the network (i.e., objects for which the selected statistics were configured for collection). You can further configure the top results graph by specifying the display condition and the maximum number of statistics, say *MaxN*, that satisfy that condition. *MaxN* is a positive integer value that determines the maximum number of statistics to be displayed. *Display condition* specifies a criterion that determines if a particular statistic can be displayed. The criterion is specified by listing a function of the statistic value to be compared (i.e., minimum, average, maximum, or standard deviation), the comparison sign (i.e., greater than ">" or less than "<"), and the real number against which the statistic value will be compared. Only those statistics that satisfy the display condition are shown in the top results graph, which means that it is possible that less than *MaxN* statistics will be displayed.

Figure 4.17 is an example of a **Top Objects** window that lists 16 objects that contain the selected statistic **IP.Traffic Sent (packets/sec)**. However, the **Top Objects** window is configured to display at most 10 statistics whose average value is greater than 0. Even though all 16 statistics have an average value greater than zero, only the top 10 statistics (whose order is based on the average value from largest to smallest) will be displayed in the top results graph as illustrated by the preview pane.

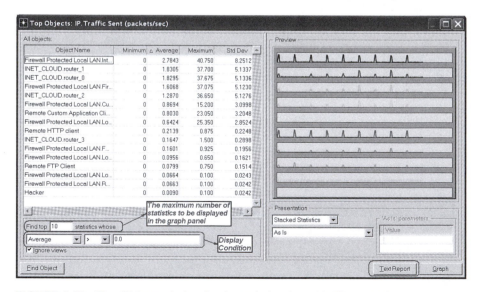

FIGURE 4.17   **Top Objects** window for the statistic selected in Figure 4.16b.

Clicking the **Graph** button in the bottom right of the **Top Objects** window will create a top results analysis panel with at most *MaxN* statistics that satisfy the display condition. If there are no statistics that satisfy the display condition, then no analysis panel will be created. Finally, you can export information about the top statistics into a text editor by clicking the **Text Report** button.

4.5.7 VIEWING RESULTS WITH THE TIME CONTROLLER

OPNET also provides an option, called the **Time Controller,** to view the progress of the simulation over time. This option allows you to view changes to the network objects (e.g., if the objects are moving) and the corresponding statistic results over a period of time.

- Display the statistics of interest in an analysis panel (see Section 4.4.4).
- Open the **Time Controller** utility:
  Approach #1: From the pull-down menu in the **Project Editor** select **View > Show Time Controller.**
  Approach #2: Press Ctrl+Alt+T.
- The **Time Controller** window will appear as shown in Figure 4.18a. You can use playback controls (i.e., go to start, scan backward, pause, play, scan forward, and to end) and speedup/slowdown buttons to control the simulation playback. You can also click the **Configure** button to open **Time Controller Settings** to further configure playback (Figure 4.18b).
- Click the **Play** button (▷) to start simulation playback. Once the playback starts, a green line, which shows the simulation progress, will appear in the displayed graphs (Figure 4.18c).

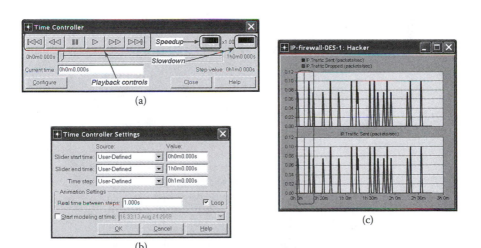

FIGURE 4.18   Example of the **Time Controller** utility. (a) The controls in the **Time Controller** window. (b) The **Time Controller Settings** window for configuring options. (c) A green line moves across the displayed graphs to show progress of the simulation with time.

- Click the **Help** button for more information regarding the controls in the **Time Controller** window.
- Once finished, click the **Close** button to exit the **Time Controller** utility.

4.6   MANIPULATING ANALYSIS PANELS

OPNET provides a set of operations to manipulate a displayed analysis panel including showing/hiding opened analysis panels, updating analysis panels with the latest simulation results, removing displayed statistic from the graph, modifying analysis panel appearance, and examining raw statistic data displayed in the graph. This section provides instructions on performing these and other operations for manipulating displayed graphs.

4.6.1   HIDING/SHOWING ANALYSIS PANELS

Once analysis panels have been displayed, you can use these operations to either hide them or show them again. This can be done on all panels or on individual panels.

The following are steps for hiding/showing all analysis panels:

Approach #1: Click the *Hide/Show Graph Panels* icon (▣) in the **Project Editor**. Clicking this icon hides or shows all the analysis panels.

Approach #2: From the **Project Editor**'s pull-down menu, select **DES > Panel Operations > Arrange Panels > Show All** to show all the analysis panels, or **DES > Panel Operations > Arrange Panels > Hide All** to hide all the analysis panels.

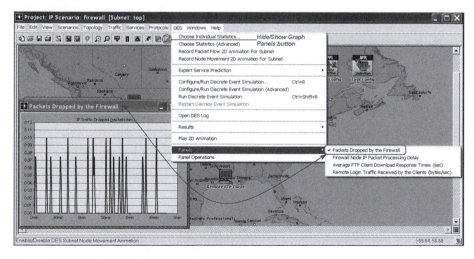

FIGURE 4.19   Showing/hiding analysis panels.

Use the following steps for hiding/showing individual analysis panels:

Approach #1: From the **Project Editor**'s pull-down menu, select **DES >
Panels** > **<analysis panel name>**, where **analysis panel name** is the name
of the panel you would like to show/hide. The **Panels** pull-down menu lists
all the available analysis panels with those panels that are currently being
displayed, containing a checkpoint next to their name. Selecting this menu
option will alternately show or hide the selected panel. Figure 4.19 illustrates a situation where the analysis panel named *Packets Dropped by the
Firewall* is being displayed.

Approach #2: Right-click within the panel space (the gray area) of the analysis panel and, from the pop-up menu, select **Hide This Panel**.

Approach #3: Click on the *Close* icon (i.e., cross) in the top-right corner of the
analysis panel window. Then click on the **Hide** button in the panel that pops up.

4.6.2 Arranging Analysis Panels

The **DES** > **Panel Operations** menu of the **Project Editor** contains a list of other
panel operations, which are shown in Figure 4.20. The **Panel Operations** > **Arrange
Panels** option allows you to arrange the currently visible panels on the screen: the
Distribute and **Tile** choices ensure that the panels are displayed without overlapping
one another and are evenly distributed over the area of the **Project Editor**'s workspace,
while the **Cascade** option shows panels one on top of the other in a cascading fashion.

4.6.3 Deleting Analysis Panels

You may delete analysis panels either individually or all together. This operation
physically removes either a selected panel or all of the created analysis panels. The
deleted analysis panels cannot be viewed anymore and must be re-created if you
want to see them again.

For deleting all of the analysis panels:

- From the **Project Editor**'s pull-down menu, select **DES** > **Panel Operations**
  > **Delete all Panels.**

For deleting individual analysis panels:

- Click the *Close* icon (i.e., cross) in the top right corner of the analysis panel
  window that you would like to delete. Then click the **Delete** button in the
  window that pops up.

4.6.4 Converting Panels into Annotation Objects

You can use these operations to convert selected analysis panels into annotation
objects and then convert them back to panels again. Converting analysis panels
into annotation objects within the workspace can be very helpful when preparing a

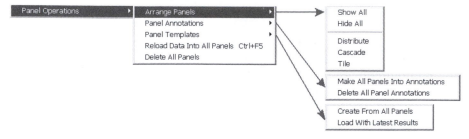

FIGURE 4.20   Available options in the **Panel Operations** pull-down menu.

presentation to describe a simulation study and the corresponding results, because annotation objects can make the appearance of the project workspace more visually appealing and easier to describe. The panel annotation object is treated the same way as any other annotation object in the **Project Editor** (see Section 2.9).

As shown in Figure 4.20, the **Panel Annotations** pull-down menu option provides two choices:

- **Make All Panels Into Annotations** embeds all analysis panels in the current scenario, including those not currently displayed, converting them into annotation objects.
- **Delete All Panel Annotations** converts all panel annotation objects back into regular analysis panels.

You can also convert selected analysis panels into annotation objects by right-clicking within the panel space (gray area) of the analysis panel and, from the pop-up menu, selecting **Make Panel Annotation In Network**. Double-clicking on the panel annotation object converts it back into the regular panel graph. The panel annotation object is treated the same way as any other annotation object in the **Project Editor**.

The following are steps for making all of the analysis panels into annotation objects:

- From the **Project Editor**'s pull-down menu, select **DES > Panel Operations > Panel Annotations > Make All Panels Into Annotations**.

For making individual panels into annotation objects, use one of the following approaches:

Approach #1: Right-click within the panel space (the gray area) of the analysis panel and, from the pop-up menu, select **Make Panel Annotation In Network**.
Approach #2: Double-click within the analysis panel space.

The steps for deleting all the annotation objects are as follows:

- From the **Project Editor**'s pull-down menu, select **DES > Panel Operations > Panel Annotations > Delete All Panels Annotations**.

For deleting individual annotation objects, use one of the following approaches:

Approach #1: Double-click on the panel annotation object. Once you move the newly restored analysis panel, you may still observe the annotation object within the **Project Editor**'s workspace with the label *Panel Open*. Select that annotation object and press the DELETE key on your keyboard.

Approach #2: From the **Project Editor**'s pull-down menu, select **DES > Panel Operations > Panel Annotations > Delete All Panels Annotations**.

4.6.5 RELOADING ANALYSIS PANELS WITH NEW RESULTS

Another useful operation is to reload existing analysis panels with results from a new simulation run. Consider a situation where you may have created several panels with various statistic graphs to analyze the results of a simulation run. Upon close examination of the collected results, you realize that one of the parameters of the simulation study is not configured properly and the simulation needs to be rerun. Once the new simulation results are collected, you would like to examine the same graphs as before but with the new data. Instead of re-creating all the analysis panels from scratch, you can convert the analysis panels into templates (i.e., analysis panels without any data) and then reload newly obtained results into these panel templates.

If some statistics have been displayed in an analysis panel, and you later change which statistics are collected during the simulation so that some of the displayed statistics are no longer chosen for collection, then reloading the analysis panel with the latest simulation results will fail. Reloading will also fail if the collected simulation results are deleted (e.g., by using the **Manage Scenarios** option), and no new results are collected for the scenario.

Use the following steps for reloading all analysis panels with the newest simulation results:

- Rerun the simulation without closing the existing analysis panels.
- Reload analysis panels with new results using one of the following three approaches:
- **Approach #1:**
    - From the **Project Editor**'s pull-down menu, select **DES > Panel Operations > Panel Templates > Create From All Panels**. This converts all the analysis panels (hidden or shown) into graph templates.
    - From the **Project Editor**'s pull-down menu, select **DES > Panel Operations > Panel Templates > Load With Latest Results**. This updates all the graph templates with the newest simulation data.
- **Approach #2:**
    - From the **Project Editor**'s pull-down menu, select **DES > Panel Operations > Reload Data Into All Panels**. This option first creates panel templates from all panels and then loads all template panels with the latest simulation results.
- **Approach #3:**
    - Press Ctrl+F5 keys on your keyboard.

Use the following steps for reloading individual panels with newest simulation results:

- Rerun the simulation.
- Right-click anywhere in the analysis panel and, from the pop-up menu, select **Make Panel Into Template**, which will change the analysis panel into a template panel.
- Right-click anywhere in the analysis panel and, from the pop-up menu, select **Load Data Into Template**, which will load the latest simulation data into the panel.

4.7   ADVANCED ANALYSIS PANEL PROPERTIES

It is often important not just to view the statistic results in the form of a graph but also to examine the raw numerical data. OPNET allows exporting statistic data into a spreadsheet or viewing the data through an internal editor program. These and other operations to change how the data is viewed in the panel are available through pop-up menus in the analysis panel.

Each analysis panel consists of two areas: graph and panel as shown in Figure 4.14. The graph area displays the actual graphs and has a white background. The panel area is gray in color and forms the boundary of the graph area on all four sides. Right-clicking on the panel and graph areas respectively generates different pop-up menus as shown in Figure 4.21. Figure 4.21a shows the pop-up menu that opens up when you right-click on the panel area, and Figure 4.21b shows the pop-up menu that appears when you right-click on the graph area. These menus are described in Sections 4.7.1 and 4.7.2.

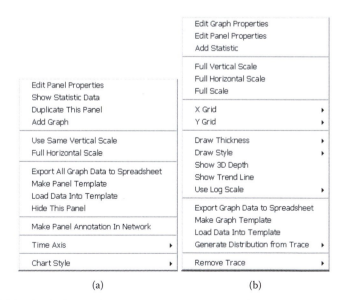

(a)   (b)

FIGURE 4.21   Analysis panel pop-up menus: (a) when you right-click on the panel area and (b) when you right-click on the graph area.

4.7.1 PANEL AREA POP-UP MENU

The *panel area pop-up menu* (Figure 4.21a) primarily deals with operations on the panel itself, which includes the following:

- **Edit Panel Properties**—allows you to change the panel properties including the panel title, the background color, the panel's location on the screen, and so on.
- **Show Statistic Data**—opens up the **OPNET editor** window, which contains the raw statistical data and its summary for the statistics displayed in this panel.
- **Duplicate This Panel**—creates another copy of the same analysis panel.
- **Add Graph**—opens up the **Results Browser** window, which allows you to select other statistic(s) to be added to the analysis panel (see also Section 4.5.5).
- **Use Same Vertical Scale**—adjusts all the graphs (if there are more than one) within the panel to have the same vertical scale.
- **Full Horizontal Scale/Full Vertical Scale/Full Scale**—restores the original horizontal/vertical/both scales of the graph.
- **Export All Graph Data to Spreadsheet**—exports the graph's raw statistical data into a space-separated text file that is automatically opened by the user-configured spreadsheet program (e.g., Excel). You can change the spreadsheet program used to open the exported data by changing the value of the **spreadsheet_prog** preference.
- **Make Panel Template/Load Data Into Template**—allows you to reload the analysis panel with the latest simulation results (see Section 4.6.5).
- **Hide This Panel**—hides the panel from view (see Section 4.6.1).
- **Make Panel Annotation In Network**—converts the panel into an annotation object and places it into the **Project Editor**'s workspace (see Section 4.6.4).
- **Time Axis**—allows you to specify the type of the simulation time axis used in the analysis panel. The possible time axis display options include **Date/Time** (if available), **seconds**, or **auto-scale** (which is the default).
- **Chart Style**—allows you to change the default appearance of the analysis panel to one of the predefined configurations (i.e., **Default**, **Classic**, **Gold Color Scheme**, or **Silver Color Scheme**).

The **Edit Panel Properties** option opens a **Panel Operations** window that allows you to change some of the panel properties. Figure 4.22 shows a graph with its corresponding **Panel Operations** window, which contains the following options:

- *Panel title:* textbox is used for specifying the title of the graph's panel.
- **Panel Coordinates** button allows you to set the panel's width, height, and location on the screen.
- **Set Color** button allows you to specify the background color of the panel, which by default is gray for the *Default* chart style.

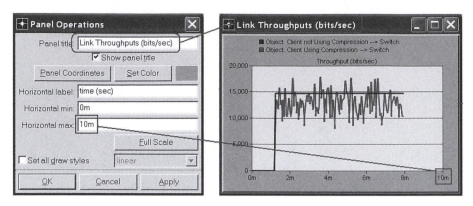

FIGURE 4.22   A **Panel Operations** window and its graph.

- *Horizontal label*, *Horizontal min*, *Horizontal max* textboxes and **Full Scale** button allows managing properties of the graph's horizontal axis.
- Checkbox *Set all draw styles* together with the pull-down list controls the draw style of the data shown in the graph. The data can be drawn using `linear`, `sample-hold`, `bar`, `discrete`, `square-wave`, or `bar chart` styles, which are described in Section 4.2.6.

The **Show Statistic Data** option opens a **Statistic Information** window that contains the raw graph data and its summary. The **Statistic Information** window (Figure 4.23) provides two options through a pull-down list:

- `General Statistic Info`—a summary of displayed graph data, which includes max, min, expected value, sample mean, variance, confidence intervals, and other statistical information.
- `Statistic Data`—a list of raw statistical values of the displayed graph.

The **Statistic Data** window is editable. You may change the raw data values in this window and then click the **Build New Statistic** button to display a panel with a new graph built using the updated statistic results values.

4.7.2   Graph Area Pop-Up Menu

The *graph area pop-up menu*, shown in Figure 4.21b, deals with manipulation of the data graph in an analysis panel. Some of the options in this menu are common to the panel area pop-up menu and were already described in Section 4.7.1. The following list provides a brief description of those options that are not available in the panel area pop-up menu:

- **Edit Graph Properties**—opens a window that allows changing the way the data is displayed in the graph, including operations such as setting the legend title, changing the draw style, and specifying vertical min and max.

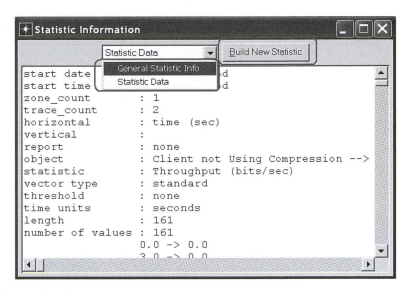

FIGURE 4.23   Statistic Information window.

- **X Grid, Y Grid**—changes how the X and Y gridlines are displayed: `Disabled` (no line), `Solid`, or `Dashed`. By default, the X axis gridline is not displayed (i.e., disabled) and the Y axis gridline is displayed using a solid line.
- **Draw Thickness/Draw Style/Show 3D Depth/Show Trend Line/Use Log Scale**—allows you to change how the actual data is displayed in the graph (e.g., the thickness and draw style of the displayed data line, inclusion or exclusion of 3D depth and trend lines, and using logarithmic scale to display the data).
- **Generate Distribution from Trace**—generates a probability distribution function (PDF) based on the graph's data and saves the result into a file. By default, the file is saved into the `op_models` directory. This file can be opened using OPNET's PDF Editor, which allows you to create, update, and view the probability density functions.
- **Remove Trace**—allows you to remove one of the statistical traces (i.e., a set of statistical data) from the graph.

Figure 4.24 shows the **Edit Graph Properties** window, which allows changing the display properties of the graph. We do not describe the options available in this window because they are very similar to other options described in this section and for the most part are self-explanatory.

Finally, it is occasionally beneficial to zoom in on the data displayed in the graph. To do that, left-click anywhere in the graph, and while holding down the mouse button, drag the mouse over the area where you would like to zoom in. Options **Full Horizontal Scale**, **Full Vertical Scale**, or **Full Scale** can restore the original graph view. Alternatively, you can use **Edit Graph Properties** and **Edit Panel Properties**

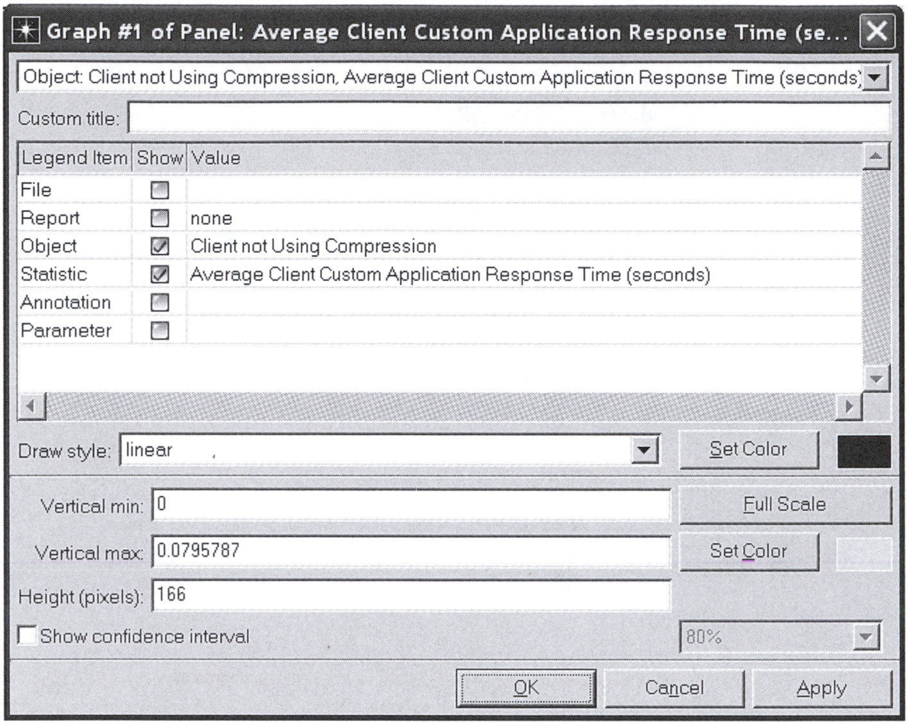

FIGURE 4.24   Edit Graph Properties window.

to change the vertical and horizontal min and max values, which will achieve the effects of zooming in or zooming out on the specified portion of the displayed graph.

4.8   DES LOG

As a simulation executes, it records all important events that occur during the simulation. Examples of such events are simulation warnings and errors, unexpected protocol behavior, and anything that may be significant or unusual. These events are stored in a plain ASCII text file called the DES Log. The file name for the DES Log is log_ info, and it is located in the project's directory in the results folder of the corresponding scenario. Usually, the results folder name starts with the project name followed by the scenario name and the DES run number, all separated by a dash. For example, in OPNET IT Guru Version 16.0, the results folder for a scenario named HTTP_simple of project HTTP is named HTTP-HTTP_simple-DES-1.olf.dir.

However, instead of peeking into the ASCII file, you are probably better off examining the DES Log through OPNET's **Log Viewer** facility. The **Log Viewer** can be accessed by selecting the **DES > Open Des Log** option from the pull-down menu in the **Project Editor**. The **Log Viewer** window, shown in Figure 4.25, consists of two tabs, several panes, and a few controls at the bottom of the window. The number of tabs in the **Log Viewer** depends on the simulation configuration and the modules (e.g., Flow Analysis) currently running together with the main OPNET software

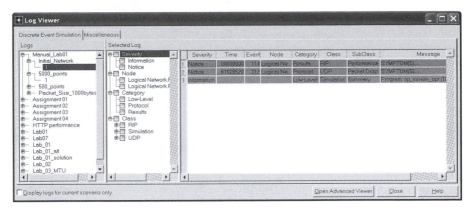

FIGURE 4.25   *Discrete Event Simulation* tab of the **Log Viewer** window.

(i.e., Modeler, IT Guru, etc.). In this section, we will only describe the *Discrete Event Simulation* and *Miscellaneous* tabs.

The DES Log is very helpful when debugging a simulation as it provides insight into what has happened during the simulation and it records any unusual events that occurred as the simulation was being executed. Each simulation run results in at least a few DES Log messages generated. When a simulation run generates three or fewer DES Log messages, it typically means the simulation was configured correctly and OPNET did not identify anything unusual. However, if there are more than three DES Log messages, then it is a good idea to examine these messages to ensure that your simulation behaves as expected.

4.8.1   DISCRETE EVENT SIMULATION TAB

The *Discrete Event Simulation* tab consists of three panes: *Logs*, *Selected Log*, and *Messages*. The *Logs* pane contains a tree-like view of the available simulation log. By default, this pane contains only the simulation logs for the current scenario. However, if desired, you can examine simulation logs of other projects and scenarios by unselecting the *Display logs for current scenario only* checkbox, located at the bottom of the **Log Viewer** window. By clicking on a simulation run number, you specify the simulation run for which you would like to examine DES Log messages. You can only examine the DES Log for a single simulation run at a time.

The *Selected Log* pane allows you to specify the type of log messages you would like to display. By default, none of the log types are selected, which results in all DES Log messages for the current simulation run being displayed. If the simulation run generated a lot of DES Log messages, then you can display only those messages that pertain to the problem you are currently examining by clicking on the desired log category. For example, you can only display log messages that were generated for a specific node in the network or that are relevant to a certain protocol; alternatively, you can choose to display those messages that record configuration errors or warnings and so on.

The *Messages* pane contains information pertaining to the DES Log messages. The *Messages* pane is structured in the form of a table where each row contains information about a single DES Log message, while the columns represent various message attributes such as the severity of the message (e.g., notice, warning, and error); the simulation time when the message was recorded; the event number that caused the DES Log message to be generated; the node for which the message was generated; the DES Log message category, class, and subclass; and the message itself. Clicking on the *Message* field in the table opens a **Log Entry** window that contains a full description of the message including information such as the simulation event that caused this message to be generated, the effects of this event on the simulation, and suggestions for possible onfiguration changes that may address the issue described by this DES Log message.

At the bottom of the **Log Viewer** window, there are four controls:

- *Display logs for current scenario only* checkbox, which controls which DES Logs will be available for examination in the *Logs* pane.
- **Open Advanced Viewer** button, which when pressed opens the **Advanced DES Log Viewer** window, which provides additional options for examining the DES Log for a selected simulation run. The **Advanced DES Log Viewer** allows you to specify custom filters for searching through the DES Log messages. We do not describe the **Advanced DES Log Viewer** in this book.
- **Close** button closes the **Log Viewer** window.
- **Help** button opens a web page, which provides a brief description of the options available in the **Log Viewer** window.

These window controls remain unchanged when you open the *Miscellaneous* tab.

4.8.2   MISCELLANEOUS TAB

The *Miscellaneous* tab, shown in Figure 4.26, consists of only two panes called *Logs* and *Selected Log*. Through the *Logs* pane, you can choose which of the two

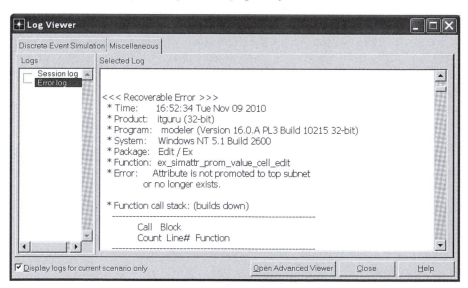

FIGURE 4.26 *Miscellaneous* tab of the **Log Viewer** window.

available log files you would like to examine: Session Log or Error Log. You can also examine these logs by opening session_log or err_log ASCII text files, respectively, both of which are located in the op_admin directory. Depending on your choice, the *Selected Log* pane will display the contents of either the session log or the error log file.

5 Standard Applications

5.1 MODELING TRAFFIC SOURCES IN OPNET

Traffic generation sources form a key aspect of parameter configuration in the modeling of network systems. The data traffic traveling through the network enables us to observe the behavior and study the performance of various network protocols in their operational environment. Without the data sources, the network is only populated with the control packets generated by certain protocols and technologies upon initialization or periodically (e.g., status update in routing protocols). However, the total amount of such control traffic is very small, and as a result, control traffic usually has very little, if any, effect on the overall network performance. Therefore, in most cases, to conduct a study of a network system, it is vital that the corresponding simulation model includes traffic sources that generate data packets to be exchanged between various nodes of the modeled network system.

OPNET provides a variety of traffic source models that may be included in a simulation. How a traffic source is deployed in a simulation model depends on the types of traffic sources required in the model. Some types of sources are configured to run on individual network nodes and can be created in a simple manner, while others require a more complex configuration process.

The rest of this chapter and Chapters 6 and 7 describe the configuration and deployment of traffic sources in OPNET. Section 5.2 begins by explaining the different types of traffic source models available in OPNET. The remaining sections of this chapter deal with standard applications only. Section 5.3 describes the general procedure by which standard applications may be included in an OPNET simulation. Section 5.4 provides a description of each of the available standard applications along with their configuration attributes. Section 5.5 explains an advanced feature of how to use symbolic names for network nodes when configuring applications. Section 5.6 describes the standard application statistics available for collection. Chapter 6 describes other advanced OPNET features for generating traffic, such as explicit packet generation sources, demands, and baseline loads. If desired, you may skip Chapter 6 and go directly to Chapter 7, which concludes our discussion of modeling traffic sources in OPNET by introducing user profiles and the methods for deploying standard and custom applications within the simulated network system.

5.2 TYPES OF TRAFFIC SOURCE MODELS IN OPNET

Typically, OPNET simulates network traffic using **explicit**, **background**, or **hybrid** (i.e., explicit and background together) traffic models. Each of the explicit and background models can be deployed in a simulation by a choice of different mechanisms. Figure 5.1 illustrates the overall hierarchy of the traffic models available in OPNET. These models are further described in Sections 5.2.1 through 5.2.3.

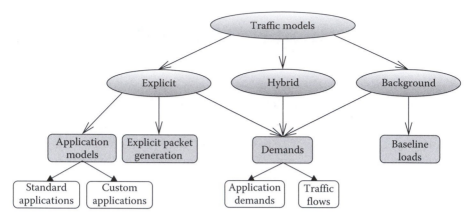

FIGURE 5.1   Summary of OPNET's traffic model hierarchy.

5.2.1 EXPLICIT TRAFFIC MODELS

Explicit traffic models provide a very accurate description of traffic behavior, which is achieved by modeling the complete life cycle of every packet generated by the traffic sources. Packet traffic is the lifeblood of an OPNET simulation as the data packets flowing through the protocol layers invoke the various functions in those protocols. Typically, the data generated by the traffic sources is split into messages or packets and is then sent to the lower protocol layers that further segment, encapsulate, fragment, queue, forward, receive, reassemble, and perform other processing, all of which are the responsibilities of various protocols in the simulated system. Without data packets, the simulation will not be able to model precisely all the features of such protocols as TCP, UDP, IP, Resource Reservation Protocol (RSVP), OSPF, and so on, and therefore, it would be difficult to accurately evaluate the performance of these protocols.

However, the high precision of packet-by-packet modeling of network traffic comes at a cost. Explicit traffic generation models may result in high memory consumption, since each packet is explicitly modeled and is represented in the main memory as a data structure that requires memory allocation. Moreover, explicit models require longer simulation execution times, because each stage of every packet's life cycle is simulated as one or more DES events, each of which contributes to the execution time.

OPNET provides the following mechanisms for modeling explicit traffic: **explicit packet generation**, **application demands**, and **application models**. **Explicit packet generation** is primarily concerned with specifying the rate at which the traffic is to be generated, the size of the packets created by individual traffic sources, and traffic destinations. This method does not model any particular application layer protocol. Defining traffic sources using explicit packet generation requires modifying attribute values of all the nodes that generate traffic. In OPNET, only certain node models (e.g., typically either traffic source or sink nodes) contain attributes for specifying the explicit packet generation characteristics. Such node models usually contain the word `_station` in their names and can be easily found by searching the **Object Palette Tree** (Section 2.3.2).

Application demand is another mechanism for specifying explicit traffic exchanged between two nodes in a simulated network system. This mechanism is also primarily concerned with representing the rate and the packet sizes of traffic, and it does not model any specific protocol behavior. Application demands are implemented as a flow of request and response messages exchanged between the application layers of two nodes. Configuration of the application demands does not require setting attribute values of individual nodes. Instead, it only involves the demand object itself, which defines the traffic flow between two nodes through such attributes as the duration of the traffic flow (i.e., start and end times), request and response parameters (i.e., packet size and traffic rate), transport protocol, and traffic mix. The **Traffic Mix (%)** configuration parameter allows modeling application demands using both explicit and background traffic models. Specifically, this attribute defines the percentage of an application demand modeled as background traffic. OPNET also contains other demand models for defining such traffic flows as voice, ATM, and IP (both generic and specific such as security and ping).

Application models also specify an explicit message exchange between the nodes in the simulated network system. However, in addition to specifying characteristics of the traffic flow (i.e., rate, packet size, destination, etc.), application models are concerned with representing the behavior of application layer protocols, or simply applications. That is why application models provide a more accurate representation of traffic sources in the simulated system than application demands or explicit packet generation. OPNET provides a set of standard applications that are already implemented to model specific application layer protocols such as FTP, HTTP, E-mail, Remote Login, and others, as well as a mechanism for defining custom applications to model other nonstandard application layer protocols.

5.2.2 BACKGROUND TRAFFIC MODELS

Background traffic models provide an analytical, and therefore less accurate, representation of data transfer over the network. The occupancy of the queues at the intermediate nodes is adjusted based on the configuration of the background traffic models, which in turn influences the delays and other performance measures of the corresponding simulated protocols and network devices. The background traffic models do not simulate the life cycle of individual packets; instead, they employ an analytical representation of traffic behavior. Therefore, a simulation study that includes background traffic models will execute faster and will consume less memory than a simulation study that uses explicit traffic models.

Background traffic is modeled through either **traffic demand objects** (i.e., traffic flows or application demands) or **baseline loads**. **Demand objects** usually specify traffic between a pair of nodes. OPNET provides a distinction between application demands and other demand types, which are referred to as **traffic flows**. Application demands represent application layer traffic in the form of request-response messages exchanged between two nodes (i.e., traffic is configured and modeled to travel in both directions). Traffic flows are used to model data generated from other nonapplication sources such as IP voice, IP ping, IP security, IP multicast, ATM, and Frame Relay. Traffic flows represent general traffic characteristics such as rate, duration, and start time for data that travels in one direction only. Demands are added to the simulation by dragging and dropping the corresponding demand objects into the network topology and then configuring them to represent the traffic flow of interest. Traffic demands are often deployed between nodes that are not directly connected by a link. Specifically, the demands need not be associated with any specific path, in which case it is up to the simulation and deployed routing protocols to determine how the traffic will travel through the network. However, specifics of demand configuration depend on the type of the demand in question and are discussed in Section 6.6. Notice that any traffic demand can be configured to be modeled as discrete or explicit traffic only, as background traffic only, or as combination of the two.

Baseline loads, on the other hand, only represent background throughput applied to a single object such as a link, node, or connection. Similarly to explicit packet generation, baseline loads do not require any additional objects placed within the project workspace and are configured directly on the object of interest. In this book, we only discuss baseline loads applied to link models. Typically, baseline loads specify such information about the background traffic traveling through that particular link as average packet size, which is assumed to remain constant throughout the simulation, and traffic rate, which could be configured to vary. The overall traffic load on the link is computed by adding together the loads generated by all traffic types that traverse the link and could include explicit traffic, demands, and baseline loads. The delays introduced by the baseline loads influence every explicitly modeled packet that travels through the corresponding link.

5.2.3 HYBRID TRAFFIC MODELS

Generally, explicit traffic models are appropriate for simulation studies that require detailed modeling of the protocols and technologies deployed in the simulated network, which in turn necessitates an accurate representation of the traffic source behavior. On the other hand, background traffic models are often used in simulation studies that have scarce computing resources and where the aggregate behavior of the protocols is sufficient for the purpose of the study. In practice, simulation studies often rely on a combination of explicit and background models to represent the traffic sources in the simulated network system. Such **hybrid traffic models** combine the advantages of both approaches by using explicit models and obtaining an accurate representation of those traffic sources that require a detailed evaluation, while employing background models for the traffic sources that do not require accurate representation. As a result, hybrid models provide greater accuracy over background models, while using fewer resources and speeding up execution as compared to explicit models.

5.3 INCLUDING APPLICATIONS IN A SIMULATION MODEL

Configuring a simulation system to run application models is a multistep process that includes configuring applications of interest and the corresponding statistics to be collected during simulation, specifying the user profile(s), and finally deploying the configured applications on nodes in the simulated network system. OPNET makes a distinction between an **application model**, which specifies the properties of the traffic generated by an application, and the **user profile**, which describes how the application will be used. Generally, the definition of an application model consists of such information as packet interarrival times and sizes, while the user profiles specify which applications are used, when each application starts, for how long each application executes, whether the applications within the profile are executed concurrently or sequentially, when the whole profile starts and ends, and so on.

In OPNET, deploying applications actually means deploying defined user profiles, which typically entails specifying the nodes on which the profiles will be executed and the nodes that will provide services for the profiles' applications. A node that runs user profiles is said to support user profiles and is labeled a **source** or a **client**. Similarly, a node that provides services for applications is said to support applications and is labeled a **server** or a **destination**. Therefore, client or source nodes support user profiles, while server or destination nodes support application services.

Note that a server may not support all the profile's applications, and therefore, the client node may need to contact different servers to obtain services for the applications within a single profile. Furthermore, a node can operate both as a client/source and as a server/destination, which means that the same node can support user profiles and applications. Such a design is very flexible and it allows specifying application configurations that closely reflect application deployment in real-life networks.

Consider the following situation where a network simulation is configured with three user profiles. Let us call them user profiles 1, 2, and 3. Each of these profiles supports a different number of applications. User profile 1 represents a clerk who only runs e-mail and print applications; profile 2 represents a student who runs web, remote login, and e-mail applications; while profile 3 represents a customer service employee who only answers the phone and therefore only runs a voice application. These profiles are deployed in the network as indicated in Table 5.1. A single client node (e.g., nodes A and C) can support multiple profiles, perhaps representing different users working on the computer at different times or one user performing multiple tasks or operating in different roles. Similarly, a single server can support multiple applications (e.g., nodes D and E), which is quite common in today's world. Finally, it is also possible for the same node to support both user profiles and applications (e.g., node F).

TABLE 5.1
Example of Application and User Profiles

Node	User Profiles Supported	Application Services Supported
Node A	2 and 3	None
Node B	2	None
Node C	1 and 3	None
Node D	None	E-mail, Web (i.e., HTTP)
Node E	None	Remote Login, E-mail
Node F	3	Voice

5.3.1 APPLICATION CONFIG UTILITY OBJECT

The first step in specifying and configuring standard applications is to include an *Application Config* utility object in the simulated system. This is usually done by opening the **Object Palette Tree** (Section 2.3) and then selecting and dragging the *Application Config* object into the project workspace. The icon for the *Application Config* object is shown in Figure 5.2. Typically, the *Application Config* object could be found in the following **Object Palette Tree** categories:

- *Node Models > Fixed Node Models > By Name > Application*
- *Node Models > Fixed Node Models > By Machine Type > utility*
- *Shared Object Palettes > applications > Node Models*
- *Shared Object Palettes > internet_toolbox > Node Models*
- *Shared Object Palettes > utilities > Node Models*
- and others

A simulated scenario should not contain multiple *Application Config* objects. Only one *Application Config* object is needed to configure all necessary standard and custom applications. Figure 5.3 shows the **Attributes** window of the *Application Config* object. As you may recall, this window may be opened by right-clicking on the *Application Config* object and selecting the **Edit Attributes** option.

As Figure 5.3 shows, the *Application Config* object contains three compound attributes called **Application Definitions**, **MOS**, and **Voice Encoder Schemes**. The latter two attributes deal with the modeling of voice applications, while we are primarily concerned with the **Application Definitions** compound attribute that defines configuration of standard and custom applications.

The **Application Definitions** compound attribute contains attribute **Number of Rows**, which specifies the number of applications to be configured within the *Application Config* object. The default value for **Number of Rows** is 0, which indicates that there will be no applications configured in the current scenario. Setting **Number of Rows** to any value greater than 0, for example A, results in A entries created within the **Application Definitions** compound attribute, each of which is responsible for providing a single definition of a standard or custom application. For example, Figure 5.4a shows that the **Number of Rows** is set to 3, which results in three application definition entries. Since none of the entries have been configured yet, each entry is named Enter Application Name...

FIGURE 5.2 *Application Config* utility object.

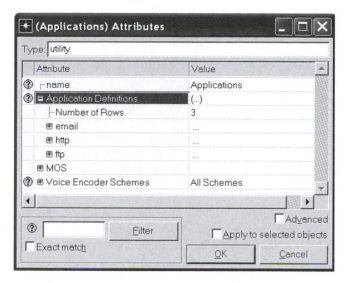

FIGURE 5.3   **Attributes** window of the *Application Config* object.

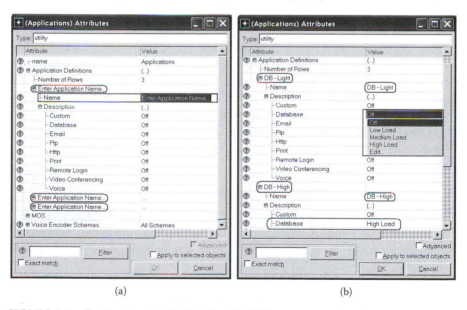

FIGURE 5.4   Configuring the **Application Definitions** compound attribute (a) with three rows, none of which has been configured yet, and (b) with the first two rows both configured to model Database applications.

An application definition entry consists of an attribute **Name**, which specifies an alphanumeric name of the current application, and a compound attribute **Description**, which consists of 10 compound attributes that specify parameters for one custom and 9 standard application definitions. Only one of these compound attributes can be configured for each single application definition entry.

Thus, even though the attribute **Description** contains 10 application configuration attributes, only one of these attributes could be configured while the rest must remain set to the value `Off`. Figure 5.4b shows the same object as in Figure 5.4a, but with the first two rows both configured to model `Database` applications.

Novice users commonly make one of the following mistakes when configuring OPNET applications:

- Placing multiple *Application Config* objects within the project workspace, one per each application definition. This is incorrect because a single *Application Config* object allows you to specify multiple application definitions.
- Attempting to configure multiple applications within a single application definition entry by setting several application definitions under the compound attribute **Description**. Such attempts will result in an OPNET warning appearing at the top or bottom of the **Application Attributes** window stating that "`Only one application type can be set per application name.`"

Additionally, all applications run over a specific default transport protocol. HTTP, FTP, E-mail, print, database, and remote login applications, by default, run over TCP, while video conferencing and voice applications, by default, run over UDP. The definition of custom applications contains an attribute that explicitly configures the transport protocol employed by that custom application (see Section 6.2). However, if needed, any of the standard and custom applications can have their default transport protocol changed at the client and server nodes. The procedure for making such a change is discussed in Section 7.5.5.

5.3.2 CONFIGURING STANDARD APPLICATIONS

The standard applications available in OPNET are described in detail in Section 5.4. That section also provides an explanation of the attributes that may need to be configured for each of these applications. The general instructions to configure standard applications in an OPNET simulation are as follows:

- Add the *Application Config* object if it is not already in the project workspace.
- Right-click on the *Application Config* object and select **Edit Attributes**.
- Expand the **Application Definitions** compound attribute.
- Specify the value of the **Number of Rows** attribute, which corresponds to the total number of different applications to be configured within the current simulation scenario.
- Expand the rows one at a time to configure each application.
  - Set the value of the **Name** attribute, which will automatically change the name of the corresponding row that contains the application definition.

- Expand the compound attribute **Description** to configure the desired application:
  - Left-click in the *Value* column of the desired application and, from the pull-down list, select one of the predefined values. Table 5.2 provides a list and a brief description of the differences between the predefined application values.
  - Alternatively, you can specify the values for individual attributes by choosing `Edit...`, which opens the *Attributes Table* for the corresponding application. The *Attributes Table* can also be accessed by double-clicking in the *Value* column of the desired application. In the *Attributes Table* specify the values of the application attributes to define the application as desired.
- OPNET allows configuring only one application within a single **Description** attribute (i.e., per row), while the rest of the application descriptions within that attribute must be set to the `Off` value. In fact, OPNET does not allow changing the `Off` setting in any row if there is another application already configured within the **Description** attribute.

TABLE 5.2
Description of Preset Application Values

Application	Preset Values	Brief Description
Database	• *Low Load* • *Medium Load* • *High Load*	Each of the preset values is configured to execute 100% of query transactions. The only difference in the settings is the size of the transaction response and the frequency of transaction arrival.
E-mail	• *Low Load* • *Medium Load* • *High Load*	Each of the preset values is configured to send and receive three e-mail messages per group. The primary differences in configuration are the size of e-mail messages and the interarrival times of send and receive messages.
FTP	• *Low Load* • *Medium Load* • *High Load*	Each of the preset values is configured to have all FTP operations evenly distributed between get and put operations. The main differences in the configuration are the sizes of transferred files and the frequency with which FTP operations are generated.
HTTP	• *Light Browsing* • *Heavy Browsing* • *Searching* • *Image Browsing*	Each of the values is configured to run HTTP 1.1. The primary differences in configuration are the frequency of page requests, definition of the web pages, and specification of the user's web-browsing behavior.
Print	• *Text File* • *B/W Images* • *Color Prints*	These preset attributes differ in the frequency and the files sizes of print job requests, with the *Text File* setting having the smallest file size and the highest frequency, while *Color Prints* have the largest size and the lowest job submission frequency.

(Continued)

TABLE 5.2   (*Continued*)
Description of Preset Application Values

Application	Preset Values	Brief Description
Remote Login	• *Low Load* • *Medium Load* • *High Load*	Each of the preset values is configured to have a different generation frequency and size of host and terminal commands.
Video Conferencing	• *Low-Resolution Video* • *High-Resolution Video* • *VCR Quality Video*	The only differences between these settings are the size of the video frames and frequency of their generation.
Voice	• *PCM Quality Speech* • *PCM Quality Speech and Silence Suppressed* • *Low-Quality Speech* • *Low-Quality Speech and Silence Suppressed* • *IP Telephony* • *IP Telephony and Silence Suppressed* • *GSM Quality Speech* • *GSM Quality Speech and Silence Suppressed*	The only differences between these preset values are the type of voice encoding scheme used and the value of the ToS field. The rest of the attribute values are the same: the lengths of silence and talk spurts are set to exponential distribution with mean outcomes of 0.65 and 0.352 seconds respectively; the number of voice frames per packet are set to 1, all voice data is modeled as discrete traffic, none of the settings use any signaling protocols, compression and decompression delays are set to 0.02 seconds, and both voice application clients are configured to place a call through a land line, located in a quiet room.

5.4   DESCRIPTION OF STANDARD APPLICATIONS

All standard applications model **two-tier**, request-response application layer proto-cols where a client node issues a request to a server and that server sends a direct response back to the client. Custom applications allow modeling applications with more than two tiers, where a client request can be forwarded through multiple servers before a response message is sent back. We will refer to such applications as **multi-tier**. Figure 5.5 illustrates the difference between two-tier and multitier applications.

The standard applications in OPNET include Database, E-mail, FTP, HTTP, Print, Remote Login, Video Conferencing, and Voice. Sections 5.4.1 through 5.4.8 describe these applications along with their configuration attributes.

5.4.1   DATABASE

The database application is modeled as a protocol that executes two types of database operations: query and entry. The **database query** operation retrieves data from the data-base. It consists of a query message that carries the database request and a response message that carries the data. For query operations, the size of the query message is

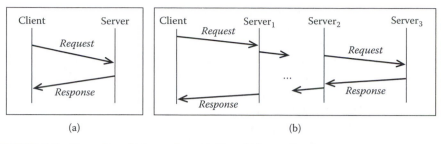

FIGURE 5.5   Examples of (a) two-tier and (b) multitier applications.

(Database) Table	
Attribute	Value
Transaction Mix (Queries/Total Transactions)	100%
Transaction Interarrival Time (seconds)	exponential (30)
Transaction Size (bytes)	constant (16)
Symbolic Server Name	Database Server
Type of Service	Best Effort (0)
RSVP Parameters	None
Back-End Custom Application	Not Used

| Details | Promote | OK | Cancel |

FIGURE 5.6   *Database Configuration Table.*

always 512 bytes, while the size of the database response is configurable through the **Transaction Size** attribute. The **database entry** operation writes data into the database. It consists of an entry message that carries the data and a response message that carries the database acknowledgement of the operation. For the database entry operation, the size of an entry message, which carries the data, is also configurable through the **Transaction Size** attribute, while the size of a response message is always set to 512 bytes.

By default, the database application runs over the TCP transport protocol. When the simulation is configured to run the database application over TCP, then a single TCP connection is established between the client node and the database server, and this connection is used to perform all database query and entry transactions. The *Database Configuration Table* contains seven attributes as shown in Figure 5.6. The first three attributes are specific to the database application, while the other four are common to many other standard application definitions. To avoid unnecessary repetition, we describe the latter attributes only once in this subsection.

- **Transaction Mix (Queries/Total Transactions)**—specifies the percentage of query transactions among all database operations. If this attribute's value is set to 100%, as shown in Figure 5.6, then all database transactions are queries. On the other hand, if the attribute's value is set to 0%, then all database operations are database entry transactions.

- **Transaction Interarrival Time (seconds)**—specifies the duration of the time interval between two consecutive database transactions. This attribute helps to determine the start time of the next database transaction. The start time of the next transaction is computed by adding the value of this attribute to the time when the previous transaction started. Database transactions are independent of one another, which means that the client can start a new transaction before receiving a response for the previous transaction.
- **Transaction Size (bytes)**—defines the size in bytes of the database transaction request. For database entry transactions, this attribute defines the size of the request messages, while the size of the responses is set to 512 bytes. For database query transactions, it is the opposite: the size of the requests is set to 512 bytes while the size of the responses is set to the value defined by this attribute.
- **Symbolic Server Name**—specifies a common symbolic name for all the nodes that operate as database servers for a given application definition (i.e., different definitions of the same application may have different symbolic server names). This attribute is part of every standard application definition (excluding the *Custom* application). During the application deployment process, the symbolic server name, defined by this attribute, is mapped to the actual names of physical nodes in the simulation. The symbolic server name is part of the so-called destination preferences definition, which specifies the servers that support this application. Specifically, the destination preferences specify which servers will be contacted by the client (i.e., the client could be configured to contact only selected servers that support this application) and the probability with which the selected servers will be contacted (i.e., the client could be configured to contact different servers with different probabilities). This topic is discussed in greater detail in Section 5.5. Typically, the value of this attribute remains unchanged.
- **Type of Service**—is the value to be assigned to the type of service (ToS) byte or the differentiated services code point (DSCP) field in the IP header in each individual packet generated by this application's client. When employed together with various quality of service (QoS) mechanisms, this value specifies the type, class, or "priority" of the packet in the IP queue and determines how the packet will be treated as it travels through the network. Packet treatment mechanisms are typically configured through the *IP Qos Attrib* object. We discuss this issue in detail in Chapter 10. This attribute is part of most application definitions, including custom applications but excluding the HTTP application. In practice, the value of this attribute often remains unchanged since it has no effect on the simulation unless the simulated system is configured to support QoS mechanisms that rely on ToS or DSCP markings.
- **RSVP Parameters**—is a compound attribute that configures the Resource Reservation Protocol (RSVP) by specifying such subattributes as **RSVP Status** (i.e., whether this application supports RSVP), **Outbound Flow** (the bandwidth and buffer size to be reserved for this application), and **Inbound Flow** (the bandwidth and buffer size to be reserved for the traffic of the peer

application). The values for flow characteristic parameters have to be defined in the *IP Qos Attrib* object first and only then they may be used for setting **RSVP Parameters**. This attribute is part of most application definitions, including custom applications but excluding the Print application. In practice, the value of this attribute often remain set to None since it has no effect on the simulation unless the simulated system is configured to support RSVP.

- **Back-End Custom Application**—specifies the name of the custom application and probability of this custom application executing on the application server upon arrival of the client's request. If this attribute is configured, then the server will execute the specified custom application upon arrival of the client's request. The server will send a response back to the client only when it completes execution of the custom application. This attributes allows combining standard two-tier applications together with multitier custom applications. This attribute is part of most application definitions, excluding HTTP, Print, Video Conferencing, and Voice. In practice, the value of this attribute often remains set to Not Used unless the simulated system requires a multitier custom application to be executed at the back-end of another application.

5.4.2 E-MAIL

OPNET models the e-mail application as a two-tier request-response message exchange between an e-mail client and an e-mail server. The e-mail response and request operations are client driven: the e-mail client periodically polls the server to retrieve its e-mail messages and it also periodically sends newly written e-mails to the server. By default, the e-mail application uses TCP as its transport protocol. When an e-mail application is configured to run over TCP, then a single TCP connection is opened between the client and the server and all messages are sent and received through that TCP connection. OPNET does not model client-to-client e-mail message exchange. Developing a multitier application model, where a client node sends the message to a server and that message is then retrieved from the server by another client, requires using custom applications.

The *E-mail Configuration Table* contains a total of nine attributes. However, we describe only the five key attributes (shown in Figure 5.7) that define the e-mail

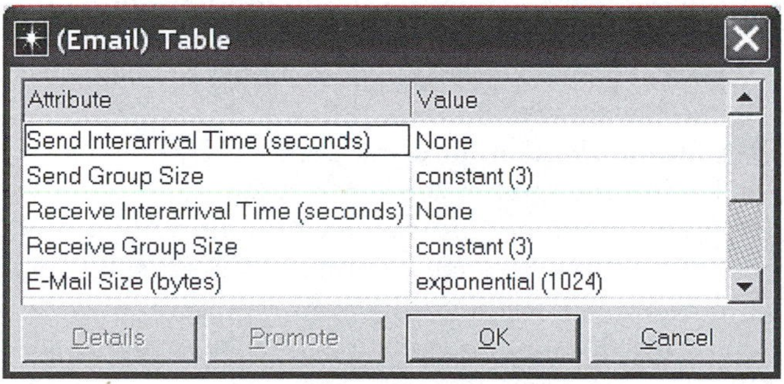

Attribute	Value
Send Interarrival Time (seconds)	None
Send Group Size	constant (3)
Receive Interarrival Time (seconds)	None
Receive Group Size	constant (3)
E-Mail Size (bytes)	exponential (1024)

FIGURE 5.7 *E-mail Configuration Table.*

application, while the other four attributes **Symbolic Server Name**, **Type of Service**, **RSVP Parameters**, and **Back-End Custom Application** are omitted to avoid repetition.

- **Send Interarrival Time (seconds)**—specifies when the next batch of e-mail messages will be uploaded to the e-mail server. The sending time of the next e-mail batch is computed by adding the value of this attribute to the time of the previous e-mail upload. The send and the receive interarrival times are independent of each other. This means that the client could be configured to frequently upload e-mail messages onto the server while seldom polling the server to retrieve its e-mails, and vice versa.
- **Send Group Size**—specifies the number of messages per batch of e-mails to be uploaded.
- **Receive Interarrival Time (seconds)**—specifies when the next batch of e-mail messages will be received from the e-mail server. The receiving time of the next e-mail batch is computed by adding the value of this attribute to the time the previous batch of e-mails arrived from the server.
- **Receive Group Size**—specifies the number of e-mail messages per batch of e-mails to be received.
- **E-mail Size (bytes)**—the size in bytes of a single e-mail message. If the client uploads 5 messages in each batch of e-mails and if the value of this attribute is 100 bytes (i.e., size of each e-mail is 100 bytes), then the total amount of data sent by the client node to the server in each upload is 5 × 100 bytes = 500 bytes.

5.4.3 FTP

The FTP standard application models the basic operation of the File Transfer Protocol (FTP). Even though the regular FTP application consists of multiple commands, this OPNET model only simulates two primary FTP operations for data transfer: put and get. The **FTP put** operation uploads a file onto the FTP server while the **FTP get** operation downloads a file from the FTP server onto the client node. Both operations consist of two message types: control and data. **Control messages** are either requests for a file (i.e., get operation) or acknowledgements of a file transfer completion (i.e., in put operation). **Data messages** carry a file that is being transferred between the client and the server. The control message size is always 512 bytes, while the size of the data message is configurable.

In OPNET, the FTP application model transfers only one file at a time. By default, FTP runs over the TCP transport protocol. When the default transport protocol is used, a separate TCP connection is opened for each file transfer operation. Unlike in real networks, this FTP model sends control and data messages of a file transfer operation over the same TCP connection. The *FTP Configuration Table* contains a total of seven attributes. However, we only describe the key three attributes, which define the FTP application, shown in Figure 5.8.

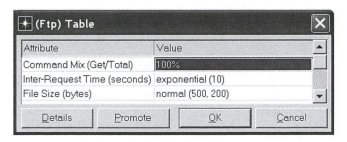

FIGURE 5.8 *FTP Configuration Table.*

- **Command Mix (Get/Total)**—specifies the ratio between the number of get operations and the total number of FTP operations. For example, if the value of this attribute is set to 40%, then 40% of all FTP operations will be get operations, while the remaining 60% will be put operations.
- **Inter-Request Time (seconds)**—specifies the amount of time between consecutive FTP operations. The start time of the next file transfer is computed by adding the value of this attribute to the time when the previous FTP operation started. FTP operations are independent of one another, which means that a new file transfer can start before the previous FTP operation has completed.
- **File Size (bytes)**—specifies the size in bytes of the file to be transferred.

5.4.4 HTTP

OPNET's HTTP application simulates Internet browsing activity where a client node periodically contacts web servers to retrieve web pages. In OPNET, each web page is modeled as a combination of text (i.e., a base HTML file) and several inline objects (i.e., referenced objects such as images and data files). The HyperText Transfer Protocol (HTTP) uses TCP as its transport protocol and it operates as follows. The client sends an HTTP request for a web page. The server receives the request and sends the corresponding web page back to the client. During the parsing of this web page, if the page contains multiple inline objects, then the client node subsequently requests these objects from the server.

The HTTP version determines how the objects are transferred over the network. For example, in HTTP version 1.0, the client node uses a nonpersistent connection where each web page object is requested and transmitted over a separate TCP connection. HTTP 1.1, on the other hand, employs persistent connections where the web page objects are transmitted over the same TCP connection. HTTP 1.1 also uses pipelining technique that combines multiple requests for inline objects into a single message, unlike HTTP 1.0 that sends each object request separately.

The *HTTP Configuration Table* consists of four key attributes that describe the properties of the HTTP application, as shown in Figure 5.9, and two additional attributes **RSVP Parameters** and **Type of Service**, which are common to many

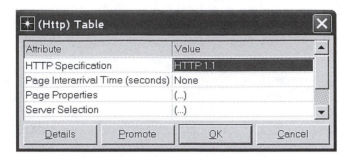

FIGURE 5.9 *HTTP Configuration Table.*

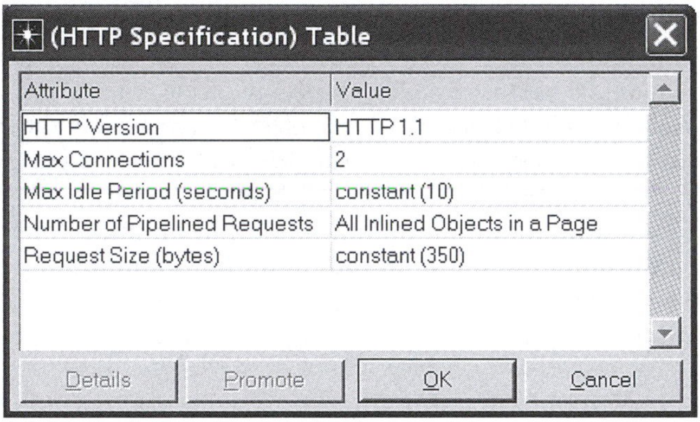

FIGURE 5.10   HTTP Specification compound attribute.

application definitions and were described in Section 5.4.1. The following list provides a description of the HTTP application attributes.

- **HTTP Specification**—defines the configuration of the HTTP protocol used by the web browser. As shown in Figure 5.10, this compound attribute consists of the following subattributes:
  - **HTTP Version**—defines the version of the HTTP protocol and has two possible values: HTTP 1.0, which supports nonpersistent connections and does not employ pipelining, and HTTP 1.1, which supports persistent connections with pipelining.
  - **Max connections**—the maximum number of parallel connections that could be established between the client and the server. This attribute is applicable only to HTTP version 1.0, which allows the opening of multiple parallel connections for transmitting requests for page objects. Notice that the number of parallel connections is different from pipelining in HTTP 1.1 where multiple web object requests are combined into a single message and are transmitted over a single connection.
  - **Max Idle Period (seconds)**—this attribute determines for how long an HTTP connection can be idle before it expires. This attribute is only

applicable to HTTP 1.1, which may keep a connection open in case there are other pages or objects to be retrieved over the same connection.

- **Number of Pipelined Requests**—specifies the number of requests that can be pipelined (i.e., combined together) into a single transport layer protocol message. This attribute applies only to HTTP 1.1.
- **Request Size (bytes)**—specifies the size of the request for a web page.
- To simplify configuration of web applications, OPNET provides four preset values for the **HTTP Specification** attribute. These preset values represent different configurations of the HTTP protocol. Each of these preset values defines the request size to be 350 bytes, while the rest of the attributes are set as follows:
  - `HTTP 1.0`—is configured to run HTTP 1.0 with the maximum number of parallel connections set to 4.
  - `HTTP 1.1`—is configured to run HTTP 1.1, has the max idle period set to 10 seconds, and allows pipelining of all inline objects within the page.
  - `Microsoft IE 5.0`—is configured to run HTTP 1.1, has the max idle period set to 10 seconds, and does not allow pipelining.
  - `Netscape Communicator 4.0`—is configured to run HTTP 1.0 with the maximum number of parallel connections set to 4.
- **Page Interarrival Time (seconds)**—specifies the time between consecutive web page requests. This attribute effectively defines the web browsing behavior of a user. For example, if you would like to model a user who reads news articles off the Web, then this attribute should be set to a relatively large value. If we assume that that it takes about 5 minutes for a regular person to read an article, then you should set this attribute's value to 300 seconds. On the other hand, if you want to model a user who searches the web for information, then the value of this attribute should be much smaller; let us say 10 seconds, which is about the average time a user spends reviewing a web page while searching. Notice that this attribute is defined through a probability distribution function that allows modeling of randomly varying values and therefore a more realistic user behavior.
- **Page Properties**—describes an OPNET representation of a typical web page. This attribute specifies such information as the size of the base HTML file (i.e., header page) together with the number and the sizes of inline objects. Each page object is represented as a row in the table as shown in Figure 5.11. The following list provides a description of the columns in the **Page Properties** compound attribute table:
  - **Object Size (bytes)**—specifies the size of a single object within the web page.
  - **Number of Objects (objects per page)**—specifies the number of objects of a certain type within the web page. The first row in the table always represents the base HTML file. Since each web page can have only one base HTML file, the number of objects in the first row is always 1, regardless of the actual value set in the *Number of Objects* column.
  - **Location**—specifies the symbolic name of the HTTP server where the inline web object resides. Even though inline objects may have the same symbolic server name, physically they can be located on different nodes,

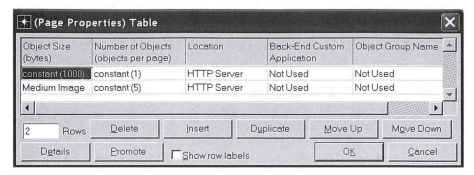

FIGURE 5.11   Page Properties compound attribute.

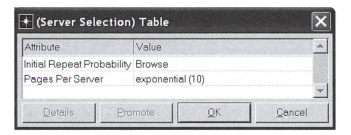

FIGURE 5.12   Server Selection compound attribute.

and therefore, the client may need to contact different servers to retrieve all inline objects of a single web page. Otherwise, this attribute has the same meaning as the **Symbolic Server Name** attribute (see Section 5.4.1).

* **Back-End Custom Application**—defines the custom application to be used at the back-end of the server (see Section 5.4.1).
* **Object Group Name**—specifies a common name for web page objects that have these characteristics. This attribute is used for collecting *Object Response* statistics that aggregate simulation results based on the name of the group of objects.

* **Server Selection**—defines whether the web pages referenced through embedded web page links will be retrieved from the same server or from another one. The web pages often contain embedded links that point to other web pages. This attribute specifies if the embedded links point to web pages located on the same server or not. Specifically, this compound attribute consists of two subattributes as shown in Figure 5.12:
  * **Initial Repeat Probability**—specifies the probability that links embedded in the web page, which is accessed for the first time, are located on the same server. The value of this attribute ranges from 0.0 to 1.0 inclusively.
  * **Pages Per Server**—specifies the number of consecutively accessed web pages from the same server.

Together, these attributes model the following server selection behavior. If the web page is the first page retrieved from the current server, then the simulation uses the value of the **Initial Repeat Probability** attribute to determine if this server will be accessed again. If the same server is chosen, then the client retrieves N consecutive web pages from that server, where N is the value specified by the **Pages Per Server** attribute. If the simulation determines that a different server should be selected or if all N web pages have been already accessed from the current server, then a new server is selected based on the client's destination preferences. OPNET provides three preset values for the **Initial Repeat Probability** attribute, which are named to represent various user behaviors:

- `Searching`—when searching the Web, the user typically accesses various pages often located on different sites. Furthermore, the user will often access a single page per site. Both these factors imply that the probability that the next web page is located on the same server should be small. That is why to represent such a behavior, the value of the **Initial Repeat Probability** attribute is set to 0.3.
- `Browsing`—when browsing the Web, the user accesses multiple sites but often explores multiple links on the same server before moving on to a different site. To represent such a behavior, the value of the **Initial Repeat Probability** attribute is set to 0.5.
- `Research`—when doing research, the user typically accesses multiple pages from the same site and visits only a small number of different web servers. That is why, to model such a behavior, the value of the **Initial Repeat Probability** attribute is set to 0.9.

5.4.5 PRINT

The Print application models the operation of submitting a printing job to a print server or a printer. OPNET even has several node models that represent network printers. The destination of the print application may be either a printer or a print server. The print application runs over the TCP transport protocol and initiates a new TCP connection for each print job request. The *Print Configuration Table* consists of four attributes as shown in Figure 5.13. We do not provide descriptions of the **Symbolic Printer Name** and **Type of Service** attributes, since one of them (**Type of Service**) was described in Section 5.4.1 while the other (**Symbolic Printer Name**) is similar to attribute **Symbolic Server Name** described in that section.

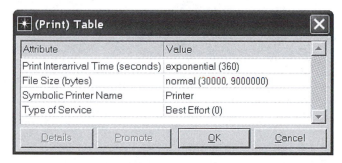

Attribute	Value
Print Interarrival Time (seconds)	exponential (360)
File Size (bytes)	normal (30000, 9000000)
Symbolic Printer Name	Printer
Type of Service	Best Effort (0)

FIGURE 5.13 *Print Configuration Table.*

- **Print Interarrival Time (seconds)**—specifies the time elapsed between two consecutive print job requests. The next print job starts after "interarrival time" seconds have passed since the previous print job request was submitted.
- **File Size (bytes)**—specifies the size in bytes of a single print job to be forwarded to a print server or a printer.

5.4.6 REMOTE LOGIN

The Remote Login application models a virtual terminal service, which is also known as a Terminal Network or Telnet. The Remote Login application allows the user to connect to a remote server and perform various operations on it by issuing commands from a local machine. Commands issued from a local system together with the responses generated by the remote server create traffic that travels through the network between the local and remote nodes.

By default, the remote login application runs over TCP, and each remote login session requires a separate TCP connection. OPNET refers to the traffic received by the client node as *Host Traffic* and calls the traffic sent by the client node (i.e., received by the remote terminal) as *Terminal Traffic*. The *Remote Login Configuration Table* contains a total of seven attributes. However, we only describe three key attributes that define the remote login application as shown in Figure 5.14. Descriptions of the other four attributes, **Symbolic Server Name**, **Type of Service**, **RSVP Parameters**, and **Back-End Custom Application**, are omitted to avoid repetition, as they were described in Section 5.4.1.

- **Inter-Command Time (seconds)**—specifies the time between two consecutive remote login commands. The next command is issued after "intercommand time" seconds have elapsed from the time the previous command was executed.
- **Terminal Traffic (bytes per command)**—defines the size in bytes of each command generated by the local host to be sent to the remote destination.
- **Host Traffic (bytes per command)**—defines the size in bytes of each response received by the local host from the remote destination.

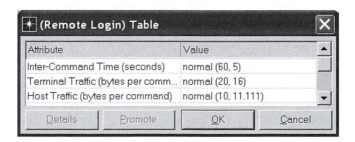

FIGURE 5.14   *Remote Login Configuration Table.*

5.4.7   VIDEO CONFERENCING

The Video Conferencing application models transmission of video traffic between two nodes in the network. OPNET represents video traffic as a sequence of data frames with the frame size being a configurable parameter. By default, the Video Conferencing application runs over the UDP transport protocol to avoid connection management and other delays associated with the TCP protocol. However, if Video Conferencing is configured to run over TCP, then each client opens a separate connection for a one-way video stream. Typically, a Video Conferencing session is established between the two client nodes without the use of a server.

The *Video Conferencing Configuration Table* contains a total of six attributes; however, we only describe three key attributes, as shown in Figure 5.15. These attributes specify the characteristics of the traffic load generated by the Video Conferencing application. Descriptions of the other four attributes, **Symbolic Server Name**, **Type of Service**, **RSVP Parameters**, and **Back-End Custom Application**, are omitted to avoid repetition.

- **Frame Interarrival Time Information**—is a compound attribute that specifies the frequency of video frame arrival for both incoming and outgoing traffic. If one of the predefined values is selected, then the frame arrival rate is the same in both directions. Expanding this attribute allows configuring the frame arrival rate independently for each direction. As shown in Figure 5.16, the definition of the compound attribute **Frame Interarrival Time Information** consists of two subattributes that specify interarrival time in seconds for the incoming and the outgoing streams of video frames. The attribute **Incoming Stream Interarrival Time (seconds)** defines

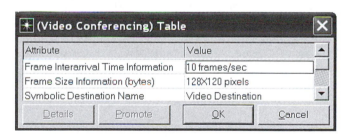

FIGURE 5.15   *Video Conferencing Configuration Table.*

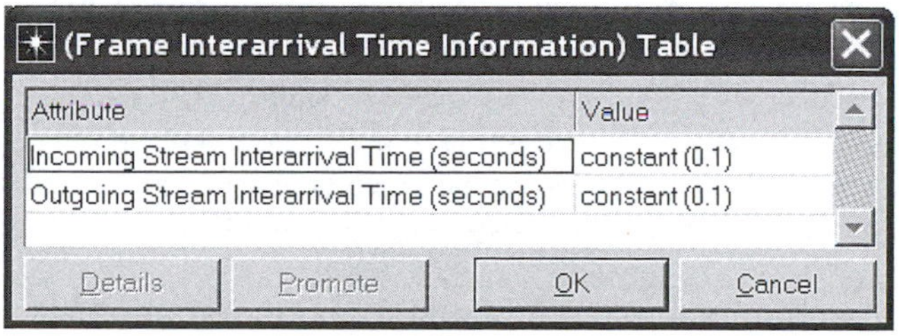

FIGURE 5.16   **Frame Interarrival Time Information** compound attribute.

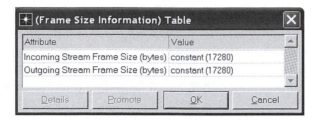

FIGURE 5.17   Frame Size Information (bytes) compound attribute.

the time between two consecutive frames in the incoming stream, while
Outgoing Stream Interarrival Time (seconds) specifies the time between
two consecutive video frames in the outgoing stream. The time when the
next frame is generated is computed by adding the "interarrival" time to
the time when the previous video frame was generated and sent from the
video application to the lower layer. The frame arrival frequency or rate is
inversely proportional to the frame interarrival time (e.g., an arrival frequency of 10 frames per second corresponds to a frame interarrival time of
0.1 seconds, meaning that one frame is generated each 0.1 seconds).
*   **Frame Size Information (bytes)**—is also a compound attribute. It specifies
    the size of each incoming (i.e., generated at the destination node) and outgo-
    ing (i.e., generated at the source node) frame in units of bytes. The predefined
    values for this attribute are specified in units of pixels (e.g., in Figure 5.15,
    the frame size is 128 × 120 pixels). However, as shown in Figure 5.17, subat-
    tributes of this compound attribute specify the frame sizes in units of bytes.
    The assumption is that each pixel requires 9 bits and therefore a 128 × 120-
    pixel frame requires 128 × 120 × 9 bits = 138,240 bits, which corresponds to
    17,280 bytes. As Figure 5.17 illustrates, the compound attribute **Frame Size
    Information (bytes)** consists of two subattributes: **Incoming Stream Frame
    Size (bytes)** and **Outgoing Stream Frame Size (bytes)**, which specify the size
    of the video frames for the incoming and the outgoing streams, respectively.

5.4.8   Voice

OPNET's Voice application models network communication between two clients using
a digitized voice signal. The definition of this application is fairly complicated, and a
complete description of all configuration features for voice applications available in
OPNET may span tens of pages. This section provides only an overview of the main
features available in OPNET for modeling a Voice application.* Typically, a voice
call consists of talk spurts followed by periods of silence. The lengths of the silence
and talk spurts can be explicitly configured in the definition of the Voice application.

Typically, the total time it takes for the voice signal to travel from one caller to
another (i.e., mouth-to-ear delay) consists of the time it takes to encode and packetize

---

* The following paper contains additional information about deployment of VoIP applications in
OPNET: Salah, K. and Alkhoraidly, A., An OPNET-based simulation approach for deploying VoIP,
International Journal of Network Management, 16(3) (May 2006), 159–183.

the voice data on one end, compress the voice packet, transmit the encoded and compressed voice data packet through the network, decompress the received voice data packet, and finally decode it for playback at the other end. Encoding and decoding delays are determined by the type of encoding scheme specified for this voice call. OPNET provides a wide range of encoding scheme models for configuring voice applications. The compression and decompression delays can also be explicitly configured through the attributes of the Voice application definition. The time to transmit the packetized voice data through the network depends on the actual network configuration and is not specified in the voice application.

By default, Voice applications run over UDP. However, internally, OPNET simulations transmit voice packets using the Real-Time Protocol (RTP), which requires no additional configuration.

The Voice application is usually run between two client nodes and it does not require the presence of a server node. OPNET allows voice traffic to be modeled as either discrete (i.e., explicit packet exchange), background (i.e., analytical modeling), or a combination of the two. Figure 5.18 shows the *Voice Configuration Table*, and the list that follows provides a brief explanation of every attribute.

- **Silence Length (seconds)**—is a compound attribute that defines the lengths of the silence periods during voice conversation for inbound and outbound traffic streams. As shown in Figure 5.19, by default, the length of the silence periods, for incoming and outgoing traffic, is computed using the exponential distribution with a mean outcome of 0.65 seconds. If needed, these values can be changed. The incoming and the outgoing periods of silence (i.e., silence on either side of the call) are independent of each other, which means that it is possible for both clients to talk or to be silent simultaneously, or for one side of the call to talk while the other side is silent.

Attribute	Value
Silence Length (seconds)	default
Talk Spurt Length (seconds)	default
Symbolic Destination Name	Voice Destination
Encoder Scheme	G.723.1 5.3K
Voice Frames per Packet	1
Type of Service	Best Effort (0)
RSVP Parameters	None
Traffic Mix (%)	All Discrete
Signaling	SIP
Compression Delay (seconds)	0.02
Decompression Delay (seconds)	0.02
Conversation Environment	(...)

(Voice) Table — Details | Promote | OK | Cancel

FIGURE 5.18 *Voice Configuration Table.*

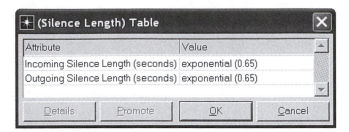

FIGURE 5.19   Silence Length (seconds) compound attribute.

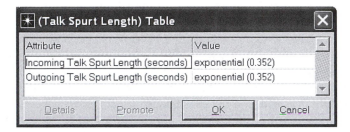

FIGURE 5.20   Talk Spurt Length (seconds) compound attribute.

- **Talk Spurt Length (seconds)**—a compound attribute that defines the lengths of the uninterrupted talk periods during the voice conversation. This attribute also specifies the lengths of the talk spurts for incoming and outgoing traffic. As shown in Figure 5.20, by default, the lengths of the talk spurts, for incoming and outgoing traffic, are computed using the exponential distribution with a mean outcome of 0.352 seconds. These values also can be changed if needed.
- **Symbolic Destination Name**—specifies the symbolic name for a destination node that participates in this voice conversation with the client node (see Section 5.4.1).
- **Encoder Scheme**—specifies the algorithm used for encoding voice into a digital signal. OPNET provides 30 preconfigured models of various encoding schemes. You can use an existing encoding scheme or create a new one by modifying the *Voice Encoder Schemes Table*, which is one of the attributes in the *Application Config* object. Voice encoding schemes are defined through attributes such as codec type (e.g., PCM, ADPCM, QCELP, VSELP, etc.), frame size, lookahead size, digital signal processing ratio, coding rate, and speech activity detection. Voice encoding is a separate topic, which we do not discuss in this book due to space limitations.
- **Voice Frames per Packet**—specifies the number of encoded voice frames placed into a single application layer packet before the voice data is sent to the lower layers. The lower layers can, and often do, fragment the application layer packets into smaller chunks.
- **Type of Service**—specifies the ToS/DiffServ field value in the IP packet header (see Section 5.4.1).

- **RSVP Parameters**—specifies the configuration of the RSVP protocol (see Section 5.4.1).
- **Traffic Mix (%)**—defines how the voice traffic is modeled during the simulation. This attribute contains the following preset values:
  - `All Discrete`—all voice data is modeled as explicit traffic only.
  - `All Background`—all voice data is represented through an analytical model with no explicit voice data packets being sent into the network.
  - `25%`, `50%`, or `75%`—indicates that the voice traffic is modeled using a mixture of discrete and background traffic. The attribute value specifies the actual percentage of voice application data modeled as background traffic.
  - `Edit...`—allows inputting the actual percentage of the voice data to be modeled as background traffic, with 0% specifying that the voice data will be represented as all discrete traffic and 100% denoting that the voice data will be modeled as all background traffic.
- **Signaling**—is a compound attribute that specifies how the voice connection is established and terminated. This attribute contains the following preset values:
  - `None`—there is no protocol for managing the voice connection.
  - `SIP`—a connection for this voice call will be set up and released using the Session Initiation Protocol (SIP) defined in RFC 2543.
  - `H.323`—the signaling protocol for connection management is based on the recommendation from the ITU Telecommunication Standardization Sector (ITU-T).
  - `Edit...`—allows setting subattribute values for this compound attribute:
    - **Protocol**—defines the signaling protocol for setting up and releasing the voice connection. Possible values for this attribute are `None`, `SIP`, or `H.323`.
    - **Traffic Modeling**—specifies what portion of the voice traffic will be modeled. This subattribute has two possible values:
      - `Control Plane Only`—means that only the control packets associated with the signaling protocol will be modeled. If selected, then no application traffic will be modeled while the connection is active (i.e., from the time the connection is established until the time the connection is released).
      - `Control and Traffic Plane`—means that both the control packets of the signaling protocol and the data packets of the voice traffic will be modeled. This is the default setting for all preset values of the compound attribute **Signaling**.
- **Compression Delay (seconds)**—this attribute specifies the time it takes to compress a voice packet.
- **Decompression Delay (seconds)**—this attribute specifies the time it takes to decompress a voice packet.
- **Conversation Environment**—configures the quality of the environment on both ends of the voice call. The quality of the conversation environment could be set to one of the following preconfigured values: `Land phone - Quiet room`, `Land phone - Noisy room`, `Cell phone in building`, `Cell phone in SUV or sedan`, and `Cell phone in convertible`.

These values are configurable through the *MOS* table, which is one of the attributes in the *Application Config* object. MOS stands for **Mean Opinion Score**, which is a measure for assessing the quality of a voice call.

5.5   USING SYMBOLIC NODE NAMES

In OPNET, when defining standard applications, the application's servers or destinations are referred to through symbolic names instead of the names of the actual nodes in the simulated network. Thus, for example, you can choose a symbolic name such as "FTP Server" and then later, while deploying the application in the simulated network, map the symbolic name to one or more specific nodes. The same idea is used for specifying custom applications where, instead of trying to predict which nodes will operate as source and destination for the custom application traffic, you only need to specify the sources and destinations through symbolic names (e.g., "Custom HTTP Login Source" and "IM Authentication Server") This feature separates application definitions from the configuration of the actual network topology and results in a flexible architecture for configuring simulation scenarios. Consider the following situation: A network architect creates a simulation study of several applications over a range of different network topologies. Using the "symbolic name" feature, OPNET is able to maintain the names of the application sources and destinations independent of the actual network topology. As a result, once the applications have been defined and configured, the architect can directly deploy the defined applications in any desired network topology by specifying the mapping between the symbolic client and/or server names and the names of the actual nodes in the new topology.

In OPNET, only the advanced node models (i.e., node models whose names include the suffix *_adv*) allow specifying a mapping between the actual and the symbolic names by explicitly changing the attribute values. Typically, this step is performed automatically by OPNET when deploying user profiles. Furthermore, OPNET also differentiates between source and destination symbolic names. The source symbolic name is typically used only by custom applications to specify a name that identifies the nodes that serve as the originators or starters for a custom application phase. The destination symbolic name is used to designate the nodes that operate as a server or traffic destination for both standard and custom applications. The instructions in Sections 5.5.1 and 5.5.2 provide the steps for mapping source and destination symbolic names respectively to actual node names. These instructions are applicable only to end node models such as workstation, server, LAN, and so on. OPNET specifies the mapping between actual node names and the corresponding source or destination symbolic names through the node attributes called **Application: Source Preferences** and **Application: Destination Preferences** shown in Figure 5.21.

5.5.1   MANUALLY CONFIGURING AN APPLICATION'S SOURCE PREFERENCES

This attribute is only applicable to custom applications. It specifies that the current node will serve as the initiator or a starter for a phase in a custom application task, whose attribute **Task Specification...Manual Configuration...Source** matches the specified symbolic name. Refer to Section 6.2 for additional details on defining

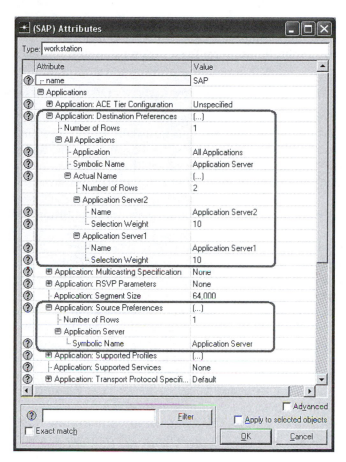

FIGURE 5.21 **Application: Source Preferences** and **Application: Destination Pre-ferences** attributes.

custom applications, their tasks, and phases. Follow these steps to map symbolic source names to actual node names:

- In the network topology of the simulated system, right-click on the node that will serve as a source for the corresponding application, and from the menu that appears, select the **Edit Attributes** option.
- Expand the attribute **Applications...Application: Source Preferences**. Notice that this attribute is only available in the advanced node models (i.e., node models that have the suffix _adv_ as part of their names).
- Set the value of the attribute **Number of Rows**, which corresponds to the number of symbolic names that may map to this node. It is possible that a single node operates as a source for multiple applications.
- For each new row that appears, expand the compound attribute **None** and change the value of the attribute **Symbolic Name** to one of the values in the pull-down list of available symbolic names. This change will cause the current

node to serve as a source node for all applications that have their attribute **Source** set to the corresponding symbolic name. If multiple nodes are mapped to the same symbolic source name, then the actual node is selected randomly.

5.5.2 MANUALLY CONFIGURING AN APPLICATION'S DESTINATION PREFERENCES

This attribute is applicable to both custom and standard applications. It provides a mapping between a symbolic destination name and the actual nodes in the network. If multiple nodes match a single destination symbolic name, then the destination node is selected based on the associated selection weight. If a mapping between the destination symbolic name and the actual node is not specified, then the destination node is chosen randomly out of all the nodes in the network that support services for the specified application. Follow these steps to map symbolic destination names to actual node names:

- In the network topology of the simulated system, right-click on the node that will serve as a source. From the menu that appears, select the **Edit Attributes** option. Notice that both source and destination preferences are configured at the source node.
- Expand the **Applications...Application: Destination Preferences** attribute. This attribute is also only available in the advanced node models (i.e., node models that have the suffix _adv_ as part of their names).
- Set the value of attribute **Number of Rows**, which corresponds to the number of applications for which you would like to provide a mapping between a symbolic server name and the actual node names. The server or destination mapping is done on a per application basis.
- Perform the following actions for each newly appeared row, which is a compound attribute. The new rows will all have the default name **All Applications** indicating that the mapping rule between the symbolic server name and the actual physical node in the simulation will be applicable to all defined applications.
  - Expand the row.
  - Set the value of the attribute **Application**, which by default is set to All Applications. This attribute specifies the name of the applications to which this mapping rule will apply. This attribute can help to differentiate between applications that have been defined with the same symbolic server name. The value field of this attribute contains a pull-down list of all applications currently defined in the _Application Config_ utility object. You can only set the attribute **Application** to one of the values in that list.
  - Set the value of the attribute **Symbolic Name**, which specifies the symbolic name to be mapped to an actual node name. Again, the drop-down list provides all the symbolic names defined in the _Task Config_ and _Application Config_ utility objects.
  - Configure the attribute **Actual Name** that specifies the number of actual nodes that map to this symbolic server name, the actual node names, and selection weights, which are the probabilities with which the server node will be selected. For example, as shown in Figure 5.21, the destination's symbolic name Application Server is mapped

to two actual nodes in the simulated network called `Application Server2` and `Application Server1`. Specifically, this compound attribute contains the following subattributes:

- **Number of Rows** that specifies the total number of nodes that can operate as a server for this application. There will be a new row created for each server.
- For each row, specify the mapping to an actual server node by setting the values of the following two attributes:
  - **Name**—the name of the actual node in the simulated system, which will operate as a server. This attribute provides a list of names for all the nodes available in the system. You must select only those nodes that support the current application; otherwise, this configuration setting will be ignored.
  - **Selection Weight**—is the server selection weight that determines the probability with which this server will be contacted during application execution.

5.6 APPLICATION STATISTICS

Every standard application in OPNET has a set of statistics available for collection during the simulation. Typically, these statistics describe the performance of the application in terms of various application delay properties and traffic arrival and departure rates. For example, the e-mail application includes such statistics as download and upload response times, and traffic received and sent rates in units of bytes per second and packets per second.

Global application statistics describe the overall application performance in the network. For example, the **Traffic Received (bytes/sec)** Global Statistic of the e-mail application collects data from the transport layer to measure the arrival rate in bytes per second averaged over all nodes (i.e., client and server) that run the e-mail application. Node application statistics differentiate between client and server nodes and collect information about application performance at one particular node. Each standard application contains a set of statistics that describe the application performance recorded at the client and at the server nodes. For example, OPNET collects statistics for the e-mail application in the following two categories:

- **Client E-mail**, which collects statistics related to application performance at the client node (e.g., delay experienced between sending a request and receiving a response) and
- **Server E-mail**, which collects statistics related to performance of the e-mail application at the server node (e.g., e-mail request arrival rate at the server or the time it takes the server to process an application request).

Client statistics are not available for collection if the simulation model does not contain an *Application Config* node or demand objects. Server statistics are not available for collection if there are no server nodes in the simulation model.

A summary of the available application statistics is presented in Tables 5.3 and 5.4. To avoid repetition, we combined several common statistic categories into a single section.

TABLE 5.3
Summary of Global Application Statistics

Category	Name	Description
DB Entry, DB Query	*Response Time (sec)*	Time it takes for a database response packet to arrive as a reply to a request packet.
	Traffic Received (bytes/sec), *Traffic Received (packets/sec)*	Average traffic arrival rate to all database applications deployed in the network. The statistic is computed based on the application data arriving from the transport layer.
	Traffic Sent (bytes/sec), *Traffic Sent (packets/sec)*	Average traffic departure rate from all database applications deployed in the network. The statistic is computed based on the application data forwarded to the transport layer.
Email	*Download Response Time (sec)*	Time it takes for e-mail messages to arrive from the server in response to a client's request. The time includes the connection setup delay.
	Traffic Received (bytes/sec), *Traffic Received (packets/sec)*	Average traffic arrival rate to e-mail applications deployed in the network. The statistic is computed based on the application data arriving from the transport layer.
	Traffic Sent (bytes/sec), *Traffic Sent (packets/sec)*	Average traffic departure rate from all e-mail applications deployed in the network. The statistic is computed based on the application data forwarded to the transport layer.
	Upload Response Time (sec)	Time it takes for an ACK to arrive from the e-mail server in response to the client uploading e-mail messages. The time includes the connection setup delay.
FTP	*Download Response Time (sec)*	Time it takes to download a file from the server starting from the time the client sends a get request until the file download completes. This time includes the connection setup and tear-down delay. The statistic is updated only after the connection is closed.
	Traffic Received (bytes/sec), *Traffic Received (packets/sec)*	Average traffic arrival rate to all FTP applications deployed in the network. The statistic is computed based on the application data arriving from the transport layer.
	Traffic Sent (bytes/sec), *Traffic Sent (packets/sec)*	Average traffic departure rate from all FTP applications deployed in the network. The statistic is computed based on the application data forwarded to the transport layer.

Application	Statistic	Description
	Upload Response Time (sec)	Time it takes to upload a file to the server starting from the time the client starts uploading the file (i.e., put request issued) until the time an ACK arrives from the server. This time includes the connection setup and tear-down delay. The statistic is updated only after the connection is closed.
HTTP	*Object Response Time (sec)*	Time it takes to retrieve a single in-lined object from the HTML page.
	Page Response Time (sec)	Time it takes to retrieve an entire HTML page including all in-lined objects.
	Traffic Received (bytes/sec), *Traffic Received (packets/sec)*	Average traffic arrival rate to all HTTP applications deployed in the network. The statistic is computed based on the application data arriving from the transport layer.
	Traffic Sent (bytes/sec), *Traffic Sent (packets/sec)*	Average traffic departure rate from all HTTP applications deployed in the network. The statistic is computed based on the application data forwarded to the transport layer.
Print	*Traffic Received (bytes/sec),* *Traffic Received (packets/sec)*	Average traffic arrival rate to all Print applications deployed in the network. The statistic is computed based on the application data arriving from the transport layer.
	Traffic Sent (bytes/sec), *Traffic Sent (packets/sec)*	Average traffic departure rate from all Print applications deployed in the network. The statistic is computed based on the application data forwarded to the transport layer.
Remote Login	*Response Time (sec)*	Time it takes for a Remote Login response packet to arrive as a reply to a request packet.
	Traffic Received (bytes/sec), *Traffic Received (packets/sec)*	Average traffic arrival rate to Remote Login applications deployed in the network. The statistic is computed based on the application data arriving from the transport layer.
	Traffic Sent (bytes/sec), *Traffic Sent (packets/sec)*	Average traffic departure rate from all Remote Login applications deployed in the network. The statistic is computed based on the application data forwarded to the transport layer.
Video Conferencing	*Packet Delay Variation*	End-to-end delay variance experienced by the video packets from the time they are created until they are received.
	Packet End-to-End Delay	The time it takes for a video packet to arrive at the destination node starting from the time the packet is created at the source node.
	Traffic Received (bytes/sec), *Traffic Received (packets/sec)*	Average traffic arrival rate to all Video Conferencing applications deployed in the network. The statistic is computed based on the application data arriving from the transport layer.
	Traffic Sent (bytes/sec), *Traffic Sent (packets/sec)*	Average traffic departure rate from all Video Conferencing applications deployed in the network. The statistic is computed based on the application data forwarded to the transport layer.

(Continued)

TABLE 5.3   (Continued)
Summary of Global Application Statistics

Category	Name	Description
Voice	*Jitter (sec)*	If $Tc(i)$ is the difference between the times when packets i and $i + 1$ were created at the source node and $Tp(i)$ is the difference between the times when packets i and $i + 1$ were played back at the destination node, then $Jitter = Tp(i) - Tc(i)$.
	MOS Value	Recorded Mean Opinion Score (MOS) of the voice traffic.
	Packet Delay Variation	End-to-end delay variance experienced by the voice packets from the time they are created until they are received.
	Packet End-to-End Delay (sec)	The total delay experienced by the voice packets (i.e., "analog-to-analog" or "mouth-to-ear" delay). It includes network, encoding/decoding, and (de)compression delays. This statistic records data collected from all the nodes in the network.
	Traffic Received (bytes/sec), *Traffic Received (packets/sec)*	Average traffic arrival rate to all Voice applications deployed in the network. The statistic is computed based on the application data arriving from the transport layer.
	Traffic Sent (bytes/sec), *Traffic Sent (packets/sec)*	Average traffic departure rate from all Voice applications deployed in the network. The statistic is computed based on the application data forwarded to the transport layer.

TABLE 5.4

Summary of Node Application Statistics

Category	Name	Description
Client DB, Client DB Entry, Client DB Query	Response Time (sec)	Time it takes for a database response packet to arrive as a reply to a DB Entry/Query request packet for all Database applications in this node. This statistic is not available for the Client DB category.
	Traffic Received (bytes/sec), Traffic Received (packets/sec)	Average traffic arrival rate to Database (Entry and Query)/Entry/Query requests for all Database applications deployed at this node. The statistic is computed based on the application data arriving from the transport layer.
	Traffic Sent (bytes/sec), Traffic Sent (packets/sec)	Average traffic departure rate from Database (Entry and Query)/Entry/Query requests for all Database applications deployed at this node. The statistic is computed based on the application data forwarded to the transport layer.
	Transaction Size (bytes)	Size of the transaction packets generated by the Database Entry/Query requests for all Database applications in this node. This statistic is not available for the Client DB category.
Server DB, Server DB Entry, Server DB Query	Load (requests/sec)	Arrival rate of Database (Entry and Query)/Entry/Query requests at the server node. These requests can belong to different Database sessions that could originate from different nodes in the network.
	Load (sessions/sec)	The number of currently active sessions at this database server. This statistic is only available for the Server DB category.
	Task Processing Time (sec)	The time it takes for this server node to process a Database (Entry and Query)/Entry/Query request. This statistic is computed by subtracting the time at which the request arrives from the time the server completes processing this request.
	Traffic Received (bytes/sec), Traffic Received (packets/sec), Traffic Sent (bytes/sec), Traffic Sent (packets/sec)	Each of these Server DB statistics has the same meaning as the corresponding Client DB statistics except they are collected on a single specific server node that supports the Database application.

(Continued)

TABLE 5.4 (Continued)
Summary of Node Application Statistics

Category	Name	Description
Client E-mail	*Download Response Time (sec),* *Traffic Received (bytes/sec),* *Traffic Received (packets/sec),* *Traffic Sent (bytes/sec),* *Traffic Sent (packets/sec),* *Upload Response Time (sec)*	Each of the Client E-mail statistics has the same meaning as the corresponding Global E-mail statistics except that these statistics are only collected on a single specific client node that runs e-mail application services.
Client FTP	*Download Response Time (sec),* *Traffic Received (bytes/sec),* *Traffic Received (packets/sec),* *Traffic Sent (bytes/sec),* *Traffic Sent (packets/sec),* *Upload Response Time (sec)*	Each of these Client FTP statistics has the same meaning as the corresponding Global FTP statistics except that these statistics are only collected on a single specific client node that runs the FTP application.
	Download File Size (bytes), *Upload File Size (sec)*	The size of the download/upload file received/sent by the client node.
Client HTTP	*Object Response Time (sec),* *Page Response Time (sec),* *Traffic Received (bytes/sec),* *Traffic Received (packets/sec),* *Traffic Sent (bytes/sec),* *Traffic Sent (packets/sec)*	Each of these Client HTTP statistics has the same meaning as the corresponding Global HTTP statistics except that these statistics are only collected on a single specific client node that runs the HTTP application.
	Downloaded Objects, *Downloaded Pages*	The total number of downloaded individual in-lined objects or complete HTML pages.

	User Cancelled Connections	This statistic collects the number of times the page download was cancelled because the user started a new page download while the previous page was still being loaded.
Client Print	*File Size (bytes)*	The average size of the job sent to the printer.
	Traffic Sent (bytes/sec), Traffic Sent (packets/sec)	Average traffic departure rate from Print applications deployed in this node. The statistic is computed based on the application data forwarded to the transport layer.
Client Remote Login	*Response Time (sec), Traffic Received (bytes/sec), Traffic Received (packets/sec), Traffic Sent (bytes/sec), Traffic Sent (packets/sec)*	Each of these Client Remote Login statistics has the same meaning as the corresponding Global Remote Login statistics except that these statistics are only collected on a single specific client node that runs the Remote Login application.
Server E-mail, Server FTP, Server HTTP, Server Remote Login, Server Print	*Load (requests/sec)*	Arrival rate of client requests at the server node. These requests can belong to different sessions that could originate from different nodes in the network.
	Load (sessions/sec)	Collects the number of currently active sessions at this server.
	Task Processing Time (sec)	The time it takes for this server node to process a client request. This statistic is computed by subtracting the time at which the request arrives from the time the server completes processing this request.
	Traffic Received (bytes/sec), Traffic Received (packets/sec), Traffic Sent (bytes/sec), Traffic Sent (packets/sec)	Each of these Server statistics has the same meaning as the corresponding Client statistics except that they are collected on a single specific server node that supports the corresponding application services. The Traffic Sent statistics are not available for the Server Print statistics category.
Video Called Party, Video Calling Party, Video Conferencing	*Packet Delay Variation, Packet End-to-End Delay (sec), Traffic Received (bytes/sec), Traffic Received (packets/sec), Traffic Sent (bytes/sec), Traffic Sent (packets/sec)*	Each of these statistics has the same meaning as the corresponding Global Video Conferencing statistics except that they are collected on a single specific node that supports the Video Conferencing application. The difference between these statistics categories is as follows: • Video Called Party—collects statistics on a per caller basis for the called party. • Video Calling Party—collects statistics on a per caller basis for the calling party. • Video Conferencing—collects statistics for all video traffic at this node.

(Continued)

TABLE 5.4   *(Continued)*
Summary of Node Application Statistics

Category	Name	Description
Voice Application, Voice Called Party, Voice Calling Party	*MOS Dejitter Delay (sec)*	Delay in seconds caused by eliminating jitter when packets are sent over the network.
	MOS Dejitter Loss Rate	The packet loss ratio computed as the ratio of the number of packets lost to the total number of packets, where loss is due to the interarrival delay being greater than the de-jitter delayl.
	MOS Network Loss Rate	The packet loss ratio computed as the ratio of the packets lost due to certain network conditions out of the total number of packets sent.
	MOS Value	Mean Opinion Score value.
	Packet Delay Variation, Packet End-to-End Delay (sec), Traffic Received (bytes/sec), Traffic Received (packets/sec), Traffic Sent (bytes/sec), Traffic Sent (packets/sec)	Each of these statistics has the same meaning as the corresponding Global Voice statistics except that they are collected on a single specific node that supports the Voice application. The difference between these statistics categories is as follows: • Voice Application—collects statistics for all voice traffic at this node. • Voice Called Party—collects statistics on a per-caller basis for the called party. • Voice Calling Party—collects statistics on a per-caller basis for the calling party.

In such situations, the statistic description applies to all statistics categories in that section unless noted otherwise. A separate statistic exists for each of these categories, and it only computes the statistic value for traffic belonging to that category alone. For example, the global statistic **Traffic Received (bytes/sec)** exists for both the **DB Entry** and the **DB Query** categories. This statistic measures the average traffic rate in bytes per second received by all database applications deployed in the network. However, the statistic is computed separately for the Database Entry and the Database Query operations.

6 Advanced Traffic Generation Features

6.1 INTRODUCTION TO CUSTOM APPLICATIONS

The standard application models are fairly easy to configure, and they provide an adequate level of detail for modeling commonly used applications such as e-mail, FTP, and remote login. However, the standard applications only model a **two-tier** communication paradigm and do not allow for modifications of the simulated application protocols. To address this issue, OPNET also provides facilities for modeling **custom applications**, which could represent nonstandard, multitier applications that follow a user-defined protocol. For example, the standard application model for e-mail only simulates a two-tier client-server message exchange, where the client periodically sends newly written messages to its e-mail server and also independently polls the server to retrieve e-mails destined for it. However, in real life, an e-mail application behaves slightly differently. The main steps in sending an e-mail message from one client to another include the following:

- Client A writes an e-mail message to client B.
- Client A uploads this e-mail message to its e-mail sever.
- The e-mail server for client A relays the message to the e-mail server for client B.
- Client B downloads the e-mail message sent by client A from its e-mail server.

The above behavior represents a **multitier** message exchange in which there is a complex sequence of interactions between clients A and B through their e-mail servers. Modifying the standard e-mail application model to simulate such an e-mail protocol is a very challenging undertaking, which probably will require the use of OPNET Modeler to change the code of the e-mail application process model. On the other hand, such behavior can be easily simulated using the custom application-modeling framework without writing a single line of code.

In OPNET, all custom applications are defined through a series of **tasks**. Each task is further divided into individual **phases**. Figure 6.1 summarizes the architectural hierarchy of the custom application-modeling framework. Each phase of a task represents either a request-response message exchange between two end nodes or processing on a single end node. For example, the first phase might represent the message exchange between an originating node and server A, the second phase might model the message exchange between server A and server B, and so on, creating a succession of phases, where each end node represents a separate communication tier.

153

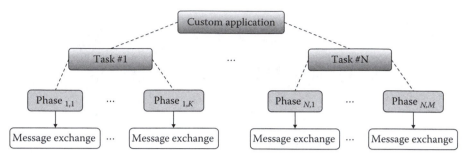

FIGURE 6.1   Custom application hierarchy.

Moreover, the end node that serves as a destination in one phase typically serves as a source of the message exchange in the following phase. However, the node that initiated the task (i.e., the source/originator of the first phase within the task) might not necessarily be the destination node of the last phase within the task, although it is fairly common for a custom application to have the source in the first phase and the destination in the last phase be the same end node.

Consider an example of an application for purchasing items over the Internet, which requires the following steps:

- **Login**
  - The client sends its credentials to the web server.
  - The web server contacts an authentication server to verify the client's credentials.
  - The authentication server processes the provided credentials and generates a response for the web server.
  - The web server sends an ACK back to the client indicating whether or not the login was successful.
- **Browsing**
  - The client sends a request for an item to the web server.
  - The web server searches its internal database and sends the results of the search back to the client.
  - This process is repeated multiple times while the client browses the inventory of items.
- **Purchase**
  - The client provides to the web server the list of items it wishes to purchase and its credit card information.
  - The web server contacts the bank server to verify the validity of the credit card information.
  - The web server contacts the database server to update the stored information about the purchased items.
  - The web server sends a receipt for the purchase back to the client.

To model the application described in the example above, each top-level bullet of the list (e.g., login, browsing, and purchase) could represent a task within the custom application definition, while each of the subbullets could represent a phase within the corresponding task. The message exchange between individual end nodes during each of the phases could be modeled by specifying such configuration parameters as message length, message generation rate, and the total number of messages exchanged, similar to the attributes used in the definitions of standard applications. Note that every end node participating in the information transfer during the corresponding custom application phase effectively denotes a separate communication tier.

As you can imagine, the process of specifying a custom application is fairly complex. For this reason, prior to starting the configuration of a custom application using OPNET, it is prudent to carefully examine and specify all the aspects of the custom application to be modeled. Specifically, it is important to identify and define the following characteristics of the application:

- The number and type of end nodes or tiers involved in this application.
- The individual tasks that compose the modeled application.
- The individual phases of each task.
- For each phase of the task:
  - The end nodes or tiers involved in the phase.
  - If the phase is a data transfer, the characteristics of the traffic exchanged between the end nodes.
  - If the phase involves processing at the server, the amount of time typically taken for processing.

For the custom application example described above, the description has already clearly identified the end nodes involved in the process (e.g., client, web server, database server, authentication server, and bank server). The application has also been divided into individual tasks, and the phases for each task have been clearly specified. The only remaining job is to define the characteristics of the traffic exchanged during each phase and the length of the authentication processing during the login task. As you can imagine, the details of such a configuration are highly application-specific and must be treated on a case-by-case basis. Hopefully, this section has provided a good overview of the key steps in defining custom applications in OPNET and you now have a basic understanding of how an application can be represented through tasks and phases.

Once the application specifications are determined, the process of configuring a custom application in OPNET consists of the following steps:

- Configure each task of the custom application using the *Task Config* utility object.
- Configure the custom application using the *Application Config* utility object.
- Specify the statistics to be collected during the simulation.
- Specify a user profile that employs the configured custom application using the *Profile Config* utility object.
- Deploy the defined user profile in the simulated system.

In this chapter, we primarily concentrate on the first three items mentioned in the list, while specifying user profiles and deploying applications are discussed in Chapter 7. Section 6.2 describes OPNET specifics for configuring individual tasks and phases. It is then followed by sections that describe how to configure custom applications using specified task definitions (Section 6.3), and then a detailed example of configuring a custom application (Section 6.4). Sections 6.5 and 6.6 provide an overview of explicit traffic generation sources and traffic demands. The chapter concludes with a description of available statistics for evaluating the performance of various advanced OPNET traffic generation mechanisms.

6.2   CONFIGURING TASKS AND PHASES FOR CUSTOM APPLICATIONS

The first step for specifying custom applications is to add the *Task Config* utility object into the project workspace.

6.2.1   *Task Config* Utility Object

The *Task Config* utility object, shown in Figure 6.2, is located in the same folders of the **Object Palette Tree** as the *Application Config* object (see Section 5.3.1). The steps for adding the *Task Config* object in the **Project Editor**'s workspace are the same as for any other object:

- Identify and select the object in the **Object Palette Tree**.
- Drag and drop the object into the workspace.

As with other utility objects, only one *Task Config* node is needed to configure all the tasks within the current simulation scenario.

As shown in Figure 6.3, the *Task Config* object consists of only two attributes: **name**, which specifies the name of the object as it appears in the workspace, and **Task Specification**, a compound attribute that allows specifying individual tasks for custom applications. As common to many compound attributes that define multiple instances of the same item, the **Task Specification** attribute contains a subattribute called **Number of Rows**, which specifies the number of tasks to be defined for all custom applications in this simulation. As usual, this attribute accepts either zero or a positive integer value. The value of zero represents a

FIGURE 6.2   *Task Config* utility object.

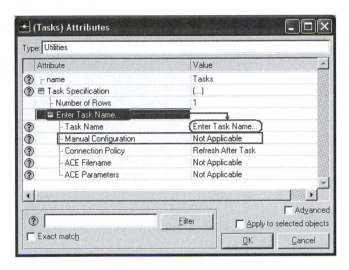

FIGURE 6.3   Attributes of the *Task Config* object.

situation when there are no tasks available for custom applications. Setting this attribute to a positive integer value, say N, will result in N compound subattributes for configuring custom application tasks appearing below the attribute **Number of Rows**.

6.2.2   SPECIFYING TASK DEFINITIONS

Each row of the **Task Specification** attribute in the *Task Config* object is used to define a separate task. When first created, each row, which we will call a task definition, is named **Enter Task Name...** by default. As shown in Figure 6.3, a task definition consists of the following attributes:

- **Task Name**—specifies an alphanumeric name of the task. Setting this attribute's value automatically changes the name of the whole row that contains the current task definition.
- **Manual Configuration**—configures the task behavior by defining the actions performed during individual phases of the task, which will be discussed later.
- **Connection Policy**—specifies the connection reuse policy. The connection could be either reused across all the tasks or reset for every task. However, the connection reuse policy defined by this attribute is superseded by the connection policies defined for individual phases of the task. We discuss this issue in more detail in Section 6.2.3.
- **ACE Filename** and **ACE Parameters**—these attributes allow configuring custom application tasks using files generated by the Application Characterization Environment (ACE) OPNET software program. This topic is beyond the scope of this book and is not discussed here.

FIGURE 6.4 *Manual Configuration Table.*

The Manual Configuration attribute is represented in the form of a table, shown in Figure 6.4, where each row specifies a single phase within the task. The number and the order of the phases in the *Manual Configuration Table* are controlled through the textbox called *Rows*. This textbox accepts a nonnegative integer value that specifies the total number of rows, or phases, to be defined within the table. The following buttons are located next to the textbox:

- The **Delete** button removes the highlighted row or phase from the table. Removing a row from the table also decrements by 1 the number of rows specified in the textbox *Rows*. The **Delete** Button is unavailable when the number of rows is set to 0.
- The **Insert** button adds a row into the table. The new row is added above the row that was selected when the **Insert** button is clicked. When a new row is inserted into the table, the number of rows specified in the textbox *Rows* is incremented by 1.
- The **Duplicate** button copies the selected row. The copy is inserted below the selected row. The number of rows specified in the textbox *Rows* is also incremented by 1 when a row is duplicated. The **Duplicate** Button is unavailable when the number of rows is set to 0.
- The **Move Up** and **Move Down** buttons allow manipulation of the order of the rows within the table by moving them up or down. Neither button is available when the number of rows is set to 0 (i.e., there are no rows to be moved up or down).
- The **OK** button saves the changes and closes the window that contains the *Manual Configuration Table*, while the **Cancel** button closes the window without saving the changes.
- The **Details** button provides a description of the selected column.
- The **Promote** button promotes the value of the selected cell to the next higher level.

6.2.3 Specifying Phase Configurations

The columns of the *Manual Configuration Table* represent the configuration attributes of the phases, while each row in the table corresponds to a complete configuration of a single phase. We now describe the meaning of each of these columns.

The first column, **Phase Name**, specifies an alphanumeric name of a phase within the task and the statistic group that the phase belongs to (optional). Clicking on a cell under the **Phase Name** column opens a window for defining the name and the statistic group. The name of the phase is primarily needed for readability (e.g., Phase #2, Transaction #7, Authentication Processing, or Submission of Credentials) and has no influence on the simulation. If the statistic group name is set, then the simulation will record the response time for the phases that share the same statistic group name. The response time is computed as the difference between the time the last phase completed its execution and the time when the first phase in the group started its execution.

The second column, **Start Phase After**, specifies when the phase will start. This attribute accepts the following values:

- `Application Starts` is a predefined value, which, if selected, will configure the current phase to begin as soon as the application starts.
- `Previous Phase Ends` is a predefined value, which, if selected, signifies that the current phase will begin right after the previous phase (i.e., the phase defined in the row above) completes its execution.
- The pull-down list of values available for this attribute also contains the names of all the phases defined so far. It is possible to set the value of the **Start Phase After** column to a specific phase name, in which case the current phase will start when the specified phase completes its execution.
- `Edit…` specifies a comma-separated list of phase names. In such a case, the current phase will start only after all the phases in the provided list complete their execution. This option is particularly useful when a single phase depends on multiple phases. For example, consider a situation where a proxy server requires login information and the name of the website to allow the client to access that site. When the proxy server receives the required information, it starts the authentication and name resolution phases. These phases may run concurrently and may complete in any order. Only after both the authentication and the name resolution phases complete can the data transfer phase begin. Specifying multiple phase names in the comma-separated list allows such behavior to be modeled. The comma-separated list should contain no spaces between the comma and the name of the following phase. OPNET does not do any error-checking at run-time, which means that this attribute will accept any typed-in value, even an incorrect one. However, in the presence of errors (e.g., an incorrect phase name or a space between the comma and the next list entry), the simulation will fail to execute and will provide an error message `Phase Error: Encountered phase configuration error`.

The third column of the *Manual Configuration Table* is called **Source**. This column specifies a symbolic name for the source node of the current phase. For each task, the actual name of the source or destination node is resolved only once, and all the phases of the task adhere to the same name resolution. A workstation, a server, or a LAN object can serve as the source or destination node. This attribute accepts the following values:

- Originating Source—the node on which a custom application that includes the current task (i.e., phase) is deployed will serve as the source node for the current phase.
- Previous Source—the node that was the source during the previous phase will serve as a source for the current phase as well.
- Previous Destination—the node that was the destination during the previous phase will serve as the source for the current phase.
- Edit...—allows specifying the symbolic source name as a string. This name must be mapped or resolved through the source or destination preferences attribute of an actual node in the simulated network. If the **Source** symbolic name is defined for the first time within the current task, then it should be mapped to the actual node name through the **Source Preferences** attribute in the corresponding node. The same mapping will then be used in all the phases throughout the task. ·

The next column is called **Destination** and it specifies the symbolic name of the destination node for the current phase. This attribute accepts the following values:

- Originating Source—the node on which a custom application that includes the current task (i.e., phase) is deployed will serve as the destination node for the current phase.
- Previous Source—the node that was the source during the previous phase will serve as the destination for the current phase.
- Previous Destination—the node that was the destination during the previous phase will serve as the destination for the current phase as well.
- Not Applicable—there is no destination node in this phase. Typically, in such a case, the phase is used to model processing at the source node. Also, if the **Destination** attribute is set to Not Applicable, then the attribute values that define the traffic sent between the source and the destination of this phase are ignored.
- Edit...—allows specifying the destination symbolic name as a string. This name also must be mapped to an actual node in the simulated network, if the symbolic name is used for the first time within the task. The same mapping will be used in all the phases throughout the task.

The next two columns specify the characteristics of the traffic generated in the phase. The column **Source->Dest Traffic** defines the characteristics of the traffic transmitted from the source node to the destination. Effectively, this compound attribute specifies how request messages are generated at the source.

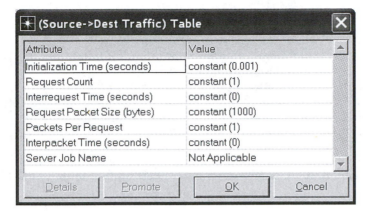

FIGURE 6.5   Table for source–destination traffic characteristics.

The source–destination traffic characteristics are defined through the following subattributes and are shown in Figure 6.5:

- **Initialization Time (seconds)**—the time that elapses before the source node starts generating requests. If there is no destination node defined for this phase, then this attribute specifies the processing time at the source node.
- **Request Count**—the number of requests sent from the source node to the destination.
- **Interrequest Time (seconds)**—the amount of time elapsed between two consecutive requests.
- **Request Packet Size (bytes)**—the size of the individual packet. Since a request may consist of multiple packets, this attribute defines the size of one of the request's packets. Thus, the size of the request itself can be computed by multiplying the packet size by the number of packets per request.
- **Packets Per Request**—specifies the number of packets in a single request.
- **Interpacket Time (seconds)**—the amount of time elapsed between two consecutive transmissions of packets that belong to the same request.
- **Server Job Name**—the name of the server job defined within the *Server Config* utility object that is associated with this phase. The topic of server configuration is outside the scope of this book and is not discussed here.

The column **Dest->Source Traffic** of the *Manual Configuration Table* defines the characteristics of the traffic transmitted from the destination node to the source. Effectively, this compound attribute defines how response messages are generated by the destination node. This attribute has one preset value called No Response, indicating that the destination node generates no response message. Also, if the symbolic name of the destination node is left undefined, then the values of this attribute are

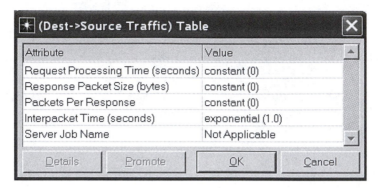

FIGURE 6.6   Table for destination–source traffic characteristics.

ignored. The destination-source traffic characteristics are defined through the following subattributes and are shown in Figure 6.6:

- **Request Processing Time (seconds)**—specifies the time spent by the destination node processing the request before it can generate a response.
- **Response Packet Size (bytes)**—specifies the size in bytes of each response
  packet generated by the destination node (there may be multiple packets in
  the response).
- **Packets Per Response**—specifies the number of packets in each response
  message generated as a reply to the request packet.
- **Interpacket Time (seconds)**—the amount of time elapsed between two
  consecutive transmissions of packets that belong to the same response
  message.
- **Server Job Name**—the name of the server job associated with this phase.
  The topic of server configuration is outside the scope of this book and is not
  discussed here.

The next column of the *Manual Configuration Table* is **REQ/RESP Pattern**.
It specifies the message exchange pattern between the source and the destination
nodes. This attribute allows only two preset values:

- REQ->RESP->REQ->RESP... (Serial) value indicates that the source
  will not generate a new request until it receives a response from the destination. Specifically, the source node will wait for the **Interrequest Time
  (seconds)** after the response message arrival before sending the next request
  message. This value is not available if the **Dest-> Source Traffic** attribute
  is set to No Response.
- REQ->REQ->...->RESP... (Concurrent) value indicates that the source
  will generate the requests according to the configuration of the **Interrequest
  Time (seconds)** attribute without waiting for the response arrival.
  Effectively, the request messages will be always generated **Interrequest
  Time (seconds)** apart, independent of destination responses.

The next column of the *Manual Configuration Table*, **End Phase When**, speci-fies the conditions that denote the completion of a phase. This attribute allows one of the following values:

- `Final Request Leaves Source`—the phase completes after the source node generates **Request Count** number of request messages (i.e., the last packet of the last request have been sent by the source).
- `Final Request Arrives at Destination`—the phase completes after the destination node receives **Request Count** number of request mes-sages (i.e., the last packet of the last request has arrived at the destination).
- `Final Response Leave Destination`—the phase completes after the destination node generates the response message to the last request from the source (i.e., the last packet of the last response message has been sent by the destination).
- `Final Response Arrives at Source`—the phase completes after the source node receives the last response message from the destination (i.e., the last packet of the last response message arrives at the source node).

The column **Timeout Properties** specifies the maximum amount of time allowed for the phase to execute. This attribute was added to handle situations where the phase connection was terminated or when the phase transaction packets have been lost. If the phase does not complete within the specified time, then it is forcefully ter-minated regardless of whether all transactions have ended or not. This attribute also allows specifying what will be executed after a forceful phase termination: the next phase (i.e., continue normal task execution) or the next task (i.e., skip the remaining phases of the task). This attribute has a preset value called `Not Used` indicating that the phase must complete all of its transactions before termination.

The final column of the *Manual Configuration Table* is called **Transport Connection** and it specifies how the transport protocol is used by this phase. Figure 6.7 illustrates the *Transport Connection Table*, which consists of two subattributes:

- **Policy**—defines how the transport connections are established within this phase. This attribute accepts two values:
  - `New Connection per Request`—means that the request from the source to the destination will be transmitted over a separate con-nection. After the source receives the response, the corresponding con-nection is closed. Only one request–response pair is allowed for each connection.
  - `As Specified per Connection`—means that the connection policy is determined through the other compound attribute called **Connections**.
- **Connections**—is a compound attribute that specifies such information as the number of connections, transport protocol, transport port, and the policy for connection reuse. Figure 6.8 illustrates the *Connections Table*. The num-ber of connections per phase is controlled through the textbox titled *Rows*. The rest of the buttons in this table provide the same functionality as in

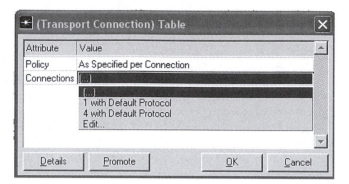

FIGURE 6.7   *Transport Connection Table.*

Tag	Transport Protocol	Transport Port	Policy
Default	Default	Default	Refresh After Phase
Default	TCP	Default	Refresh After Phase
Default	TCP	Default	Refresh After Phase
Default	TCP	Default	Refresh After Phase

4 Rows   Delete   Insert   Duplicate   Move Up   Move Down
Details   Promote   ☐ Show row labels   OK   Cancel

FIGURE 6.8   *Connections Table.*

other similar tables (e.g., the *Manual Configuration Table*). The meaning of the subattributes in the *Connections Table* is as follows:

- **Tag** is the symbolic name of the connection used to identify shared connections.
- **Transport Protocol** is the name of the transport protocol used by this connection. If set to Default, then this connection will run over the transport protocol defined for this custom application through the *Application Config* utility object.
- **Transport Port** specifies the port number on which the connection will connect. If set to Default, then this connection will connect to the port number defined for this custom application through the *Application Config* utility object.
- **Policy** specifies the policy for reusing the connection. If the **Policy** attribute of the *Transport Connection Table* is set to New Connection per Request (see Figure 6.8), then the configuration of this attribute is ignored. This attribute can take one of two values:
  - Refresh After Phase means that all message transfers of this phase will be conducted over defined connections. Once the phase

completes, all opened connections will be closed. If the phase is configured with multiple connections, then the connections are chosen for message transfer in a round-robin fashion.

– Reuse Across Phases means that the connection belongs to the whole task and the actual policy for reusing the connection is defined through the **Connection Policy** attribute of the *Task Config* object (see Figure 6.3). Once the phase completes, the connections will remain open until the last phase of the task is reached. After that, the connections will either be closed, if the **Connection Policy** attribute value is set to Refresh After Task, or be reused, if it is set to Reuse Across Tasks. Specifically, in the latter case, the connections will remain open until the application, to which the task belongs, terminates.

Note that connection sharing is only allowed across the same application instance. For example, if two nodes execute the same application, then each node creates a new application instance, and therefore, the connections cannot be shared among these applications. In practice, if a new phase that transmits traffic is about to start, then it first searches for any shared connection that has the same connection tag value. If such a connection is found, then that connection is reused; otherwise, a new connection is initiated. Once the phase completes, it checks the value of the **Policy** attribute in the *Connections Table* (e.g., Reuse Across Phases or Refresh After Phase) to determine if the connection is to be terminated. The above situation is not applicable to the case when the **Policy** attribute in the *Transport Connection Table* is set to New Connection per Request, in which case a new connection is opened for every request within the phase.

The **Connections** attribute in the *Transport Connection Table* accepts one of the following two preset values:

- 1 with Default Protocol—configures a single default transport protocol, connected to a default port, with the connection being refreshed after the phase.
- 4 with Default Protocol—configures four connections totally. The first connection runs over the default transport protocol, connects to the default port, and is refreshed after the phase. The other three connections are the same except that they are explicitly configured to run over TCP. Such a configuration is shown in Figure 6.8.

6.2.4 Summary: Configuring Tasks for Custom Applications

Sections 6.2.2 and 6.2.3 have described the various attributes that are used to configure tasks and phases for custom application definition. We conclude this section by providing a summary of the steps to specify a task for a custom application:

- Add the *Task Config* object into the simulation scenario's workspace.
- Right-click on the *Task Config* object and select the **Edit Attributes** option.

- Expand the **Task Specification** attribute.
- Specify the total number of tasks to be configured by setting the value of the attribute **Rows** to the corresponding value.
- For each row:
  - Expand the row and set the name of the task.
  - Edit the *Manual Configuration Table* by selecting Edit... from the pull-down list for the **Manual Configuration** attribute:
    - Specify the number of phases in the current task by setting the value of the textbox *Rows*.
    - Configure each individual phase of the task.
    - When all the phases of the task have been completed, press the **OK** button to save the phase configuration.
  - Set the value of the **Connection Policy** attribute.
- When all the tasks have been configured, press the **OK** button to save the changes and exit the *Task Config* **Edit Attributes** window.

6.3   DEFINING CUSTOM APPLICATIONS IN OPNET

After all the tasks for a custom application have been specified, the custom application itself can be defined through the *Application Config* object. The process for defining custom and standard applications is very similar and it consists of the following steps:

- Add the *Application Config* object if it is not already in the project workspace.
- Right-click on the *Application Config* object to open the **Edit Attributes** window.
- Expand the **Application Definitions** compound attribute and specify the total number of applications to be configured (i.e., set the value of the **Number of Rows** attribute).
- Expand the rows one at a time to configure each application:
  - Set the value of the **Name** attribute.
  - Expand the **Description** attribute and configure the *Custom* application.
- Repeat the process for the rest of the applications.
- Click the **OK** button to close the window and save the changes.

The only difference between these steps and those for standard applications (Section 5.3.2) is the set of attributes available for defining *Custom* applications, which are shown in Figure 6.9.

Unlike standard applications, the definition of custom applications does not include such information as packet size and data rate. All these attributes have been already defined within the corresponding tasks and phases. The definition of a custom application primarily deals with specifying which tasks are to be used and how

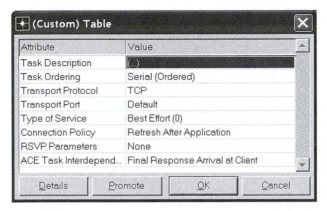

FIGURE 6.9   Attributes available for defining custom applications.

these tasks are executed. Specifically, the *Custom* application definition consists of the following attributes:

- **Task Description** defines the tasks to be used in this custom application. This compound attribute consists of two subattributes:
  - **Task Name** specifies the name of the task. Only those tasks that have been already defined in the *Task Config* object can be selected as a value for this attribute. In fact, the pull-down menu for this attribute provides the list of available tasks.
  - **Task Weight** controls the probability that the current task will be selected. This attribute is only used when there are multiple tasks within the application and the tasks are configured to run in a random fashion. In such a case, the selection probability of a task is computed as the ratio between the task's weight and the sum of all weights for all the tasks within this application.
- **Task Ordering** specifies the order of task execution. This attribute can be set to one of three preset values: `Serial(Ordered)`, `Serial(Random)`, or `Concurrent`. Each of these values is configured to execute all the tasks within the custom application using the ordering scheme that corresponds to the name of the value. The meaning of these values is self-explanatory. This compound attribute also consists of two subattributes, one of which has the same name:
  - **Task Ordering** subattribute can be also set to either `Serial(Ordered)`, `Serial(Random)`, or `Concurrent`. Notice that if the parent attribute **Task Ordering** is set to one of its values, then its subattribute **Task Ordering** will be automatically set to the corresponding value with the same name.
  - **Number of Used Tasks** specifies the number of application tasks that will be executed. The default value for this attribute is `Number of Specified Tasks`, which means that the simulation will attempt to

execute all the tasks defined for this application. If this attribute value is set to a number, say N, then only N tasks within the application will be executed. Specifically, if the `Serial(Ordered)` or `Concurrent` task ordering is selected, then only the first N application tasks will be executed. However, if the application is configured to run tasks using the `Serial(Random)` ordering, then the simulation will select N tasks randomly.

- **Transport Protocol** specifies the transport protocol used by the custom application. This value is used only when the phase of the application's task has its **Transport Protocol** attribute value set to `Default`. Otherwise, the transport protocol defined for the application phase takes precedence. One of the predefined values for this attribute is `Direct Delivery`. If the transport protocol value for a custom application is set to `Direct Delivery`, then the message exchange is simulated directly between the application instances, and all delay and other characteristics associated with transport, network, medium access, and other layers are ignored.

- **Transport Port** specifies the port number used by the custom application. This value is used only when the phase of the application's task has its **Transport Port** attribute value set to `Default`. Otherwise, the port number defined for the application phase takes precedence.

- **Type of Service** specifies the *ToS/DiffServ* field value in the IP packet header. For more details, refer to Section 5.4.1.

- **Connection Policy** specifies the connection reuse policy. This attribute can only be set to the value `Refresh After Application`, which means that the connection will be terminated upon application completion and that the connection reuse policy defined in the *Task Config* object governs the application's message exchange.

- **RSVP Parameters** specifies the configuration of the RSVP protocol. For more details, see Section 5.4.1.

- **ACE Task Interdependency** deals with defining custom applications using ACE. This topic is outside the scope of this book and is not discussed here.

6.4   EXAMPLE OF CONFIGURING CUSTOM APPLICATIONS IN OPNET

Let us use an example shown in Figure 6.10 to illustrate the procedure for configuring custom applications in OPNET. This is the same example discussed in Section 6.1. The custom application illustrated in Figure 6.10 can be configured as a single task with several phases. However, to illustrate the additional features available in OPNET, we will model this application using three separate tasks with the following phases:

Task *Login*:
- **Phase *Submit Credentials*:** The client submits its credentials as a single message of size 1024 bytes. The web server will not send a response to the client until it verifies the credentials with the authentication sever.

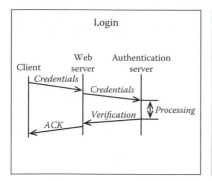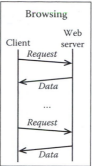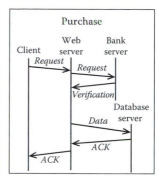

FIGURE 6.10   Example of a multitier application.

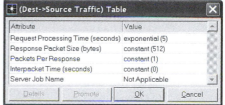

FIGURE 6.11   Traffic configuration for the *Verify Credentials* phase.

- **Phase** *Verify Credentials*: The web server forwards the client's credentials (a single 1024-byte message) to the authentication server, which spends about 5 seconds processing the data before sending the response, which consists of a single 512-byte packet. Figure 6.11 illustrates the traffic configuration in both directions for this phase. Notice that we simulated the processing at the authentication server by setting the value of the **Request Processing Time (seconds)** attribute to 5 seconds.
- **Phase** *Send ACK*: The web server sends an ACK back to the client after it receives verification from the authentication server. The ACK message is 1024 bytes.

Task *Browsing*:
- **Phase** *Browse*: The client periodically submits a request to the web server. Each client request is 1024 bytes long and consists of a single message. The client takes about 5 minutes to process the received information before it issues another request. The client sends about 40 requests before purchasing an item. For each request, the web server sends a response message that is 10,000 bytes long. Each response message is split into five packets. It takes about 2 seconds for the web server to retrieve the information about the requested item. Figure 6.12 illustrates the traffic configuration in both directions for this phase. The interrequest time for generating request messages is set to 300 seconds, which represents the 5 minutes the client spends processing

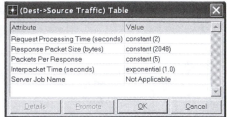

FIGURE 6.12 Traffic configuration for the *Browse* phase.

information from the web server. The response packet size is set to 2048 bytes. Since there are 5 packets per response message, the total size of the response is 5×2048 bytes $= 10,000$ bytes.

Task *Purchase*:

- **Phase *Submit Request*:** The client sends a request to the web server to purchase an item. The request is 6,000 bytes long and is split into three messages. The web server does not send a response until it receives an ACK from the other servers.

- **Phase *Validate CC*:** the web server sends the client's credit card information to the bank server. The credit card information is sent as a single 2000 bytes packet. The bank server verifies the correctness of the provided information and sends back a response message with a confirmation code. The size of the response message is 512 bytes.

- **Phase *Update Database*:** This phase starts simultaneously with the *Validate CC* phase. The web server sends information about the purchased item to the database server. The request size is 2000 bytes and it consists of a single message. The database server sends an ACK back to the web server. The size of the response packet is 128 bytes.

- **Phase *Complete Purchase*:** Once the web server receives the response messages from the bank and database servers (i.e., both the *Validate CC* and the *Update Database* phases have completed), it generates a message back to the client. The confirmation message is 10,000 bytes long and it is split into five packets.

Due to space considerations, the above example does not model situations such as failure to login (i.e., invalid login information) and failure to purchase (i.e., invalid credit card information). The summary of task configurations is provided in Figure 6.13. Here are some observations regarding the above configurations:

- For simplicity, we keep the transport connection policy set to default values.
- The client node is denoted as *Originating Source*, which is the node on which the custom application is deployed.
- The *Submit Credentials* and the *Submit ACK* phases of the *Login* task have no responses generated by the destination. These phases represent the request message sent by the client to the web server and a response by the web server after it completes verification of credentials.

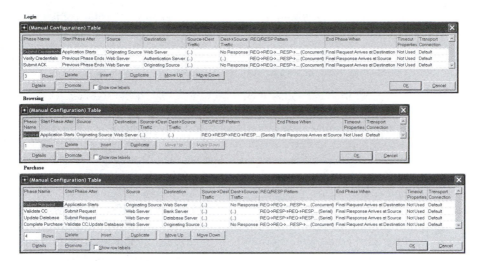

FIGURE 6.13 Summary of task configurations.

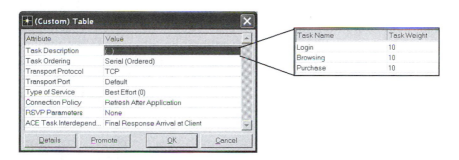

FIGURE 6.14 Configuration of the custom application.

- The *Verify Credentials* phase is configured to have the concurrent REQ/RESP pattern. However, the actual setting of this attribute for the *Verify Credentials* phase is irrelevant because the source generates a single request message.
- The *Browse* phase is configured to have the serial REQ/RESP pattern, because the client should not generate a new request before it receives a response to its previous message.
- In the *Purchase* task, both the *Validate CC* and the *Update Database* phases start after the *Submit Request* phase. This configuration models the simultaneous execution of these phases. As a result, the *Complete Purchase* phase is configured to start only after both the *Validate CC* and *Update Database* phases have completed.

Once all the phases have been specified, the only thing that is left is to define the custom application itself. Figure 6.14 illustrates the configuration of the custom application described in this example. This application is configured to execute the tasks in a serial order: *Login, Browsing*, and *Purchase*. The weight values of the task are irrelevant since the execution order of the tasks is configured as serial.

6.5   EXPLICIT PACKET GENERATION SOURCES

Explicit packet generation is another method for modeling traffic sources. As mentioned in Chapter 5, this technique does not model any particular application layer protocol. Instead, it simulates packet-by-packet traffic generated by an end node. Only the end node models that contain the word *_station* in their names include attributes for defining the explicit packet generation sources. However, different node models often define packet generation parameters differently.

For example, the *ethernet_ip_station_adv* model, shown in Figure 6.15, defines packet generation parameters under a compound attribute called **Traffic Generation Parameters**, which is a child of the compound attribute **IP**. This model specifies the packet generation configuration through such attributes as **Packet Inter-Arrival Time (seconds)**, **Packet Size (bytes)**, **Destination IP Address**, **Start Time (seconds)**, and **Type of Service**. We do not describe these attributes in detail here because their purpose is self-explanatory. Notice that the attribute **Destination IP Address** requires an actual IP address for the destination node. By default, all the nodes in the simulated network system are assigned IP addresses dynamically during the simulation execution. However, you can set the IP address of any node in the network before the simulation execution. In end nodes, the IP address is set through the attribute **IP...IP Host Parameters...Interface Information...Address**. Alternatively, you can set the IP addresses for all nodes in the network by selecting the option **Protocols > IP > Addressing > Auto-Assign IPv4 Addresses** from the pull-down menu in the **Project Editor** (see Chapter 9). Once the destination end nodes have been assigned IP addresses, you need to retrieve those values by examining the value of each node's **IP...IP Host Parameters...Interface Information...Address** attribute and then assigning that value to the **Destination IP Address** attribute in the end node station that explicitly generates packets.

The traffic source definition configured at a node only specifies traffic in one direction from this node to a destination. To configure traffic in the reverse direction requires defining the traffic source at the destination node also. The number of traffic sources originating from a node is controlled through the attribute **Number of Rows**, where each row represents a single traffic flow between this node and a destination.

On the other hand, the *ethcoax_station* model, shown in Figure 6.16, defines ON-OFF traffic sources through the top-level compound attribute **Traffic Generation Parameters**. The ON-OFF traffic source model consists of a series of ON-OFF periods (i.e., ON, OFF, ON, OFF, ON, OFF, etc.), where the source generates traffic during the ON period and is idle during the OFF period. The **Traffic Generation Parameters** attribute consists of the following subattributes:

- **Start Time (seconds)**—the time when the node starts transmitting the traffic.
- **ON State Time (seconds)**—the duration of the ON period.

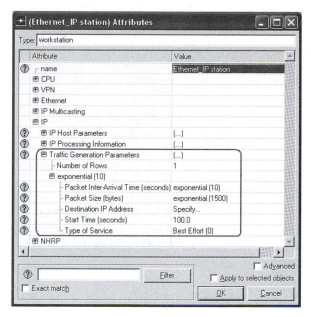

FIGURE 6.15 Explicit packet generation parameters of the *ethernet_ip_station_adv* model.

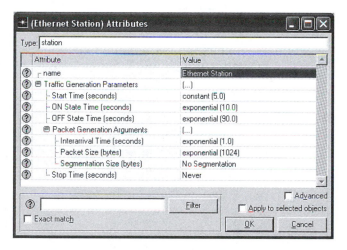

FIGURE 6.16 Explicit packet generation parameters of the *ethcoax_station* model.

- **OFF State Time (seconds)**—the duration of the OFF period.
- **Stop Time (seconds)**—the time when the node stops transmitting traffic. This attribute accepts the preset value `Never`, which configures the node to transmit traffic until the end of simulation.
- **Packet Generation Arguments**—a compound attribute that specifies the characteristics of the generated traffic. This attribute consists of **Interarrival Time (seconds)**, **Packet Size (bytes)**, and **Segmentation Size**

(bytes) subattributes. In certain situations, the traffic source packets are segmented into smaller data chunks before they are sent to the lower layers. The attribute **Segmentation Size (bytes)** specifies the size of such a segment. If this attribute is set to the value No Segmentation, then the packets are sent to the lower layers without being segmented.

Packet generation in the *atm_station* model is defined in yet another manner. The ATM node models contain a compound attribute called **ATM Application Parameters**, which consists of the following subattributes: **Arrival Parameters**, **AAL Parameters**, **Traffic Contract**, and **Destination Address**, as shown in Figure 6.17. The actual traffic characteristics are defined through the compound attribute **Arrival Parameters**, which includes subattributes **Packet Interarrival Time (seconds)**, **Packet Size (bytes)**, **Call Start Time (seconds)**, **Call Duration (seconds)**, **Transmission Size (Mbytes)** (the total amount of traffic transmitted during a single call), and **Call Interarrival Time (seconds)**. In addition, ATM node models that include the suffix *uni_src* in their name also allow configuration of packet generation traffic sources.

Despite these differences, configuring explicit packet generation is fairly straightforward. First, you need to ensure that the nodes of interest actually allow configuring explicit packet generation traffic. The next step is to identify the location of the attributes that configure explicit packet generation. Typically, such attributes are grouped together under a compound attribute called **Traffic Generation**

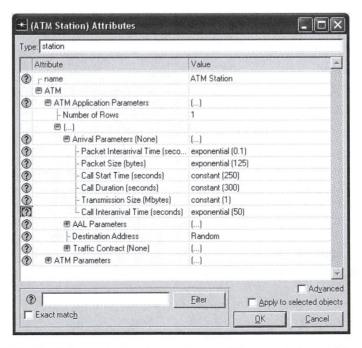

FIGURE 6.17   Explicit packet generation parameters of the *atm_station* model.

Parameters, **Application Parameters**, or a similar name. The final step is to specify the attribute values that define the characteristics of the explicit packet generation source. Typically, such traffic sources are defined through attributes that specify the time when packet generation starts and ends, packet interarrival times, the packet sizes, and the addresses of the destination nodes.

6.6 APPLICATION DEMANDS AND TRAFFIC FLOWS

Demand is another mechanism for specifying traffic flows between two nodes in the network. OPNET differentiates between application demands and traffic flows. An **application demand** is a demand object that models an application layer message exchange between two end nodes.

Traffic flow is another type of demand that simulates end-to-end traffic without representing the details of the traffic flow pattern. Traffic flow demand only specifies the traffic generated at the network, or more specifically the IP layer. Only nodes that contain the IP layer are capable of deploying traffic flow demands. Each demand model has a different set of configuration attributes that primarily depends on the traffic type being modeled. Demands only represent characteristics of a certain type of traffic, and they do not model any specific protocol. Demands are defined between two nodes that are not necessarily connected through a direct link. It is up to the simulation to determine how the demand traffic is to be routed through the simulated network. Describing all available traffic demand models would take a lot of time and may require a separate chapter in itself. That is why we only provide a summary of the characteristics for application and traffic flow demands. If you have some familiarity with the properties of the modeled traffic type (e.g., IP ping, VoIP, and ATM PVX), then you should be able to easily discern the meanings of the corresponding demand attributes.

The process of specifying demands in a simulated system is similar to that of connecting nodes in a network topology with a link. The only difference is that the links join two directly connected nodes, while demands join any two nodes in the network. Figure 6.18 illustrates the difference between demands and links. A demand object is typically represented as a dotted line, while a link is represented as a solid line in the

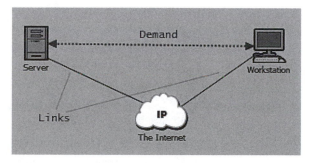

FIGURE 6.18 Example of network topology with demands.

network topology. The process of adding a demand object into a simulation scenario consists of the following steps:

- Identify the demand model of interest in the **Object Palette Tree**.
- Drag the demand model object into the **Project Editor**.
- Connect any two desired nodes.

6.6.1 Application Demands

Application demands model application traffic that flows in both directions between source and destination nodes. By default, application demands are denoted as dotted lines with arrowheads on both ends indicating bidirectional traffic. As shown in Figure 6.19, application demands are configured through the following attributes:

- **Duration** is a compound attribute that specifies the start and end times of the application demand.
- **Request Parameters** is a compound attribute that specifies the size of each request message, the generation rate of the request messages (i.e., the number of requests generated each hour), and the ToS value of each request packet.
- **Response Parameters** is a compound attribute that consists of a single subattribute **Size (bytes)**, which specifies the size of the response message.
- **Traffic Mix (%)** specifies how the application demand will be modeled. In particular, it defines the percentage of demand traffic modeled as background traffic. This attribute can take any of several preset values: All Discrete, 25%, 50%, 75%, All Background, and Edit…. The All

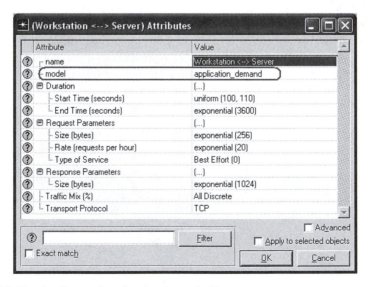

FIGURE 6.19 Attributes of application demand objects.

`Discrete` setting corresponds to a value of 0% while `All Background` corresponds to a value of 100%. Setting `Edit...` allows you to type in any desired value.

- **Transport Protocol** specifies the transport protocol over which the request and response messages of the application demand are to be transmitted.

6.6.2 Traffic Flow Demands

Traffic flows model a certain type of traffic, which could be unidirectional or bidirectional. Traffic flows that model a unidirectional flow are represented by a dotted line with an arrowhead on one end. The arrow points in the direction of traffic flow, where the node pointed to is the destination. Figure 6.20 shows attributes of the IP traffic flow demand. The attributes that are circled in this figure are common to most traffic demands, while the remaining attributes are specific to the type of traffic flow being modeled.

The attributes **Traffic (bits/second)** and **Traffic (packets/second)** specify how the transmission rate of the traffic flow changes over the course of time. These attributes are also known as profiles and they have the following possible values:

- `NONE`, which means that the profile is not defined.
- `Select...`, which provides a list of traffic profiles that you can choose from. As shown in Figure 6.21, a traffic profile consists of three sections: *Profile Library*, which specifies the location of profile definitions; *Profile*, which provides a list of the available profile definitions; and *Preview*, which

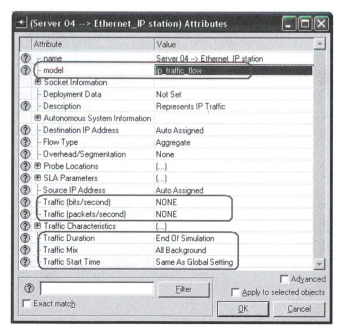

FIGURE 6.20   Attributes of IP traffic flow objects.

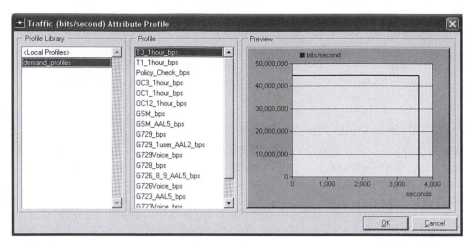

FIGURE 6.21   Selecting preset traffic profiles.

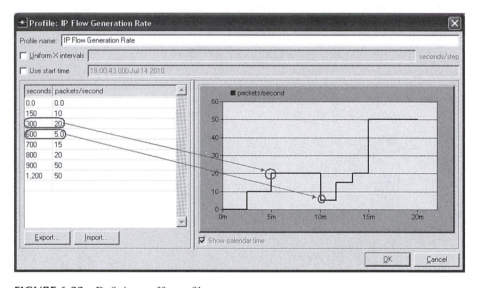

FIGURE 6.22   Defining traffic profiles.

illustrates the selected traffic profile as a graph. For example, in Figure 6.21, the *T3_1hour_bps* profile was selected from the *demand_profiles* library and, as the preview window illustrates, the selected profile starts transmitting traffic at the rate of 44.736 Mbits/seconds (i.e., the rate of a T3 line) at time 0 and continues transmitting traffic at the same rate for the next 3600 seconds.

Selecting Edit... in the value field of the attribute opens a window in which you can define a new traffic profile, as shown in Figure 6.22. The actual traffic profile definition is specified through a table with two columns: *seconds* (i.e., *x*-axis) and *bits/second* or *packets/second* (i.e., *y*-axis), depending on which type of traffic profile has been selected. Figure 6.22 shows the profile configuration for the attribute **Traffic (packets/second)**, which is why the second column in the table is titled *packets/second*. The table itself specifies how the transmission rate of the traffic profile changes with time. Each entry in the table designates the start of a new transmission rate. For example, as shown in Figure 6.22, at time 300 seconds, transmission rate of the traffic flow is set to 20 packets/second, and at time 600 seconds, the transmission rate is changed to 5 packets/second, which means that from time 300 seconds until time 600 seconds, the traffic flow will transmit data at the rate of 20 packets/second, and at time 600 seconds, the rate will decrease to 5 packets/second. The traffic flow will continue transmitting data at a rate of 5 packets/second for an additional 100 seconds until the time 700 seconds when the traffic flow rate changes to 15 packets/second. The preview window to the right of the table provides a graphical representation of the transmission rate changes.

The top of the traffic profile window has a textbox titled *Profile Name*, which allows you to define a name for the given traffic profile. Right below it is a checkbox, *Uniform X intervals*, which, when selected, forces the data points that define the transmission rate of the traffic flow to be uniformly distributed. The length of the *x*-axis update interval can be specified through the corresponding textbox. For example, if the *x*-axis update interval is set to 100 seconds, then the transmission rate of the traffic flow will be defined every 100 seconds. However, if this checkbox is left unselected, then the transmission rate of the traffic flow can be specified at intervals of any length as shown in Figure 6.22, where the length of the first two intervals is 150 seconds, the length of the second interval is 300 seconds, while the fourth interval is only 100 seconds long (i.e., $700 - 600 = 100$ seconds). Another checkbox, *Use Start Time*, specifies the time when the profile will begin. Finally, OPNET allows exporting and importing traffic profiles. When the button **Export…** in the bottom left corner of the traffic profile window is clicked, the defined traffic profile is exported into a tab-delimited text file that can be opened with any text or spreadsheet editor. Clicking the **Import…** button allows importing a defined traffic profile from a tab-delimited text file.

Continuing with our discussion of traffic flow demand attributes illustrated in Figure 6.20, the compound attribute **Traffic Characteristics** contains such configuration parameters as packet size, packet interarrival time, packet markings, and others that describe properties of the packets generated for this demand (Figure 6.20 does not show these configuration parameters). Even though this attribute is common to many traffic flow definitions, the actual configuration parameters that describe the packet properties often vary depending on the traffic type. The attribute **Traffic Duration** controls the duration of the traffic flow, while the attribute **Traffic Mix** specifies how this traffic flow is to be simulated (see Section 6.6.1). The attribute **Traffic Start Time** specifies the time when this traffic flow will start execution. It has several preset values including the following:

- Never—indicates that this demand will not generate any traffic.
- **Same As Global Setting**—indicates that the traffic flow will start at the time specified through a global input attribute **Background Traffic Start Delay (seconds)** shown in Figure 6.23. The traffic flow will start executing

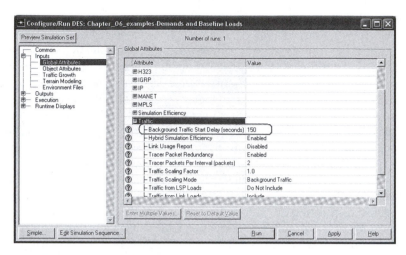

FIGURE 6.23   Background Traffic Start Delay (seconds) attribute.

its specified traffic profile this many seconds after the start of the simulation. The attribute **Background Traffic Start Delay (seconds)** is a global input attribute for the entire simulation specified through the **Configure/ Run DES** window.

6.6.3   BASELINE LOADS

Baseline load is a mechanism for modeling the background traffic of a link. Baseline loads are defined on a per-link basis, and therefore, configuring background traffic on a link requires changing the attribute values of that link. As you may recall, background traffic models a data exchange using a purely analytical approach without generating any explicit messages in the network. The effects of the background traffic are modeled as additional delay on every explicitly generated packet that travels through that link.

Each link model includes an attribute called **Traffic Information** for configuring background traffic in the form of baseline loads. Multiple instances of this attribute can be created (by creating multiple rows) to define many baseline loads for the same link. Each baseline load is defined through the following subattributes:

- **Traffic Class** specifies the traffic type of this baseline load. This information is used by the simulation engine to model the QoS handling of the background traffic.
- **Average Packet Size (bytes)** defines the average size of a packet in the background traffic. By default, the average packet size is 576 bytes.
- **Traffic Load (bps)** specifies the rate of the background traffic in bits/ second.

The subattributes **Average Packet Size (bytes)** and **Traffic Load (bps)** are specified separately for each direction of traffic flow on the link. Figure 6.24, for example,

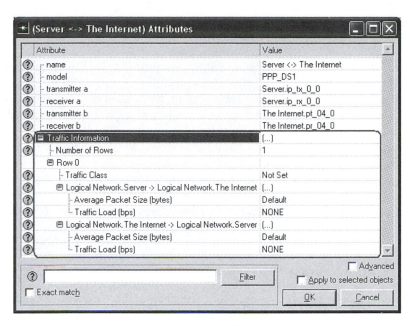

FIGURE 6.24   Attributes for configuring baseline loads.

illustrates the configuration of a baseline load for the link between the nodes *Server* and *The Internet*. This link is called *Server <-> The Internet*. The row that configures the baseline load on this link consists of two subattributes: one that defines the traffic from *Server* to *The Internet* (called *Logical Network.Server -> Logical Network. The Internet*), and another for the traffic from *The Internet* to *Server* (called *Logical Network.The Internet -> Logical Network.Server*). Both of these have separate instances of the **Average Packet Size (bytes)** and **Traffic Load (bps)** subattributes.

The value of the **Traffic Load (bps)** subattribute is also referred to as traffic intensity, and it is specified through the **Traffic Intensity Attribute Profile** window that defines how that traffic intensity varies over time. The **Traffic Intensity Attribute Profile** window is identical to that of the demand traffic profile (Figure 6.22). Overall, the baseline load's traffic intensity attributes are configured the same way as a demand's traffic profiles. OPNET refers to this type of traffic load as "static throughput" because it does not model explicit traffic and instead simply serves as an indication of how heavily the link is occupied. Unlike traffic demands, baseline loads defined traffic rate in bits/second only instead of also using units of packets/second. This is primarily because the configuration of the baseline loads includes the packet size, which remains unchanged throughout the simulation, and therefore, together with traffic rate in bits/second, can be used to compute the traffic rate in packets/second:

$$\text{Traffic rate}\left(\frac{\text{packets}}{\text{second}}\right) = \frac{\text{Traffic rate}\left(\dfrac{\text{bits}}{\text{second}}\right)}{\text{Packet size}\left(\dfrac{\text{bits}}{\text{packet}}\right)}$$

Generally, the process of configuring the link baseline loads is fairly straightforward and it consists of the following steps:

- Right-click on the link object of interest and select the **Edit Attributes** option.
- Expand the attribute **Traffic Information**. Recall that double-clicking on the value field of the attribute opens a window with the corresponding subattributes presented in the form of a table.
- Set the value of the attribute **Number of Rows**, which represents the total number of baseline loads to be defined on this link.
- To define baseline loads, expand one row at a time and specify the subattribute values for each row.
- Click the **OK** button to save the changes and close the **Edit Attributes** window.

6.7   CUSTOM APPLICATION STATISTICS

Similarly to standard applications, custom applications also have a set of predefined statistics to be collected during the simulation. A simulation scenario is configured to collect custom statistics the same way as any other type of statistic, that is, by right-clicking anywhere in the **Project Editor**'s workspace and selecting the **Choose Individual DES Statistics** option. Often collected statistic values are grouped based on the application, phase, or task name. A separate statistic vector is collected for each application, task, or profile. The name of the corresponding identifier is attached to the end of the statistic name. The identifiers for applications, tasks, and phases are typically defined according to the following conventions, where each of the corresponding names is replaced with the actual name of the profile, custom application, task, or phase:

- Application: *<Profile Name/Application Name>*
- Task: *<Profile Name/Application Name/Task Name>*
- Phase: *<Profile Name/Application Name/Task Name/Phase Name>*

For example, consider a simulation configured to run a custom application named **Banking**, which is part of a profile named **Online Customer**. The **Banking** application consists of **Login, Money Transfer**, and **Logout** tasks, and task **Login** consists of several phases, one of which is a phase named **Verification**. Based on the above configuration, upon completion, the simulation may have collected the following application and phase statistics:

- **Application Response Time (seconds) <Online Customer/Banking>**, which collects the response time for the **Banking** application defined within the profile **Online Customer**.
- **Phase Response Time (seconds) <Online Customer/Banking/Login/ Verification>**, which collects the response time for the **Verification** phase

of the **Login** task that is part of the **Banking** application defined within the **Online Customer** profile.

Generally, OPNET provides four different types of custom application statistics listed as follows:

- Global Statistics called *Custom Application* contain the results for all instances of the custom applications deployed in the simulated network. Depending on the type of statistic, these results are aggregated or averaged over the data collected from all the nodes that participate in the data transfer of the particular custom application.
- Node Statistics called *Custom Application* contain the results about the custom applications deployed at a particular node.
- Node Statistics called *Requesting Custom Application* contain the simulation results that pertain to situations in which the current node serves as a source for the custom application phases. Specifically, only statistics for those phases that originate from the current node will be available for viewing when accessing this node. To view these statistics for all the phases of a custom application requires examining results collected at every node that serves as a source for a phase within the custom application of interest.
- Node Statistics called *Responding Custom Application* contain the simulation results that pertain to situations in which the current node serves as a destination for the custom application phases. Specifically, only statistics for those phases that have the current node as their destination will be available for viewing when accessing this node. To view these statistics for all the phases of a custom application requires examining results collected at every node that serves as a destination for a phase within the custom application of interest.

A summary of all available custom application statistics is provided in Table 6.1.

6.8   STATISTICS FOR APPLICATION AND TRAFFIC DEMANDS

OPNET also provides statistics for application and traffic demands. Application demand statistics are available as part of generic Node Statistics, while there is a separate category called *Demand Statistics* with statistics for traffic demands. If no traffic demands are deployed in the simulation, then the Demand Statistics category is not available in the **Choose Results** window. After simulation execution, if configured, application demand statistics can be viewed by examining the results for the nodes that serve as source or destination for application demands. On the other hand, traffic demand statistics are displayed as a separate category under Object Statistics, and they are named according to the following conventions: *<Name of the Source Node> -> <Name of the Destination Node>*. Table 6.2 summarizes the available Demand Statistics.

TABLE 6.1
Summary of Custom Application Statistics

Category	Name	Description
Global Statistics **Custom Application**	*Application Response Time (second)*	Time taken to complete execution of all tasks within a custom application. This statistic is aggregated over all the instances of custom applications active in the network. If the simulation ends before all custom application tasks complete their execution, then this statistic will not report any values.
	Packet Network Delay (seconds)	The network delay experienced by request and response packets of this custom application.
	Phase Response Time (seconds)	The time to complete a phase within a task. The values for this statistic are recorded based on the value of the **End Phase When** attribute. In certain situations (e.g., phase completes when Final Request Leaves Source), it is possible for a phase to end while some of its messages are still traversing the network. This statistic is reported on a per-phase basis.
	Task Response Time (seconds)	The time to complete all phases within a single task of a custom application. The values for this statistic are averaged for all the nodes in the network that support this custom application. This statistic is reported on a per-task basis.
	Traffic Received (bytes/sec), *Traffic Received (packets/sec),* *Traffic Sent (bytes/sec),* *Traffic Sent (packets/sec)*	The traffic arrival and departure rates averaged over all the nodes that support this custom application.
Node Statistics **Custom Application**	*Active Custom Application Instance Count*	The total number of instances of this custom application active at the current node.
	Application Response Time (seconds)	Time taken to complete execution of all the tasks within the custom application deployed at this node.
	Phase Response Time (seconds)	The time to complete a phase within a task. Similarly to the corresponding global statistic, this statistic is also recorded based on the **End Phase When** attribute value. This statistic is recorded for all the phases of the custom application deployed at this node.
	Task Response Time (seconds)	The time to complete all phases within a single task of a custom application. This statistic is reported on a per-task basis.

Node Statistics Requesting Custom Application

Request Generation Time (seconds)	The time taken to generate a single request message. As you may recall, each request message of a phase within a custom application consists of several packets. This statistic records the total time to send a request message starting from the time the first packet of the request message was sent out until the time the last packet of the request message is transmitted. This statistic value is computed only for the request messages generated from this node and not for all the request messages of the custom application. This statistic is recorded on a per-phase basis.
Request Response Round Trip Time (seconds)	The round trip time for a request–response message exchange. The value for this statistic is computed by subtracting the time the first packet of the request message was sent out from the time the last packet of the response message arrives. This statistic is recorded on a per-phase basis.
Request Size (bytes)	The size of a request message generated during a custom application phase. This statistic is computed by adding the sizes of all packets that belong to a particular request message. This statistic is recorded on a per-phase basis.
Response Packet Network Delay (seconds)	The network delay experienced by the response packets arriving at this node as a reply to the requests originating from this node. This statistic is also recorded on a per-phase basis. If a phase is configured to generate no response messages, then this statistic will record no values.
Traffic Received (bytes/sec), Traffic Received (packets/sec), Traffic Sent (bytes/sec), Traffic Sent (packets/sec)	The traffic arrival and departure rates. These statistics are computed for the traffic generated by or arriving at this node only. These statistics are collected on a per-application basis, that is, traffic received/sent by an application deployed at the current node.

(Continued)

TABLE 6.1 (Continued)
Summary of Custom Application Statistics

Category	Name	Description
Node Statistics **Responding Custom Application**	Load (requests/sec)	The arrival rate of request messages at this node. This statistic is recorded on a per-application basis.
	Load (sessions/sec)	The rate at which new phases are created at this node. Typically, a new phase starts at a responding node when it receives a packet from the first request message of a phase. Therefore, this statistic collects the arrival rate of the first request messages at this node. This is different from the **Load (requests/sec)** statistic, which collects the arrival rate of ALL request messages. This statistic is recorded on a per-application basis.
	Request Packet Network Delay (seconds)	The network delay experienced by the request packets arriving at this node. This statistic is also recorded on a per-phase basis.
	Request Processing Time (seconds)	The time taken to process a request message. This statistic is computed by subtracting the time the last packet of the request message arrived at this node from the time the last packet of the corresponding response message is sent out from this node. This statistic is also recorded on a per-application basis.
	Response Size (bytes)	The size of a response message generated during a custom application phase. This statistic is computed by adding the sizes of all packets that belong to a particular response message. This statistic is recorded on a per-phase basis.
	Traffic Received (bytes/sec), Traffic Received (packets/sec), Traffic Sent (bytes/sec), Traffic Sent (packets/sec)	The traffic arrival and departure rates. These statistics are computed for the traffic generated by or arriving at this node only. These statistics are collected on a per-application basis, that is, traffic received/sent by an application deployed at the current node.

TABLE 6.2

Summary of Application and Traffic Demand Statistics

Category	Name	Description
Node Statistics **Application Demand**	*Response Time (seconds)*	The time for a response message to arrive after a request is sent. This statistic is computed by subtracting the time the source node generated a request from the time the source node receives a response message, which is a reply to its request.
	Traffic Received (bytes/sec), *Traffic Received (packets/sec),*	The traffic arrival rate of an application demand.
	Traffic Sent (bytes/sec), *Traffic Sent (packets/sec)*	The traffic departure rate of an application demand.
Demand Statistics	*Packet End-to-End Delay (sec)*	The time between the creation of a packet at the source node and the packet arrival at the destination node. This statistic is only available for nonbackground traffic, i.e., if traffic demand is configured as All Background traffic, then no values for this statistic will be recorded.
	Packet Jitter (sec)	The absolute value of the difference between end-to-end delays of two consecutive packets arriving at the destination node. This statistic its also only available for nonbackground traffic.
	Traffic Received (bytes/sec), *Traffic Received (packets/sec)*	The traffic arrival rate of a traffic demand.
	Traffic Sent (bytes/sec), *Traffic Sent (packets/sec)*	The traffic departure rate of a traffic demand.

6.9   STATISTICS FOR EXPLICIT PACKET GENERATION SOURCES AND BASELINE LOADS

There are no statistics defined for explicit packet generation sources and baseline loads. Therefore, to examine the influence of these traffic sources on the network performance, you may want to inspect other statistics. IP layer, Medium Access Control (MAC) layer, and Link Statistics can provide feedback about the performance of the traffic created by the explicit packet generation sources. Baseline loads model background traffic and do not generate any explicit packets. Therefore, IP and MAC layer statistics will not provide any information about baseline loads traffic. However, *point-to-point* Link Statistics such as queuing delay, throughput, and utilization will show how the configured baseline loads influence the traffic transmitted over the corresponding link.

The only exception to this rule is the *ethcoax_station* model that simulates ON-OFF explicit packet generation sources. Adding this model into the simulation causes two new categories of Global and Node Statistics to become available for collection: **Traffic Sink** and **Traffic Source**. These new additional statistics allow collecting such information as end-to-end delay as well as arrival and transmission traffic rates. The meaning of these statistics is self-explanatory and you should be able to discern their purpose based on the description of similar statistics provided above.

7 Specifying User Profiles and Deploying Applications

7.1 USER PROFILES

A **user profile** is a mechanism for specifying how standard and custom applications are used by an end user during a simulation. While we sometimes simply refer to them as profiles, user profiles should not be confused with other types of profiles (e.g., traffic profiles). Recall that the process of configuring standard and custom application traffic sources consists of the following steps: (1) define the applications, (2) configure the application statistics to be collected during the simulation, (3) define the user profiles, and (4) deploy the defined applications (which actually involves deploying the user profiles). Chapters 5 and 6 discussed the process of configuring standard and custom applications. This chapter focuses on defining user profiles and deploying the applications within the simulated network system.

While application definitions specify the characteristics of generated traffic, user profiles define how these applications are executed by the end user. This includes specifying such information as the applications' start times and duration, the number of applications executed by a single user, the order in which these applications are to be executed, and so on. For example, a definition of an e-mail application might specify that the e-mail client receives five e-mail messages of size 1024 bytes every 10 seconds, which yields 0.5 kilobytes traffic arrival rate. However, the definition of the e-mail application does not specify how many users in the simulated network run this e-mail application, when each user starts and ends executing this application, how many times the user reruns the application, whether any user runs multiple applications, do multiple applications executed by a single user run concurrently or sequentially, and so on. Such information is specified through user profiles and the application deployment process.

7.2 SPECIFYING USER PROFILES

Normally, user profiles are configured only after all the applications employed within the simulation have been defined because certain attributes of the user profiles are set to the names of the defined applications. However, if needed, an existing user profile can be reconfigured to include new application definitions. In this section, we describe how to specify and configure user profiles.

7.2.1 *Profile Config* Utility Object

Similar to application definitions, configuring user profiles also requires the presence of a utility object. Figure 7.1 shows the icon of a *Profile Config* object, which allows configuring user profiles. Typically, the *Profile Config* object is located in the same folder of the **Object Palette Tree** as the *Application Config* object (see Section 5.3.1). Similarly, only one *Profile Config* object is needed per simulation scenario to configure all the necessary user profiles.

The user profiles are defined by specifying the attribute values of the *Profile Config* object. The attributes of the *Profile Config* object are accessed the same way as the attributes of any other node: right-click on the *Profile Config* object and select the **Edit Attributes** option. Figure 7.2 shows the attributes of the *Profile Config* object.

As Figure 7.2 shows, the *Profile Config* object contains two attributes: **name** and **Profile Configuration**. The first attribute specifies the name of the object as it appears in the project workspace and the second attribute contains definitions of the user profiles configured for the current simulation scenario. The **Number of Rows**

FIGURE 7.1   *Profile Config* model icon.

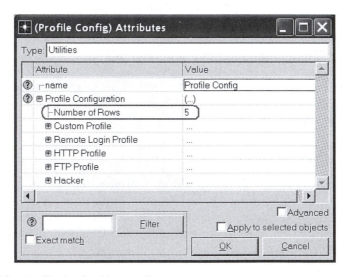

FIGURE 7.2   *Profile Config* object attributes.

subattribute of the compound attribute **Profile Configuration** controls the number of user profiles that could be configured for this simulation study. In the example shown in Figure 7.2, the **Number of Rows** attribute is set to 5 corresponding to five user profiles (i.e., `Custom Profile`, `Remote Login Profile`, `HTTP Profile`, `FTP Profile`, and `Hacker`) defined within the *Profile Config* object as separate compound attributes.

7.2.2 DEFINING A USER PROFILE

A user profile is defined by giving it a name, specifying the applications to be included in the profile, and describing the overall behavior of the profile. Let us examine one of the profile definitions of Figure 7.2 to better understand the configuration of user profiles. Figure 7.3 shows an expanded view of the `Remote Login Profile`.

As shown in Figure 7.3, the first attribute in a profile definition is **Profile Name**, which corresponds to the name of the whole profile definition row. By default, each new profile row gets the name "Enter Profile Name...", but setting the value of the attribute **Profile Name** causes the name of the profile definition to change as well.

The next attribute in a profile definition is called **Applications**. This is a compound attribute that specifies the applications, which are part of the profile; it also describes the behavior of each of those applications and how the applications are to be used within the profile. It is common to have a single profile execute multiple applications, since in real life, users frequently run multiple applications during a single computer session. A detailed description of the **Applications** attribute and how it is used to specify application behavior is given in Sections 7.2.4 and 7.2.5.

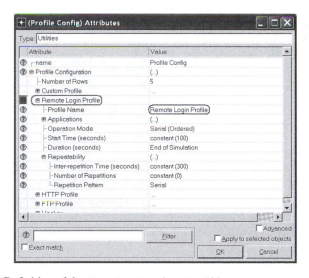

FIGURE 7.3 Definition of the `Remote Login Profile`.

Finally, the profile definition contains a set of attributes that are used to specify the overall behavior of the profile. These attributes do not describe how the individual applications within the profile will behave (which is specified through the compound attribute **Applications**), but instead they specify how the whole profile is managed during the simulation. The profile behavior attributes are as follows:

- **Operation Mode**: This attribute controls the manner in which applications are executed within the profile. Some of these applications are executed sequentially (e.g., in the morning, an office worker first reads the company announcements through the Web and then logs in to the remote server to do regular work), while others may be run in parallel (e.g., while working with the database application, the worker might also have an e-mail application running to periodically check for important messages). If the profile contains only one application, then the value of **Operation Mode** is irrelevant.
- **Start Time (seconds)**: Specifies when the user profile will start with respect to the simulation start time.
- **Duration (seconds)**: Specifies how long the profile will run.
- **Repeatability**: This is a compound attribute that describes how the user profile will be repeated if it completes before the end of the simulation.

We illustrate how the above attributes are used to define a simple user profile in Section 7.2.3. A more detailed description of the profile behavior attributes is provided in Section 7.2.6.

7.2.3 AN EXAMPLE OF A SIMPLE USER PROFILE

Consider the following example of a simple user profile: Suppose Lisa, an office secretary, normally runs only two applications during her workday, web browsing and e-mail. Lisa starts her day in the office at about 8:00 AM and goes home at about 5:00 PM. When she comes in to her office, suppose it takes her about 2 minutes to boot up her computer before she can do any work. After that, she works for two 4-hour periods, with a 1-hour break in-between. Such behavior could be easily mapped to the *Profile Config* attributes as described in the following paragraphs.

The first step is to name the user profile. For readability, let us name this profile Lisa. Next, we configure the **Application** attribute to run web browsing and e-mail. We assume that the corresponding application definitions have been created in the *Application Config* object.

After that, we set **Operation Mode** to Simultaneous indicating that Lisa runs these applications at the same time. Lisa's typical workday is 9 hours, which includes 1 hour for lunch. However, in our simulation, we also need to account for the 2 minutes it takes to boot up Lisa's computer. Therefore, the total length of the simulation should be set to 9 hours and 2 minutes or 542 minutes. We set the **Start Time (seconds)** of the profile to 120 seconds representing the 2 minutes it takes to start the computer and applications. The **Duration (seconds)** of the profile is set to 14,400 seconds (i.e., 4 hours), which is the uninterrupted period of time that Lisa spends working, both before and after lunch.

Finally, we configure the **Repeatability** attribute as follows:

- **Inter-repetition Time (seconds)** is set to 3600 seconds (i.e., 1 hour), indicating a 1-hour lunch break.
- **Number of Repetitions** is set to 1, indicating that Lisa will work for one more uninterrupted period of time (i.e., from about 1:00 PM until 5:00 PM).
- **Repetition Pattern** is set to `serial` indicating that the profile instances are repeated one after another.

Figure 7.4 shows a summary of this profile configuration, and Figure 7.5 shows a timeline of the profile execution. Notice that if the simulation time runs out before the end of the profile, the simulation will terminate without waiting for the profile to end.

7.2.4 CONFIGURING APPLICATION BEHAVIOR WITHIN A PROFILE

Now that we have a basic understanding of the relationship between the simulation and profile timelines, we add another wrinkle to this configuration process: specifying application behavior within the profile. The **Applications** compound attribute in a profile definition contains a subattribute **Number of Rows**, which controls the number of

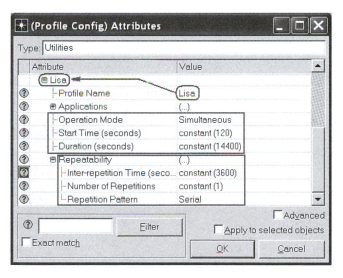

FIGURE 7.4 Example of a profile configuration.

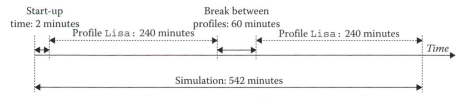

FIGURE 7.5 Execution timeline for the profile of Figure 7.4.

applications contained within the current profile, and a set of rows each describing how a single application will be used within this profile. As always, the number of application description rows corresponds to the value of the subattribute **Number of Rows**.

First, let us try to understand the relationship between the **Applications** attribute and its surrounding profile. Similar to the profile definition, the **Applications** compound attribute also contains such subattributes as **Name**, **Start Time Offset (seconds)**, **Duration (seconds)**, and **Repeatability** as shown in Figure 7.6. The main difference is that, within the profile definition, these attributes describe the profile's behavior with respect to the simulation start time, while in the case of **Applications**, these attributes describe the application behavior with respect to the current profile's start time. Therefore, the applications' attribute **Start Time Offset** specifies the time when the application will begin execution after the current profile has started. For example, consider a situation in which the profile's **Start Time (seconds)** is set to 120 while its application's **Start Time Offset (seconds)** is set to 300. In this situation, the profile and the application will begin execution 120 seconds and 420 seconds after the start of the simulation, respectively. Similarly, **Repeatability** describes the application's repetition pattern within its profile.

The above application behavior attributes are described in more detail in Section 7.2.5.

7.2.5 APPLICATION BEHAVIOR ATTRIBUTES

The attributes that describe the behavior of each application in the profile are part of the compound attribute **Applications**. The first subattribute of the **Applications** attribute is **Number of Rows**, which specifies the number of applications to be configured within this profile. A separate compound attribute is created for each

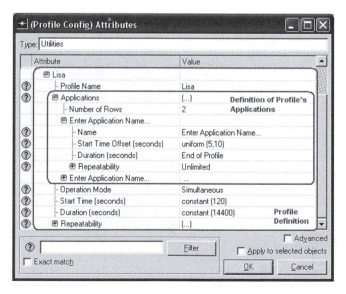

FIGURE 7.6   Definition of the **Applications** attribute within the profile.

application. This attribute is initially called **Enter Application Name…** and it contains the complete usage description of an application within the profile.

The components of an application definition are as follows:

- **Name**—the name of the application to be used within this profile. This attribute contains a list of the applications that have been already configured through the *Application Config* object. You can only select an application from the provided list. If the desired application is not present, then you need to define it in the *Application Config* object and only then return to configuring user profiles. Once the desired application has been selected, the name of the application row will change from **Enter Application Name…** to the name of the selected application.
- **Start Time Offset (seconds)**—the meaning of this attribute depends on the value of the **Operation Mode** attribute (Section 7.2.6), which specifies how the applications are executed within the profile: simultaneously (i.e., in parallel) or serially (i.e., one after another). If the applications are executed simultaneously, then the **Start Time Offset (seconds)** attribute specifies the length of the time period from the start of the profile until this application begins execution. If the applications are executed serially, then this attribute defines the length of the time period from the end of the previous application within the current profile until this application starts execution. If this is the first application in the list, then this application will begin execution **Start Time Offset (seconds)** after the start of the profile. If the previous application does not terminate before the end of the profile, then the current application will never start. More specifically, if the sum of the time at which the previous application completes and the value of the **Start Time Offset (seconds)** attribute exceeds the time when the current profile completes, then this application will never execute. Also, this attribute has two predefined values:
  - No Offset—the application will start at the same time as the profile (if the applications within the profile are executed simultaneously or if this is the first application in the list) or as soon as the previous application completes (if the applications are executed serially and this is not the first application in the list).
  - Never—the application will not start.
- **Duration (seconds)**—specifies the duration for which a single instance of this application will execute. Notice that an application may be repeated multiple times within the profile, creating new instances of itself at a later time. This attribute may be set to an explicit numeric value computed using a specified probability distribution function (PDF) or it may be selected from two preset options:
  - End of Profile—means that this application will terminate at the same time as this profile. If applications within the profile are executed serially, then none of the applications defined after the current application will execute.

- End of Last Task—means that this application will terminate only after its last task is completed. However, if the last application task completes after the end of the profile, then the application will terminate at the same time as the profile. This attribute value is primarily applicable to custom applications since they usually consist of one or more tasks (i.e., transactions between the nodes). The duration of a custom application is controlled by the time it takes to complete all of its tasks. Therefore, setting the application to terminate at the End of Last Task implies that you want to terminate the application right after it ends all of its network transactions or activities. For standard applications, which do not have explicitly defined tasks, the End of Last Task setting has the same meaning as End of Profile.

- **Repeatability**—a compound attribute that describes how the current application is to be repeated within the surrounding profile (i.e., while the profile is being executed). Specifically, OPNET defines **Repeatability** through the following three subattributes:

  - **Inter-repetition Time (seconds)**—this attribute specifies the duration of the time period between application sessions. This attribute is specified through a PDF, which is used during the simulation to compute the inter-repetition time. The meaning of this attribute depends on the repetition pattern that could be either Serial or Concurrent. If this application is being repeated serially (i.e., the next application session only starts after the previous session has ended), then **Inter-repetition Time (seconds)** specifies the length of the time period from the previous application session *completion* until the next session start. On the other hand, if the application is being repeated concurrently (i.e., multiple application sessions may be active simultaneously), then **Inter-repetition Time (seconds)** specifies the length of the time period from the previous application session *start* until the beginning of the next session. Setting **Inter-repetition Time (seconds)** to 0 when the application is repeated concurrently will cause the application sessions to be created at an infinitely high rate, which may lead to undesirable simulation outcomes.

  - **Number of Repetitions**—specifies the number of times this application will be repeated. The number of repetitions is different from the number of application sessions created during the lifetime of the profile. The number of repetitions only counts the sessions after the first application session started. For example, if the number of repetitions is set to 1, then during the lifetime of the profile, at most two applications sessions would be executed (i.e., it is possible that only one session will execute if the first session does not complete before the end of the profile and the applications are executed using the serial repetition pattern). This attribute is specified through a PDF but also has a predefined value called Unlimited, which means that the application will continue repeating until the end of the current user profile.

- **Repetition Pattern**—specifies how the current application will be repeated within this profile. This attribute has only two possible values: Serial, which indicates that the next application will only start after the previous application has completed, and Concurrent, which means that multiple application sessions can be executed at the same time and the next application can start before the previous session has completed. Figure 7.7 illustrates the differences between concurrent and serial application repetition patterns.

Finally, the compound attribute **Repeatability** has two preset values:

- Once at Start Time—meaning that the application will execute only once throughout the duration of the profile. The same effect can be easily achieved by setting the subattribute **Number of Repetitions** to 0, which is in fact how this value is set.

- Unlimited—meaning that the application will continue repeating until the end of the profile. This preset value is configured to have the application repeat in a serial order with inter-repetition time computed using an exponential distribution with a mean outcome of 300 seconds. Effectively, when this value is selected, the application will be repeated until the end of the profile with the next application session starting on average 300 seconds after the previous session completion.

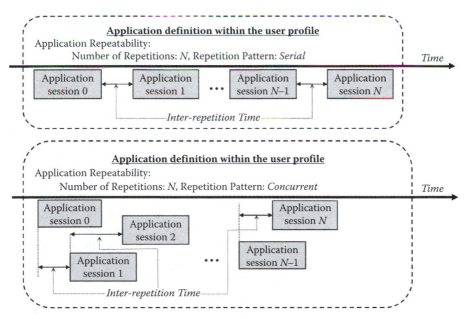

FIGURE 7.7 Application repeatability within the profile.

7.2.6   PROFILE BEHAVIOR ATTRIBUTES

As briefly introduced in Section 7.2.2, the profile behavior attributes are part of a profile definition and are used to describe the overall behavior of a profile. We now describe these attributes in more detail:

- **Operation Mode**—this attribute describes how the defined profile applications will be started within the profile. This attribute does not describe the behavior of any one individual application; rather, it configures the execution pattern of all applications within the profile. This attribute has three possible values:
  - Serial (Ordered)—the applications will be executed one after another in the order they are specified in the **Applications** attribute, that is, the application defined in the first row executes first, then the application defined in the second row, and so on. If an application is configured to be repeated an unlimited number of times or its duration is set to End of Profile, then none of the subsequent applications will start because the preceding application never finishes within the span of the profile.
  - Serial (Random)—applications will execute one after another in a random order. The danger of not executing one or more applications specified within the profile remains if at least one of the applications is configured to unlimited repeatability or has its duration set to End of Profile. Since applications will execute in random order, there is no way to predict which applications will not execute. If an application that has unlimited number of repetitions is chosen to be executed last, then all the profile's applications will run; on the contrary, if that application is chosen to execute at the start of the profile, then it will be the only application to run during that profile.
  - Simultaneous—applications will execute simultaneously. The applications are started in the order in which they are specified in the **Applications** attribute. The start of each application is determined by the **Start Time Offset (seconds)** attribute for that application. For the first application in the profile, the start time is computed by adding its start offset value to the time when the current profile was launched. The start time of each subsequent application is computed by adding its start time offset value to the start time of the previous application.
- **Start Time (seconds)**—specifies when the user profile session will start. This attribute accepts a probability distribution function as a value. For example, when set to uniform(100,110) value, the user profile will begin execution **T** seconds after the start of simulation, where **T** is a value between 100 and 110 seconds randomly selected using uniform probability distribution function.
- **Duration (seconds)**—specifies how long the profile will execute. This value supersedes the application durations, meaning that even if the application did not complete before the end of the profile, both the profile and the

application will terminate according to the value specified in the profile's **Duration** attribute. This attribute has two predefined values:

- End of Simulation — the profile will continue execution until the end of the simulation. With such a setting, it is possible that all the applications will complete before the end of the profile and will not generate any more traffic; however, the profile will remain active until the simulation ends. It is advised not to set the profile duration to End of Simulation if the profile is configured to repeat using the serial repetition pattern, because none of the subsequent repetitions will execute since the first instance of the profile will not end until the simulation terminates.

- End of Last Application—this value indicates that the profile will terminate as soon as the last instance of its applications ends. If the applications are configured to repeat an unlimited number of times, then the profile will not terminate until the end of the simulation. However, this is a very useful preset value because it ensures that if the profile has completed all its work and will not generate any more traffic then the profile will be deactivated.

- **Repeatability**—this compound attribute specifies how the profile will be repeated within the simulation. This attribute is almost identical to the application's **Repeatability** attribute. The only difference is that this attribute describes *the profile's repeatability within the simulation* while the latter attribute describes *the application's repeatability within the current profile*. This attribute also consists of three subattributes:

  - **Inter-repetition Time (seconds)**—specifies when the next profile session will start. If the profile is configured to repeat serially, then the start time of the next profile session is computed by adding the time at which the previous profile session *ended* to the profile's inter-repetition time. If the profiles are repeated concurrently, then the start time of the next profile session is computed by adding the time at which the previous profile session *started* to the profile's inter-repetition time. As before, setting the inter-repetition time to zero when the profiles are repeated concurrently will result in an infinitely large profile generation rate.

  - **Number of Repetitions**—specifies the number of times this profile will be repeated. As with applications, the number of repetitions only counts the sessions after the first profile session started. For example, if the number of repetitions is set to 1, then during the lifetime of the simulation, at most two profile sessions would be executed. This attribute is specified through a PDF but also has a predefined value called Unlimited, which means that the profile will continue repeating until the end of the current simulation.

  - **Repetition Pattern**—specifies how the current profile will be repeated within this simulation. This attribute has only two possible values: Serial, which indicates that the next profile session will only start after the previous profile has ended, and Concurrent, which means that multiple profiles can be executed at the same time, and the next profile session can start before the previous session has completed.

7.2.7 CONFIGURING USER PROFILES

To end this section, we summarize below the complete step-by-step instructions for configuring user profiles:

- If *Profile Config* is not already in the project workspace, then place the *Profile Config* object onto the project workspace.
- Right-click on the *Profile Config* object and select **Edit Attributes** from the pop-up menu.
- Expand the **Profile Configuration** compound attribute.
- Specify the value of the **Number of Rows** attribute, which corresponds to the total number of different user profiles to be configured in the current simulation scenario.
- Expand the rows one at a time to configure each user profile:
  - Set the value of the **Name** attribute, which will automatically change the name of the application's row.
  - Expand the compound attribute **Applications** to configure the applications to be used in this profile:
    - Specify the value of the **Number of Rows** attribute, which corresponds to the total number of applications used in this profile.
    - For each application, expand its corresponding compound attribute and configure each of the application's attributes. Recall that the attribute **Name**, which specifies the name of the application, can only accept one of the predefined values that are the names of applications that have been configured already through the *Application Config* object. Also, setting the value of the attribute **Name** will automatically change the name of the row to the provided value.
  - Configure the profile's **Operation Mode**, **Start Time (seconds)**, **Duration (seconds)**, and **Repeatability** attributes.
- Repeat the steps until all required user profiles have been configured.

7.3   EXAMPLES OF CONFIGURING USER PROFILES

Let us now consider a more detailed example of configuring user profiles. This example constructs two user profiles to correspond to the work schedule of two employees at a worksite.

First, consider a typical day of an employee named Jon. Jon starts his day by remotely logging in to his company's computer. After about 30 minutes of work, Jon starts periodically checking news on the Web, for example, every 30 minutes. Usually, Jon spends about 10 minutes browsing the Web in each of these sessions. In addition, every hour Jon uses the FTP application to backup his files, which takes him about 2 minutes. Jon keeps the remote login application active while browsing the Web and while transferring the backup files. Jon usually works in three shifts of 3 hours each with a break of about 45 minutes in-between. Let us assume that before taking the break, Jon closes all his applications and then restarts them after coming back from the break.

Such a user profile can be configured as shown in Figure 7.8. First, let us examine the profile execution pattern that captures Jon's overall behavior:

a. Since Jon does not terminate the remote login application while transferring files and browsing the Web, all applications are configured to execute simultaneously.
b. The profile runs for 10,800 seconds, which corresponds to Jon's 3-hour shift.
c. The profile is repeated two times serially (i.e., the profile will be executed three times totally, one after another) with 2700-second breaks (i.e., 45 minutes) in-between.

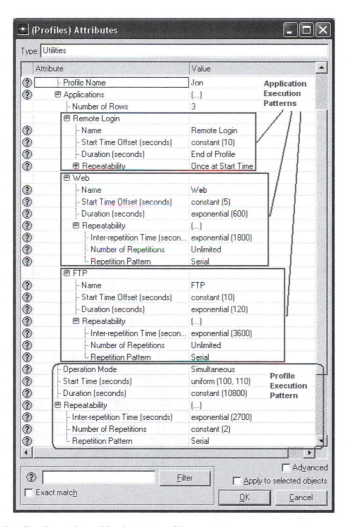

FIGURE 7.8 Configuration of Jon's user profile.

The profile start time is set to a value between 100 and 110 seconds computed using a uniform distribution. Usually, it is a good idea to set the profile start time to a nonzero value (i.e., preferably around 100 seconds), which allows other devices (e.g., routers) in the simulated system to initialize. Notice that the profile itself only describes a single 3-hour shift, while the profile repetition feature allows simulating a complete workday.

Now, let us examine how individual applications have been configured within the profile. Since Jon runs a single instance of the remote login application during each shift period, this application is configured to start only once at the beginning and to run until the end of the profile. Notice that such a configuration does not prevent other applications from starting because the profile is configured to execute applications simultaneously. We set the remote login's start time offset to 10 seconds to represent the time it takes Jon to start the application, although this value could have been set to 0. Since Jon spends 10 minutes browsing the Web every 30 minutes, we configured the web application to have a duration of 600 seconds (or 10 minutes) and repeat every 1800 seconds (or 30 minutes) an unlimited number of times, which ensures that the web application will continue executing until the end of the profile (i.e., the end of one shift period). Since Jon runs only one instance of the web browser at a time, we set the repetition pattern to serial. Notice that the start time offset of the web application is set to 1800 seconds. This simulates Jon's behavior where he first works for 30 minutes using the remote login application and only then he starts browsing the Web. Using the same logic, we configured the FTP application that runs for 120 seconds (or 2 minutes) and repeats in serial fashion, an unlimited number of times every 3600 seconds (or 60 minutes). The FTP start time offset is set to 3600 seconds to indicate that Jon does the first file transfer after 1 hour of work.

Now consider a second employee, Alice, who starts working an hour after Jon. Alice runs one application at a time in the following order: first, she spends 1.5 hours remotely connected to her company's server, then she spends about 10 minutes transferring files to and from the company's server, then for the next 30 minutes, she browses the Web, and then again she spends 1 hour and 50 minutes using the remote login application. After that, she takes a 1-hour break and then repeats the same process again for another 4 hours.

Figure 7.9 illustrates Alice's profile configuration. To enforce the ordered application execution, the profile has its attribute **Operation Mode** set to Serial (Ordered), which means that the applications will be executed in the order they are defined: *Remote Login, FTP, Web*, and *Remote Login*. The profile start time is set to a value between 3700 and 3710 seconds, which simulates Alice starting her job 1 hour later than Jon whose profile start time is set to a value between 100 and 110 seconds.

Alice's profile duration is set to 14,435 seconds, which corresponds to 4 hours and an extra 35 seconds to account for the time between application starts, that is, the first *Remote Login* application starts 10 seconds after the start of the profile, *FTP* starts 10 seconds after the first *Remote Login* application ends, *Web* starts 5 seconds after the *FTP* application ends, and the second *Remote Login* application starts 10 seconds after the *Web* application terminates. This start offset time represents additional time that may be needed to start an application, but may not be necessary in many situations.

The profile's inter-repetition time is set to 3600 seconds or 1 hour to represent the break Alice takes between her 4-hour shifts. The profile is set to repeat one more

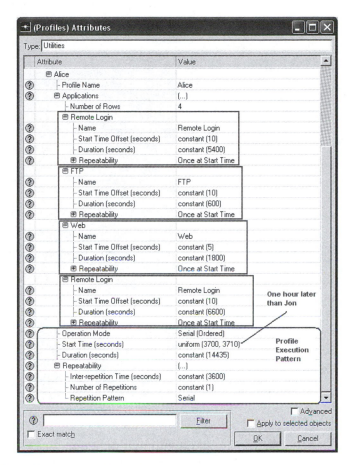

FIGURE 7.9 Alice's user profile configuration.

time after the 1-hour break, having a total of two profile sessions executed during the simulation. All the applications are configured to repeat only once at the start of the profile, ensuring that each application does not repeat within the profile. The first *Remote Login* application is set to last 5400 seconds or 1.5 hours, the *FTP* is set to run for 600 seconds or 10 minutes, the *Web* application is configured to execute for 1800 seconds or 30 minutes, and the second *Remote Login* application is configured to run for 6600 seconds or 1 hour and 50 minutes.

Finally, to ensure that both profiles execute all their configured applications during the duration of the simulation, we set the length of the simulation to 10 hours and 35 minutes (i.e., Jon's profile executes for 10 hours and 30 minutes and we give an extra 5 minutes for initialization and other possible delays). Figure 7.10 illustrates a summary of Jon's and Alice's profile configurations, while Figure 7.11 shows the relationship between the simulation and the two profiles. Figures 7.10 and 7.11 are not drawn to scale and just illustrate the relationship between the applications, the profiles, and the simulation.

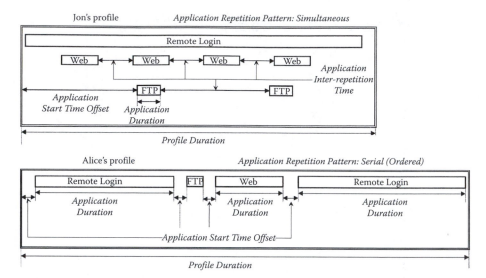

FIGURE 7.10   Jon's and Alice's profiles.

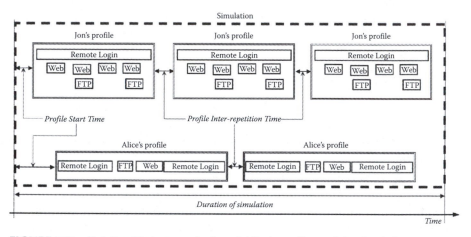

FIGURE 7.11   Relationship between Jon's and Alice's profiles and the simulation.

Note that some more complex application usage behaviors may need to be represented through multiple user profiles. For example, consider a situation where a user simultaneously runs several applications for a certain period of time and then executes a few more applications in sequence. Such a pattern cannot be represented as a single user profile because applications within the profile can only run either simultaneously or sequentially and more complex behavior is not allowed. To configure this type of user behavior, you may want to create two separate user profiles: the first one to represent applications executed in parallel, and the second one to represent applications executed in sequence. Nevertheless, the user profile utility offered by OPNET is quite flexible and can represent a variety of application usage behaviors, which is more than enough for most simulation studies.

7.4 USING THE APPLICATION DEPLOYMENT WIZARD FOR DEPLOYING USER PROFILES

Once all desired applications and user profiles have been defined, the next step is to deploy the defined applications within the simulated system. Typically, in the client–server paradigm, application deployment entails specifying the nodes that will initiate application execution (i.e., clients) and the nodes that will provide application services (i.e., servers). Using OPNET terminology, deploying applications in the simulated network system requires the following:

1. Specifying the nodes (i.e., clients or sources) that support the defined user profiles.
2. Specifying the nodes (i.e., servers) that support the application services.

In OPNET, workstation, server, LAN, and load balancer objects contain configuration attributes that allow specifying supported profiles and applications services within that object. Typically, workstation, LAN, and load balancer objects act as the client or source nodes that support configured user profiles, although in some situations, it is possible to deploy user profiles in the server nodes. Similarly, server objects usually act as the servers, while in certain situations (e.g., when running voice or video applications), it is possible to have workstation objects configured to support specific application services. LAN objects can also operate as servers because each LAN object is assumed to consist of several workstations and a server node. If a client node supports a certain profile, then every application within that profile could be executed during the simulation. There is no such restriction on server nodes. Servers are not directly tied to a specific profile. Instead, servers are responsible for supporting specific applications. Generally, it is desirable to have every application defined in the deployed profiles supported in at least one server node. If there are no servers that support a particular application service, then that application will never execute.

Typically, application deployment requires individually configuring nodes to support desired profiles and applications. Configuring one node at a time could be an error-prone and time-consuming task. That is why newer versions of OPNET products contain a separate software utility called the **Application Deployment Wizard** that allows collectively deploying specified applications at the nodes within the system. To open the **Application Deployment Wizard** from the pull-down menu in the **Project Editor**, select **Protocols > Applications > Deploy Defined Applications**. When started, the **Application Deployment Wizard** opens a **Deploy Applications** window, which consists of four distinct panels as shown in Figure 7.12:

- *Network Tree Browser* panel controls the display of objects from the simulated system in the **Deploy Applications** window.
- *Application Deployment Operations* panel provides controls for adding or removing a selected node as a source or a server within the profile, selects

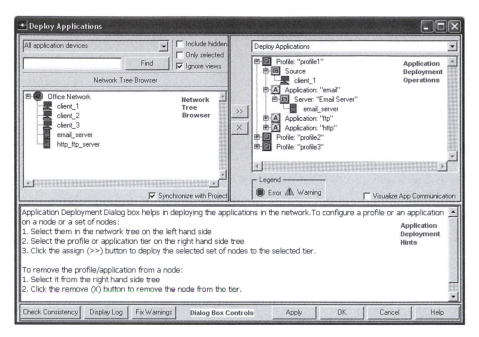

FIGURE 7.12   Deploy Applications window.

the type of application deployment operation being executed, and displays the corresponding deployment view.

* *Application Deployment Hints* panel provides helpful hints about how to configure application deployment based on the currently selected application deployment operation.
* *Dialog Box Controls* panel manipulates the **Deploy Applications** window and provides features for debugging application deployment.

7.4.1 *Network Tree Browser* Panel

The *Network Tree Browser* panel displays the objects available within the simulated network system while the controls in this section allow customization of which objects are being displayed.

* A pull-down menu, at the top of the panel, specifies the type of nodes to be displayed. By default, all application devices are shown in the *Network Tree Browser*. However, you can choose to limit the display to only *workstation*, *servers*, *mainframes*, *LAN*, or *load balancer* objects.
* A textbox with the **Find** button next to it allows searching for nodes that contain specified text in their name. The identified nodes will be selected in the **Project Editor**, and if the checkbox *Synchronize with Project* is checked, then these nodes will also be highlighted in the *Network Tree Browser*.

- In addition, there are four checkboxes that implement the following features:
  1. *Include hidden*—when checked, the *Network Tree Browser* will also display objects that are not shown within the **Project Editor**.
  2. *Only selected*—when checked, the *Network Tree Browser* will only display objects that are highlighted within the **Project Editor**. This feature only works if the *Synchronize with Project* checkbox is also checked.
  3. *Ignore views*—when checked, the *Network Tree Browser* will display all the objects in the simulation model regardless of the configured views of network topology.
  4. *Synchronize with Project*—when checked, the *Network Tree Browser* and the **Project Editor** are constantly updated so that their views are consistent, that is, when nodes are selected in the *Network Tree Browser*, then the same nodes will be selected in the **Project Editor** and vice versa.

7.4.2 APPLICATION DEPLOYMENT HINTS PANEL

The *Application Deployment Hints* panel is very simple and it contains no operational controls. This panel displays help information for performing various application deployment tasks. The text in the *Application Deployment Hints* panel changes based on the current selection in the **Deploy Applications** window (see Section 7.4.4).

7.4.3 DIALOG BOX CONTROLS PANEL

The *Dialog Box Controls* panel consists of seven buttons as described in the following:

- The **Check Consistency** button verifies that the profiles have been deployed correctly. If the consistency check identifies any mistakes in the application deployment, then the *Application Deployment Hints* panel will display an explanation of why the application deployment consistency check failed and it will also suggest possible solutions. For example, the consistency check will identify a situation where there are no server nodes supporting applications of the profile deployed at the client nodes.
- The **Display Log** button opens an **Application Deployment Logs Viewer** application that displays a list of error/warning messages and a description of operations to resolve configuration issues for the current application deployment. This log viewer application includes features for exporting log messages into a spreadsheet, comma-separated text file, web report, or XML file, as well as for managing how the log messages are displayed.
- The **Fix Warnings** button resolves possible configuration problems in the current user profile deployment and saves the performed operations into a log file.
- The **Apply** button performs the application deployment consistency check. If the profiles have been deployed correctly, it saves the configuration changes without closing the **Deploy Applications** window.
- The **OK** button also performs the application deployment consistency check. If the profiles have been deployed correctly, then it saves the configuration changes and also closes the **Deploy Applications** window.

- The **Cancel** button closes the **Deploy Applications** window without saving the configuration changes.
- **Help** opens a window in your web browser with a help page that provides a brief description of the **Deploy Applications** window features.

7.4.4 APPLICATION DEPLOYMENT OPERATIONS PANEL

The middle of the *Application Deployment Operations* panel displays the current client/server assignment for the configured profiles, which we will call the *Profile Definition Tree*. The *Profile Definition Tree* consists of several tiers. However, the actual appearance of the *Profile Definition Tree* depends on the value in the pull-down menu at the top of the *Application Deployment Operations* panel. This pull-down menu controls the type of deployment operation to be performed and has five possible options:

- **Deploy Applications** allows deploying applications in the simulated network by specifying client and server nodes for each profile. When this option is selected, the *Profile Definition Tree* will display the current application deployment in the network organized by the user profiles. The profile tier is the top-level tier and it contains application deployment for a single profile. Each profile tier contains a single source tier and one or more application tiers, one for every application defined within the profile. Finally, each application tier also includes a server tier.
- **Edit Destination Preferences** allows configuring server selection weights. When this option is chosen, the *Profile Definition Tree* will display application deployment organized by the client nodes. The client tier is the top tier in this view. It contains a separate tier for each application defined in the profiles deployed at this client node. Finally, each application tier contains the list of servers that support services for this application.
- **Edit LAN Configuration** allows specifying the number of profile clients deployed within the LAN object. When this option is chosen, the *Profile Definition Tree* will display application deployment organized by the LAN objects. The top tier of the tree will list all the LAN objects available in the simulated system, and each LAN object tier contains another tier that lists profiles deployed at the corresponding LAN object.
- **Edit Service Registration** and **Convert Ace Traffic (Flows <-> Discrete)** options deal with OPNET's ACE software and are not discussed here.

Figure 7.13 shows the appearance of the *Profile Definition Tree* for different values of the pull-down menu. In addition, the *Application Deployment Operations* panel contains a *Visualize App Communication* checkbox that when checked opens an **Application Communication Visualization** window. There are also two buttons between the *Network Tree Browser* and the *Application Deployment Operations* panels: ">>" called **Deploy Node to a Tier** and "X" called **Remove Node from a Tier**.

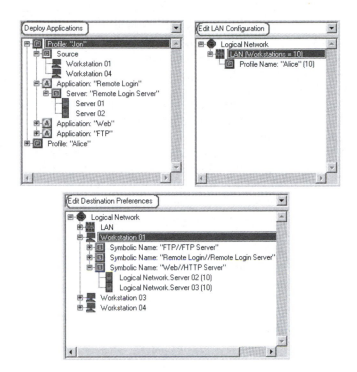

FIGURE 7.13   *Profile Definition Tree.*

The **Deploy Applications**, **Edit Destination Preferences**, and **Edit LAN Configuration** options are described in more detail in Sections 7.4.5–7.4.7.

7.4.5   DEPLOY APPLICATIONS OPTION

When the **Deploy Applications** option is selected, you can assign nodes to act as clients and servers in defined user profiles (i.e., deploy applications) by performing the following actions:

- From the pull-down menu in the *Application Deployment Operations* panel, select the *Deploy Applications* option (it is the default option when the **Deploy Applications** window is opened).
- Expand the *Profile Definition Tree* so that the source, application, and server tiers are visible.
- To assign a client or source node:
  - Select a node in the *Network Tree Browser.*
  - Select the source tier of the desired profile in the *Profile Definition Tree.*
  - Click the ">>" button to assign the node to the selected source tier. This will cause the icon of the node in the *Network Tree Browser* to appear in the source tier. Notice that the assigned node will operate as a source or an initiator for all applications within the current profile.

- Repeat this process as necessary. A single profile can have multiple client nodes, which means that each assigned node will run the corresponding profile during the simulation.
- To assign a server node:
  - Select a node in the *Network Tree Browser.*
  - Select the server tier of the desired profile in the *Profile Definition Tree.* Notice that each server tier is associated with a specific application within the profile.
  - Click the ">>" button to assign the node to the selected server tier. This will cause the icon of the selected node to appear below the server tier icon. The added node will operate as a server for the selected application in all the profiles that contain the application. Notice that a given node can operate as a server for multiple applications.
  - Repeat this process as necessary. A single application can be supported by multiple server nodes, that is, the client node that runs an application could connect to any of the servers that support services for that application.

To remove a node from its current assignment as a client or a server in the *Profile Definition Tree*, perform the following actions:

- Select a node under a source or a server tier of the desired profile in the *Profile Definition Tree.*
- Click the "X" button to remove the node from the selected source or server tier.

Once all client and server nodes have been assigned, the application deployment is complete and you can click the **Apply** or **OK** buttons to save the application deployment configuration.

A client node can connect to any of the servers that support an application on the client node (i.e., part of the profile supported by the client node). OPNET controls the probability of a particular server selection through a server selection weight attribute. The server selection weights are attributes of the client node, that is, they define the probability with which the client node connects to a particular server during application execution. That is why the *Profile Definition Tree* is organized by the client nodes when the **Edit Destination Preferences** option is selected. Therefore, different client nodes can set different server selection weight values for the same server node. By default, all servers are assigned the same weight of 10 units, and as a result, the probability of selecting a particular server is the same. Larger value of the server selection weight increases the probability that the server will be selected by the client node.*

Consider the following example. Let us assume that we have three server nodes, called *S1*, *S2*, and *S3*, supporting the FTP application. Also, let us assume that client nodes *C1* and *C2* support a user profile that includes the FTP application. Let us

---

* The external C code file **tpal_app_support.exe** contains a function called *app_server_name_select()* that randomly selects a server based on its weight.

assume that node *C1* sets server selection weights for servers *S1*, *S2*, and *S3* to be 10, 15, and 25, respectively, while node *C2* configures the same selection weights to be 25, 40, and 35. Therefore, during simulation, node *C1* will connect to server *S1* with probability $10/(10 + 15 + 25) = 0.20$, to server *S2* with probability 0.30, and to *S3* with probability 0.50. Similarly, node *C2* will select servers *S1*, *S2*, and *S3* with probabilities 0.25, 0.40, and 0.35, respectively.

Finally, the *Visualize App Communication* checkbox is only visible when the **Deploy Applications** option is selected from the pull-down menu in the *Application Deployment Operations* panel. When checked, it opens the **Application Communication Visualization** window, which provides a visual map of the application configuration as shown in Figure 7.14. This window contains three pull-down menus:

1. The first menu contains a list of applications defined within the corresponding user profile.
2. The second menu specifies the list of profiles defined in the simulated system.
3. The third menu provides a list of client nodes that support the currently selected user profile and an option to display logical tiers only.

Varying the selections in the pull-down menus changes the communication pictures displayed in the **Application Communication Visualization** window. For example, Figure 7.14 shows that the *FTP* application that is part of Jon's profile deployed at an object named *LAN* has a one-to-one client–server logical communication, and during simulation, the FTP application executed on the *LAN* node connects to two server nodes called *Server 04* and *Server 03*.

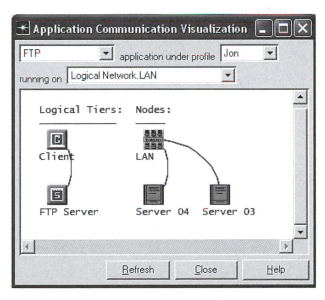

FIGURE 7.14   **Application Communication Visualization** window.

7.4.6   EDIT DESTINATION PREFERENCES OPTION

Choosing the **Edit Destination Preferences** option from the pull-down menu in the *Application Deployment Operations* panel allows changing the server weights for each client node. When this option is selected, you can only change server weights on those nodes that have been configured to support the profile's applications and you are not allowed to add new server nodes. The steps for changing the server selection weights are as follows:

- From the pull-down menu in the *Application Deployment Operations* panel, select the **Edit Destination Preferences** option.
- Expand the *Profile Definition Tree* so that the clients and their corresponding server tiers are visible.
- Select a desired server tier or an actual server node located in that tier.
- Click the **Edit** button or double-click on the server or server node tier to open the **Edit Destination Preferences** window, which contains a table of all the servers that support the current application.
- In the **Edit Destination Preferences** window, click on the *weight* column to change the server selection weight value. If all server weights for a particular application are set to 0, then all servers will be selected with the same probability. However, if at least one server has a nonzero weight, then all the servers that have a weight value of 0 will not be selected. As shown in Figure 7.13, the server weights appear in parenthesis next to each server tier when the **Edit Destination Preferences** option is selected.
- Click the **OK** button to close the **Edit Destination Preferences** window and save the changes or click the **Cancel** button to close the window without saving the changes.
- Repeat this process for the rest of the client nodes if desired.

7.4.7   EDIT LAN CONFIGURATION OPTION

The **Edit LAN Configuration** option allows specifying the number of profile clients deployed within a LAN object. When this option is selected, only those LAN objects that support user profiles will be visible in the *Profile Definition Tree*. All other types of objects and LAN objects that do not support any user profiles will not be visible. The steps for changing the number of clients deployed within a LAN object are as follows:

- From the pull-down menu in the *Application Deployment Operations* panel, select the **Edit LAN Configuration** option.
- Expand the *Profile Definition Tree* so that the tier containing LAN objects and their profiles is visible.
- Select the desired LAN object or its profile tier.
- Click the **Edit** button or double-click on the LAN object or its profile tier to open the **Edit LAN Profile Configuration** window, which contains a table of all the profiles deployed within the current LAN object.

- In the **Edit LAN Profile Configuration** window, click on the *Number of Clients* column to change the number of profile clients in the LAN. OPNET advises against configuring the number of clients to be larger than the number of workstations in the LAN although it does not prevent such a configuration. As shown in Figure 7.13, the number of clients and the number of workstations appear in parentheses next to the profile and LAN object tiers, respectively, if the **Edit LAN Profile Configuration** option is selected.
- Click the **OK** button to close the **Edit LAN Profile Configuration** window and save the changes or click the **Cancel** button to close the window without saving the changes.
- Repeat this process for the rest of the LAN objects if desired.

7.4.8 CLEARING APPLICATION DEPLOYMENT

Finally, in some situations, it is easier to completely wipe out the current application deployment instead of trying to identify problems with the configuration. OPNET provides a utility that clears all application deployments from the current simulation scenario. To clear application deployments, select the **Protocols > Applications > Clear Application Deployments** option from the pull-down menu in the **Project Editor**. As a result, the **Clear Application Deployments** window will appear. This window consists of three checkboxes and standard control buttons: **OK**, **Cancel**, and **Help**. The checkboxes allow configuring the types of application deployments to be cleared:

- *ACE traffic flow deployment*—clears all traffic deployment defined through the ACE OPNET software product, which we do not discuss here.
- *DES application demand deployment*—clears all traffic deployments defined as application demands (see Section 6.6.1).
- *DES application deployment*—clears all profile and application deployment on client and server nodes (i.e., workstation, sever, LAN, load balancer, and mainframe objects) by resetting the attributes of the corresponding objects to their default values. The application deployment will be cleared regardless of whether the applications were deployed using the **Application Deployment Wizard** or not. This action will NOT modify or remove profile, application, and application task definitions.

The control buttons have the same standard meaning:

- **OK**—clears application deployment based on the checkbox selection and closes the current window.
- **Cancel**—closes the current window without performing any actions.
- **Help**—opens a help page with a brief description of the features available in this window.

In summary, perform the following actions to clear the application deployments in the current scenario:

- Select the **Protocols > Applications > Clear Application Deployments** option from the pull-down menu in the **Project Editor**.
- In the **Clear Application Deployments** window that appears, check the checkboxes that correspond to the type of application deployment to be cleared. By default, all checkboxes are checked, which is adequate and which can be left unchanged since OPNET clears only those application deployment types that have been configured.
- Click the **OK** button to clear the selected application deployment or click the **Cancel** button to rescind this action and close the corresponding window.

7.5 DEPLOYING USER PROFILES WITHOUT APPLICATION DEPLOYMENT WIZARD

Applications can also be deployed within the simulated network system without using the **Application Deployment Wizard**. Typically, this process involves specifying which end nodes in the simulated network operate as clients by supporting defined user profiles and which end nodes operate as servers by providing services for defined applications. This section discusses these and other issues related to deploying defined applications without the **Application Deployment Wizard**.

7.5.1 CONFIGURING CLIENT NODES

Typically, the client nodes are configured to support a defined user profile by performing the following steps:

- Right-click on the client node of interest and select the **Edit Attributes** option.
- Expand the attribute **Applications**.
- Expand the attribute **Application: Supported Profiles.**
- Set the value of the attribute **Number of Rows** to the number of profiles you would like to deploy at the current client node.
- For each row that you have created:
  - Expand the row.
  - Set the value of the **Profile Name** attribute. This attribute allows you to select one of the profiles already configured in the simulation. No other values are permitted. If the current simulation scenario has no defined profiles, then the value of the **Profile Name** attribute will be None and profiles cannot be deployed. Once the value of **Profile Name** is set, the title of the corresponding row also changes to the specified value.
  - You can leave the rest of the attributes unchanged. Notice that the attribute **Traffic Type** is not configurable through the **Edit Attributes** menu. You can change the value of this attribute through the **Application Deployment Wizard**. This attribute deals with OPNET's ACE software and is not discussed in this book. The attribute **Application Delay Tracking** configures a mechanism for identifying the source of application delay in discrete event simulation and is also not discussed here.
  - Click the **OK** button to save the configuration changes.

This process should be repeated for all the client node objects that support defined user profiles. In a simulation that consists of numerous client objects, configuring one node at a time could be a tedious and error-prone task. To speed up application deployment, you can either use the **Application Deployment Wizard** or if multiple clients run the same set of profiles then they can be configured as one (i.e., select desired objects, edit attributes in one of the selected objects, and check the *Apply to Selected Objects* checkbox at the bottom of the **Edit Attributes** window before clicking the **OK** button). Figure 7.15 provides an example of profile deployment at a client node.

7.5.2 CONFIGURING SERVER NODES

Deploying application services at the server nodes is also quite simple and can be achieved by executing the following steps:

- Right-click on the server node of interest and select the **Edit Attributes** option.
- Expand the attribute **Applications**.
- Edit the attribute **Application: Supported Services** that opens the **Application: Supported Services Table** window shown in Figure 7.16.
  - Specify the number of applications supported at this node by setting the value of the **Rows** attribute at the bottom of the table window. Once the value is set, the corresponding number of rows will appear in the table. Notice that setting the number of rows to a value that is larger than the total number of applications defined in the *Application Config* node of this scenario will result in some of the rows not being set because the table cannot contain two entries with the same application name.
  - Specify the application supported at this server by selecting an application name from the pull-down menu. Once you have selected a

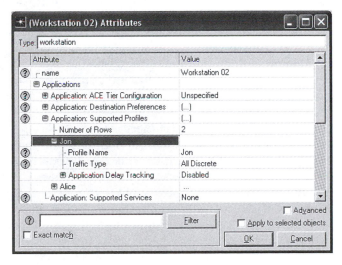

FIGURE 7.15 Profile deployment at a client node.

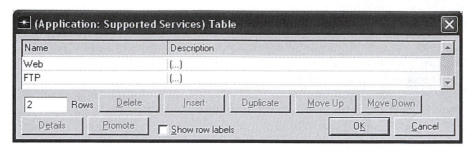

FIGURE 7.16   Application: Supported Services Table window.

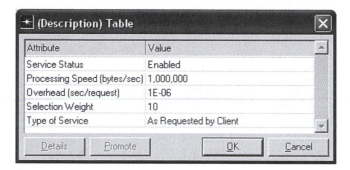

FIGURE 7.17   Default configuration of supported application services.

particular application, its name will not appear again for selection in subsequent rows of that table.

- In the description column, select `Supported`, `Not Supported`, or `Edit...` When the value `Supported` is chosen, then the application will be supported by the current server using the default setting. If the value `Not Supported` is chosen, then the application will not be supported by this node and the client nodes that run this application will not be able to connect to the current server for that application's services. If the value `Edit...` is chosen, then a table that allows configuring the server's application support attributes (i.e., processing speed, overhead delay per request, ToS value, etc.) is opened. Figure 7.17 shows the default setting of such a table.
- Click the **OK** button to save the changes.
- In addition, **Application: Supported Services** has two preset values: `None`, which means that the current server does not support any application services, and `All`, which means that this server supports all application services. OPNET advises against setting server nodes to support all application services because if the configuration changes and new applications are defined at a later time, then this node may end up supporting applications that the simulation did not intend to. In fact, the **Application Deployment Wizard** treats such a configuration as a warning, marking the corresponding node with a warning sign.

7.5.3   SPECIFYING DESTINATION PREFERENCES

Recall that each application definition (i.e., in the *Application Config* node) contains an attribute that specifies the symbolic server name. The attribute **Application: Destination Preferences**, available in all client nodes, configures the server selection weights and provides a mapping between the symbolic server name and the actual name of the node object in the simulation. With the help of this attribute, each individual client node can specify a different set of possible server destinations and server selection weights for each of its applications. If **Application: Destination Preferences** remains unchanged then, by default, OPNET configures every server node that supports application services as one of the possible destinations with a server selection weight of 10 (i.e., the same probability of selection). Refer to Section 5.5.2 for detailed instructions for specifying destination preferences.

As you can imagine, this process is error-prone because it is easy to either set an incorrect symbolic server name or configure a server node that does not support the current application as one of the application destinations (i.e., the pull-down menu lists all the nodes in the system and does not provide any hints regarding which nodes support the current application and which do not). If destination preferences have been incorrectly configured, the OPNET simulation reverts to the default setting in which each server node that supports the current application is chosen as a destination with the same probability (i.e., by default, the server selection weight is set to 10). This might be a bit confusing since OPNET uses the default setting but does not change any of the mistakes in the configuration. For this reason, it is advisable to do all the application deployment using the **Application Deployment Wizard** and only configure application destination preference attributes if they are not available for configuration in the **Application Deployment Wizard**.

7.5.4   SPECIFYING NUMBER OF CLIENTS IN A LAN OBJECT

As you may recall, the **Application Deployment Wizard** contains a feature that allows specifying the number of clients in LAN objects. You can set the number of clients without using the **Application Deployment Wizard** by simply executing the following steps:

- Right-click on a LAN object of interest and select the **Edit Attributes** option.
- Expand the attribute **Application**.
- Expand the attribute **Application: Supported Profiles** and configure the user profiles supported at this LAN object.
- Expand the attribute that describes the profile of interest.
- Set the value of the attribute **Number of Clients**, which specifies the number of users that run the current profile in the LAN. By default, this value is set to `Entire LAN`, which means that every node in the LAN will run one instance of this profile. OPNET advises against setting the number of clients to be larger than the number of workstations in the LAN.
- Click the **OK** button to close the window and save the changes.

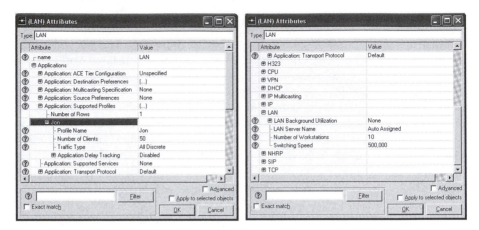

FIGURE 7.18   Configuration attributes for specifying the number of clients and the number of workstations in a LAN object.

You can also specify the number of workstations in a LAN object as follows:

- Right-click on a LAN object of interest and select the **Edit Attributes** option.
- Expand the attribute **LAN**.
- Set the value of the attribute **Number of Workstations**, which specifies the total number of workstations in this LAN. It is assumed that each LAN model consists of a number of workstations and a single server.
- Click the **OK** button to close the window and save the changes.

Figure 7.18 shows the attributes for configuring the number of workstations and the number of clients in a LAN object.

7.5.5   Specifying the Transport Protocol Used by an Application

Finally, in certain situations, it might be necessary to change the underlying transport protocol used by an application. By default, the standard applications Database, E-mail, FTP, HTTP, Remote Login, and Print use TCP, while Voice and Video Conferencing rely on UDP. The transport protocols employed by applications can only be changed in LAN, load balancer, mainframe, and advanced node models (i.e., node models that have the suffix _adv in their name) of server and workstation objects. To accomplish this task, perform the following steps:

- Right-click on a node model of interest and select the **Edit Attributes** option.
- Expand the attribute **Application**.
- Expand the attribute **Application: Transport Protocol Specification**.
- Change the transport protocol for the desired application. OPNET provides the following transport protocol options: TCP, UDP, AAL5, X25, FR, and NCP, although the available protocols vary by the application type.

You must change the application's transport protocol on both client and server nodes for the simulation to execute properly (i.e., without errors).

7.6 COMMON MISTAKES IN PROFILE CONFIGURATION AND APPLICATION DEPLOYMENT

When configuring user profiles, you must be aware of interactions between applications, profiles, and the simulation to avoid various mistakes that may lead to unpredictable results or failures to execute the simulation. First, let us examine configuration mistakes associated with the profile and simulation timing.

a. One of the most common mistakes is to set the duration of the profile to be longer than the length of the simulation. In this case, the application will terminate at the same time as the simulation. For example, suppose the length of the simulation is set to 300 seconds while the profile is configured to run for 500 seconds. If the profile is configured to run multiple applications, then it is possible that some applications within the profile will not start. As a result, there will be less traffic traveling through the network than expected.

b. Similarly, setting the profile inter-repetition time to be longer than the time left until the end of the simulation will prevent the profile from repeating and again might give unexpected results. For example, suppose the lengths of simulation, profile, and profile inter-repetition times are set to 500, 300, and 200 seconds, respectively. If the profile is configured to start at time 100 seconds, then the profile will not repeat because the first instance of the profile will end execution at time 400 seconds while the second instance of the profile will be scheduled for execution at time 600 seconds, which is after the simulation terminates. Figure 7.19 illustrates this scenario.

c. You must ensure that the profile start time is smaller than the duration of the simulation. For example, if the profile is configured to start at time 400 seconds while the duration of the simulation is only 300 seconds, then the profile will never execute.

d. Setting the profile duration to `End of Simulation` will result in the profile never repeating.

The second type of profile configuration mistake is associated with incorrect configuration of applications within the profile as described in the following:

a. Setting the application start offset time to be larger than the duration of the profile will result in that application never starting. Also, if the profile is configured to execute applications in a serial order, then the application start offset time plus start offset times and durations of preceding applications should not exceed the length of the profile, otherwise this application will never start. Figure 7.20 illustrates such a situation.

b. Similarly, if the sum of the profile start time and the application offset time is larger than the duration of the simulation, then the application will never start. OPNET documentation suggests that this problem may commonly

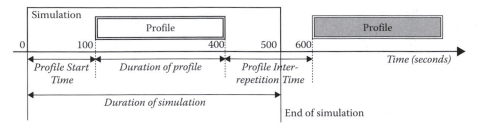

FIGURE 7.19   Example of incorrectly configured **Profile Inter-repetition Time**.

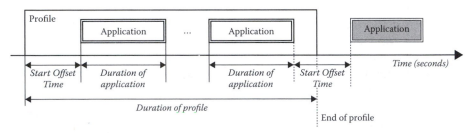

FIGURE 7.20   Example of poorly configured **Application Start Offset Time**.

occur when the values of application offset time or profile start time are configured using an exponential distribution.

c. Setting the duration of the application to be longer than the length of the profile may result in lower than expected amount of traffic generated in the network and will prevent any subsequent applications from starting if the profile is configured to execute applications in a serial order.

d. Similarly, if the profile is configured to run applications in a serial mode, then setting the duration of an application to End of Simulation will prevent all subsequent applications from starting.

e. Setting an application's repeatability to unlimited will prevent any subsequently configured applications from starting if the profile is configured to execute applications in a serial order.

In general, one should be careful when configuring an application's inter-repetition time, start offset time, and duration. If the applications within the profile are configured to be executed in a serial order, then it is important to ensure that, after the application completes its execution (i.e., including all application repetitions), it leaves a sufficient amount of time for all subsequent defined applications to execute. Figure 7.21 illustrates such a situation. On the other hand, if the applications within the profile are configured to be executed simultaneously, then for any configured application, the sum of the application duration and its inter-repetition time should not exceed the duration of the profile because it will prevent an application from repeating. This situation is similar to that depicted in Figure 7.19.

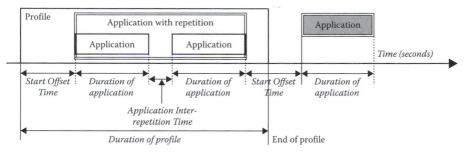

FIGURE 7.21   Example of application configured to execute so long that it prevents subsequently defined applications from starting.

Other configuration problems include deploying a profile on a client but not on a server and vice versa or, more generally, leaving some of the application or profile configuration attributes undefined. Such errors may prevent the simulation from executing. Finally, it is a good idea to configure the profiles to start a little later than the beginning of the simulation execution. This additional delay provides an opportunity for other protocols in the network to initialize. For example, configuring a profile to start at time 0 and setting the application offset time to 0 may result in no traffic traveling through the network because the routing protocols may not have had enough time to update and setup their routing tables. In this situation, when application packets arrive at the IP layer of a node, there is no route to the destination causing these packets to get discarded. Typically, setting the profile start time to 100 seconds provides a sufficient amount of time for all the protocols to initialize and converge to their stable states.

Generally, when debugging application and profile configurations or any other portion of the simulated system, you may want to employ one or more of the following suggestions:

- Use the DES Log, which provides a list of time-stamped warnings and error messages together with an explanation for the possible cause of the problem. The DES Log can be accessed from the **Project Editor's** pull-down menu by selecting **DES > Open DES Log**. See Section 4.8 for more details.
- Reduce randomness of the simulation by using the constant distribution function when specifying various attribute values.
- Consider collecting statistics such as traffic sent and traffic received using the *All Values* mode to obtain more accurate feedback regarding application performance.
- Try to isolate the problem by reducing the size of the simulated network system and simplifying the overall configuration.
- Consider using OPNET generated reports available through the pull-down **Scenarios** menu (e.g., **User-Defined Reports** and **Object/Attribute Difference Report**).

8 Transport Layer
TCP and UDP Protocols

8.1 INTRODUCTION

The Internet's transport layer (Layer 4) contains two widely used protocols: **Transmission Control Protocol (TCP)** and **User Datagram Protocol (UDP)**. TCP is a connection-oriented reliable protocol that provides ordered data delivery between processes on the end node systems. In TCP, a **segment** is a unit of data with a TCP header that is transmitted between two TCP peer entities. TCP keeps track of the segments that have been sent but have not been successfully acknowledged yet and retransmits any segments that were lost. Typically, TCP is used by such applications as e-mail, remote login, and FTP, which require reliable data delivery. UDP, on the other hand, is a connectionless and unreliable protocol that provides no guarantees regarding the success of data delivery. UDP simply hands data to the underlying network layer in the hope that it will be delivered to the destination process. UDP does not track transmitted datagrams and does nothing in the case of datagram loss. UDP is primarily used by applications for which fast data delivery (i.e., low delay) outweighs occasional packet loss and lack of guaranteed delivery. Voice over IP (VoIP) and video conferencing are typical examples of applications that run over UDP.

OPNET supports both the TCP and UDP protocols. However, UDP is a very simple protocol and does not have any configuration attributes. TCP, on the other hand, is fairly complex and has a large number of attributes configurable on the end node models such as servers, workstations, and LAN objects. Typically, the end node models contain the words *wkstn*, *server*, or *LAN* in their name. Core node models such as hubs, switches, and routers do not contain TCP configuration attributes because these node models do not support Layer 4 functionality. Even though UDP does not contain any configuration attributes, you can specify whether an applications runs over TCP or UDP. By default, OPNET runs all standard applications over TCP. The only exceptions are video conferencing and voice, which are configured to run over UDP. You can change the default transport protocol settings by modifying the value of the node attribute **Applications...Application: Transport Protocol Specification**, as shown in Figure 8.1. This attribute is only available in the advanced node models (i.e., the model name contains the suffix *_adv*). See also Section 7.5.5 for how to change the transport protocol used by an application.

The rest of this chapter is dedicated to a discussion of the supported features and available configuration parameters of TCP. Specifically, in Section 8.2, we discuss OPNET-supported features and flavors of TCP, followed by an overview of commonly used TCP configuration parameters in Section 8.3. We conclude with Section 8.4, which describes the available statistics for monitoring the performance of the TCP and UDP transport protocols.

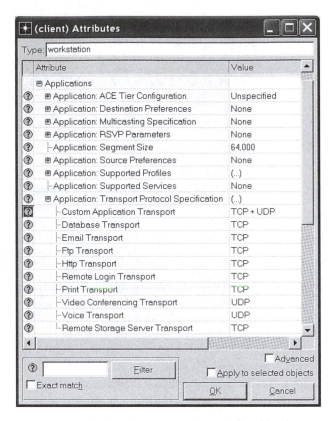

FIGURE 8.1   Attribute **Application: Transport Protocol Specification**.

8.2   SUPPORTED TCP FEATURES

The OPNET TCP model implements all basic TCP features including three-way handshake algorithms for connection establishment, acknowledgment of successfully received segments, retransmission of the segments considered to be lost, dynamic window control, in-order data delivery (i.e., segments that arrive out of order are reordered before they are sent to the upper layer), various congestion management mechanisms, four-way connection closing procedures, and others. OPNET also supports TCP's Slow Start and Congestion Avoidance mechanisms, Nagle's algorithm, Karn's algorithm, Fast Retransmit and Fast Recovery, window scaling, selective acknowledgments (SACK), Explicit Congestion Notification (ECN), persistent timeouts, and so on. These features are configurable through the **TCP...TCP Parameters** attribute, as shown in Figure 8.2. **TCP Parameters** is a compound attribute that contains all the TCP-related attributes. To specify the TCP configuration on a node, you can either configure the individual TCP configuration attributes or select one of the preset values for the **TCP Parameters** attribute. Each preset value represents a TCP configuration that corresponds to a particular TCP flavor or

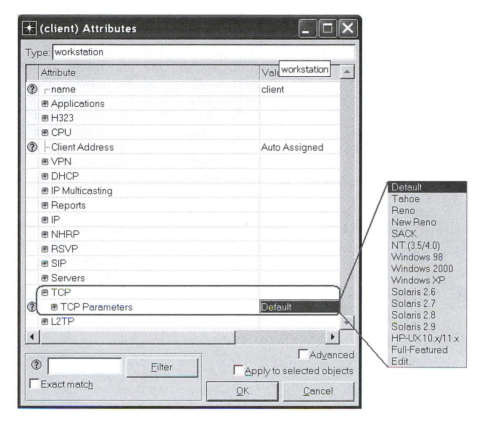

FIGURE 8.2   Attribute **TCP Parameters** and its preset values.

architecture-specific TCP setting. Figure 8.2 lists all the preset values for the **TCP Parameters** attribute.

Each TCP flavor (e.g., Tahoe, Reno, New Reno, SACK, etc.) defines a specific configuration of TCP's congestion control mechanisms. A preset value that represents an architecture-specific setting defines a TCP configuration corresponding to the TCP implementation on a particular platform. You can implement the behavior of these TCP flavors and settings by modifying individual TCP configuration attributes. However, in practice, it is more efficient to first choose your TCP flavor or setting and then tweak individual configuration parameters to your liking.

Before we describe the individual configuration attributes of TCP, let us briefly examine the meaning of the preset values for the **TCP Parameters** attribute shown in Figure 8.2.

- Tahoe—This is the most basic version of TCP, which consists of two main phases: **Slow Start** and **Congestion Avoidance**. Before we discuss these phases, recall that the TCP transmission rate is controlled through a congestion window (*cwnd*), a value that defines the number of unacknowledged segments a TCP source can send. Initially, when a TCP connection begins transmitting data, it goes into the Slow Start phase. During Slow Start, the size of the congestion window grows exponentially until it reaches the Slow Start Congestion Threshold (*ssthresh*). Once that happens, TCP switches over to the Congestion Avoidance phase in which *cwnd* grows linearly. Typically, the initial value of *cwnd* is a configurable parameter, and in OPNET, it is specified through the attribute **Slow-Start Initial Count (MSS)**. The initial value of Slow Start Congestion Threshold is set to the smaller of 64 KBytes and the Receive Buffer size at the destination host. In TCP, a timeout occurs if the source TCP process does not receive the segment acknowledgment (ACK) within a certain time period. When a timeout occurs, TCP resets the value of *cwnd* to its initial value, sets *ssthresh* to one-half of the *cwnd* value when the timeout occurred, and enters the Slow Start phase again. TCP Tahoe employs a **Fast Retransmit** mechanism that uses the arrival of a certain number (usually 3) of duplicate ACKs as an indication of segment loss. In such a case, TCP Tahoe behaves the same way as if a timeout has occurred: it retransmits the missing segment, updates the *cwnd* and *ssthresh* values, and enters the Slow Start phase. TCP Tahoe does not support Fast Recovery.

- Reno, Default—Both Reno and Default settings of **TCP Parameters** attributes correspond to the same TCP configuration that implements the Reno flavor of TCP. TCP Reno behaves the same way as TCP Tahoe initially and when a timeout occurs. However, unlike TCP Tahoe, upon the occurrence of a Fast Retransmit, TCP Reno initiates a **Fast Recovery** phase during which TCP Reno (1) sets *cwnd* to one-half of its previous value, (2) sets *ssthresh* to the new value of *cwnd*, (3) performs temporary inflation and deflation of *cwnd*, and (4) upon a new ACK arrival, enters the Congestion Avoidance phase. In step (3) of the Fast Recovery phase, TCP Reno inflates *cwnd* each time a duplicate ACK is received, and only when an ACK for a new segment arrives, TCP Reno deflates *cwnd* to the value it had before inflation. Therefore, the TCP Reno flavor has both Fast Retransmit and Fast Recovery mechanisms enabled.

- New Reno—TCP New Reno is an improvement over TCP Reno because it removes an inefficiency related to multiple losses within a single window of data. A TCP source determines that loss of multiple segments has occurred when it receives a partial ACK, that is, an ACK that acknowledges some but not all segments that were outstanding at the start of Fast Recovery. If a partial ACK is received, then the next lost segment is retransmitted immediately and *cwnd* is deflated. If more duplicate ACKs are received, then *cwnd* is inflated. The sender remains in Fast Recovery until the TCP process receives ACKs for all data that was outstanding when Fast Recovery was initiated. As a result, *cwnd* is not decremented multiple times when multiple losses occur within the same window, as would have happened with Reno.

- SACK—This flavor uses TCP's **Selective Acknowledgment (SACK)** option along with Reno for congestion control. A selective ACK is used to acknowledge only a specific segment, contrasted with a regular ACK, which is cumulative.

When a receiver decides to buffer out-of-order segments and acknowledges them selectively, the sender can skip retransmission of these segments when a timeout occurs and it only needs to retransmit those segments that are still unacknowledged. Use of the SACK option makes TCP behave more like the Selective Repeat protocol and improves protocol efficiency.

The remaining preset values of the **TCP Parameters** attribute configure various architecture-specific TCP configurations and are not described here. You can examine the TCP attribute setting that corresponds to a particular preset value by first setting the **TCP Parameters** attribute to the desired value and then expanding it to view the actual setting.

8.3   TCP CONFIGURATION ATTRIBUTES

In this section, we describe the TCP configuration attributes that are most commonly used. As we mentioned before, any node model that supports TCP contains all TCP configuration attributes under **TCP...TCP Parameters**. In general, to configure TCP on an end node, you need to perform the following steps:

- Open the **Edit Attributes** dialog box of the end node (e.g., workstation or server).
- Expand the compound attribute **TCP**.
- Set the attribute **TCP Parameters** to one of its preset values shown in Figure 8.2.
- If you would like to specify a custom TCP configuration, then you need to expand the attribute **TCP Parameters** and specify the value for the attributes of interest. Figure 8.3 illustrates the available TCP configuration attributes. Clicking on the icon in the form of a question mark inside a circle located to the left of the attribute opens an **Attribute Description** window that provides a brief description and meaning of the corresponding attribute.
- Once all the changes have been made, click the **OK** button to save the introduced changes.

⊟ TCP Parameters	(...)	⊢Segment Send Threshold	MSS Boundary
⊢Version/Flavor	Unspecified	⊢Active Connection Threshold	Unlimited
⊢Maximum Segment Size (byt...	Auto-Assigned	⊢Nagle Algorithm	Disabled
⊢Receive Buffer (bytes)	65535	⊢Karn's Algorithm	Enabled
⊢Receive Buffer Adjustment	None	⊞ Timestamp	Disabled
⊢Receive Buffer Usage Thre...	0.0	⊢Initial Sequence Number	Auto Compute
⊢Delayed ACK Mechanism	Segment/Clock Based	⊞ Retransmission Thresholds	Attempts Based
⊢Maximum ACK Delay (sec)	0.200	⊢Initial RTO (sec)	1.0
⊢Maximum ACK Segments	2	⊢Minimum RTO (sec)	0.5
⊢Slow-Start Initial Count (MSS)	2	⊢Maximum RTO (sec)	64
⊢Fast Retransmit	Enabled	⊢RTT Gain	0.125
⊢Duplicate ACK Threshold	3	⊢Deviation Gain	0.25
⊢Fast Recovery	Disabled	⊢RTT Deviation Coefficient	4.0
⊢Window Scaling	Disabled	⊢Timer Granularity (sec)	0.5
⊢Selective ACK (SACK)	Disabled	⊢Persistence Timeout (sec)	1.0
⊢ECN Capability	Disabled	⊢Connection Information	Do Not Print

FIGURE 8.3   TCP configuration attributes in the attribute **TCP Parameters**.

Now, let us examine some of the basic TCP attributes that are used to specify a custom TCP configuration:

- **Version/Flavor**—specifies the name of the TCP flavor or setting. This attribute can only be set by selecting one of the preset values of the **TCP Parameters** attribute. This is a read-only attribute and it cannot be changed once the flavor or setting in the parent attribute **TCP Parameters** has been specified. This attribute has no effect on the simulation other than the fact that its value determines the values used for some of the other TCP attributes, which may in turn affect the simulation.

- **Maximum Segment Size (MSS)**—specifies the maximum amount of data in bytes that will be transmitted in one TCP segment. Usually, the MSS value is set based on the characteristics of the underlying network so that transmitted TCP segments are not fragmented in the IP layer. To avoid fragmentation by the lower layers, set the MSS to the Maximum Transmission Unit (MTU) size of the corresponding interface minus the size of the TCP and IP headers (usually 20 bytes each). For example, if the interface MTU is 1500 bytes, then MSS should be set to 1460 bytes. This attribute also accepts `Auto-Assigned` value, which will result in the simulation automatically calculating the MSS value based on the MTU size of the node's interface. If the node contains multiple interfaces, then the MSS value is calculated based on the MTU size of the first interface (i.e., `IF0`).

- **Receive Buffer**—specifies the size of the buffer in bytes used to hold received data before it is forwarded to higher layers. The size of the window advertised by TCP is the amount of free space in this buffer. Its value can be set to `Default` in which case the attribute value is set to at least four times the negotiated MSS with a maximum value of 64 KBytes.

- **Delayed ACK Mechanism**—specifies how delayed "dataless" ACKs are transmitted by TCP. Normally, ACKs are piggybacked on data segments traveling in the reverse direction. However, if no data is being sent in the reverse direction, TCP will wait for a while before returning a "dataless" ACK. This attribute accepts two values:
  - `Clock Based`—TCP will wait for a configured number of seconds (controlled by the attribute **Maximum ACK Delay (sec)**) before transmitting a "dataless" ACK.
  - `Segment/Clock Based`—TCP will wait for a configured number of seconds or until a specified number of segments (controlled by the attribute **Maximum ACK Segments**) are received before transmitting a "dataless" ACK.

- **Maximum ACK Delay (sec)**—specifies the maximum duration of time in seconds that a TCP process will wait before sending a "dataless" ACK. By default, this attribute is set to 0.2 seconds.

- **Maximum ACK Segments**—specifies the maximum number of segments that TCP must receive before sending a "dataless" ACK. This attribute is used only if the **Delayed ACK Mechanism** attribute is set to `Segment/Clock Based`. By default, this attribute is set to 2 segments.

- **Slow-Start Initial Count (MSS)**—specifies the initial *cwnd* value at the beginning of the Slow Start phase. This attribute is specified in multiples of the MSS and by default is set to 2.
- **Fast Retransmit**—specifies whether the TCP process uses Fast Retransmit or not. If it is set to `Enabled`, then TCP will retransmit a segment on the receipt of a certain number (controlled through the attribute **Duplicate ACK Threshold**) of duplicate ACKs.
- **Duplicate ACK Threshold**—specifies the number of duplicate ACKs that triggers the start of the Fast Retransmit phase. By default, this attribute is set to 3.
- **Fast Recovery**—specifies whether the Fast Recovery mechanism is enabled on this node. This attribute accepts the following values: `Disabled`, Fast Recovery is not in use at the node; `Reno`, Reno variation of Fast Recovery is used; or `New Reno`, New Reno variation is used.
- **Window Scaling**—specifies whether, during connection establishment, a TCP SYN packet will have the window scaling option enabled or not. For window scaling to be enabled (RFC 1323), the client node must have its **Window Scaling** attribute set to `Enabled` and the server must have **Window Scaling** set either to `Enabled` or to `Passive`.
- **Selective ACK (SACK)**—determines whether or not the Selective ACK option (RFC 2018) is used. If the SACK option is to be used, this attribute must be set to `Enabled` in the client host and to either `Enabled` or `Passive` in the server host. Instead of directly setting this attribute's value, it is recommended that the **TCP Parameters** attribute be set to `SACK`.
- **ECN Capability**—specifies if TCP is configured to support **Explicit Congestion Notification** (RFC 3168). If the client node has this attribute set to `Enabled`, then it will advertise that it is ECN-capable. The server node will advertise support for ECN only if it has the **ECN Capability** attribute set to `Enabled` or `Passive` and it has received an ECN advertisement from the client node. When both the client and server have agreed to use ECN, all packets on this connection will have header bits set to indicate this fact. Normally, a router experiencing congestion will drop packets passing through it. But when an ECN-capable packet passes through a router that is experiencing congestion, the router will not drop the packet but will instead mark the header bits in the packet to indicate that it encountered congestion. The receiver of the packet echoes the congestion indication to the sender, which must react by performing congestion control just as though the packet was dropped.
- **Active Connection Threshold**—specifies the maximum number of active TCP connections allowed on the node.
- **Nagle Algorithm**—specifies whether the Nagle Algorithm (RFC 1122) is enabled. The Nagle Algorithm prevents segments smaller than an MSS from being transmitted if there are outstanding ACKs for previously sent data. This attribute accepts the following values: (1) `Disabled`, Nagle's Algorithm is not used in this node; (2) `Enabled`, Nagle's Algorithm is used in this node; and (3) `Per-Send`, implements an optimized version of this algorithm in which the check for outstanding ACKs is applied on a per-send

instead of a per-segment basis. Here, "send" refers to a single send operation from the higher-layer application.

* **Karn's Algorithm**—specifies whether Karn's Algorithm (RFC 2988) for calculating retransmission timeout (RTO) values is being used at this node or not. This attribute accepts only two possible values: `Disabled` and `Enabled`. Karn's Algorithm ignores retransmitted segments when updating the round-trip time estimate used for the RTO computation. Round-trip time estimation is based only on unambiguous ACKs, which are ACKs for segments that were sent only once.

* **Timestamp**—indicates whether the TCP Timestamp option (RFC 1323) is supported. **Timestamp** is a compound attribute that consists of subattributes **Status**, which specifies whether the option is enabled, and **Clock Tick (millisec)**, which specifies the value used to increment the timestamp value. For this option to be enabled, the client node must have its **Status** attribute set to `Enabled` and the server node must have this attribute set either to `Enabled` or to `Passive`.

* **Initial Sequence Number**—specifies the initial sequence number used for outgoing segments on all connections at this node. This attribute accepts the value `Auto Compute` in which case the initial sequence number is randomly chosen for each connection.

* **Initial RTO (sec), Minimum RTO (sec)**, and **Maximum RTO (sec)**— specify initial, minimum, and maximum values for the RTO over the lifetime of a connection in accordance with Karn's Algorithm. By default, these attributes are set to 3 seconds, 1 second, and 64 seconds respectively.

The remaining TCP configuration attributes describe various advanced TCP features and are not discussed in this book. For more information about the OPNET implementation of TCP, refer to the user guide available through the **Protocols > TCP > Model User Guide** option in the pull-down menu of the **Project Editor**. In addition, OPNET software includes a standard example project called **TCP**. This project contains several scenarios that provide an excellent illustration of various TCP features and can serve as a great reference of how to develop simulation scenarios for evaluating TCP performance.

8.4 COMMON TRANSPORT LAYER STATISTICS

Transport layer statistics are available as both Global Statistics and Node Statistics. Global Statistics contain a single category called **TCP**, which monitors the performance of TCP across all nodes. There are no Global Statistics for UDP. However, Node Statistics contain three categories for evaluating transport layer performance within individual nodes: **TCP, TCP Connection**, and **UDP**. The **TCP** and **UDP** categories of Node Statistics collect data for all TCP and UDP traffic respectively at each node. The **TCP Connection** category collects and maintains statistics separately for each TCP connection at a node. Table 8.1 provides a brief description of all the available transport layer statistics.

TABLE 8.1
Summary of Transport Layer Statistics

Category	Name	Description
Global Statistics		
TCP	*Delay (sec)*	This statistic records the average network-wide delay in seconds experienced by all TCP packets. It is measured from the time an application layer packet is sent from the source TCP layer until it is completely received by the TCP layer in the destination node.
	Retransmission Count	This statistic records the total number of TCP segment retransmissions in the network.
	Segment Delay (sec)	This statistic records the average network-wide delay in seconds for all TCP segments. It is measured from the time a TCP segment is sent from the source TCP layer until the segment is received by the TCP layer in the destination node. This is different from Delay (packet delay) since one application layer packet can be transmitted as multiple segments by TCP if the packet size is larger than the MSS.
Node Statistics		
TCP	*Active Connection Count*	This statistic records the total number of active TCP connections at this node.
	Connection Aborts	This statistic records the total number of TCP connections aborted by the higher layer at this node.
	Delay (sec), *Retransmission Count,* *Segment Delay (sec)*	These statistics have the same meaning as the corresponding Global Statistics except that, unlike Global Statistics that aggregate results over all nodes in the system, these statistics are recorded for each node separately.
	Load (bytes), *Load (bytes/sec),* *Load (packets),* *Load (packets/sec),*	These statistics record the total amount or rate of data submitted by the application layer to the TCP layer at this node. It includes data for all TCP connections at the node. These statistics are available in units of bytes, bytes per second, packets, and packets per second.

(Continued)

TABLE 8.1   (Continued)
Summary of Transport Layer Statistics

Category	Name	Description
	Traffic Received (bytes), *Traffic Received (bytes/sec),* *Traffic Received (packets),* *Traffic Received (packets/sec)*	These statistics record the total amount or rate of data received and forwarded to the application layer by the TCP layer at this node, including data for all TCP connections at the node. These statistics are available in units of bytes, bytes per second, packets, and packets per second.
Node Statistics **TCP Connections**	*Congestion Window Size (bytes)*	This statistic records the size of the congestion window for each TCP connection.
	Delay (sec), *Retransmission Count,* *Segment Delay (sec),* *Load (bytes),* *Load (bytes/sec),* *Load (packets),* *Load (packets/sec),* *Traffic Received (bytes),* *Traffic Received (bytes/sec),* *Traffic Received (packets),* *Traffic Received (packets/sec)*	These statistics have the same meaning as the corresponding Node Statistics, except that, unlike Node Statistics that collect results on a per-node basis, these statistics are recorded for each TCP connection separately.
	Flight Size (bytes)	This statistic records the amount of unacknowledged (i.e., in-flight) data per connection. This statistic is updated each time a TCP connection transmits a segment or receives an ACK.
	Received Segment ACK Number, *Received Segment,* *Sequence Number*	These statistics record the received ACK and the segment numbers for a connection respectively.

Statistic	Description
Remote Receive Window Size (bytes)	This statistic records the advertised size in bytes of the receive window at the remote node for a connection.
Retransmission Timeout (sec)	This statistic records the RTO value for a connection.
Segment Round-Trip Time (sec), *Segment Round-Trip Time Deviation*	These statistics record the mean segment round-trip time in seconds and its deviation for a connection.
Selectively ACKed Data (bytes)	This statistic records the total number of bytes sent by a connection and acknowledged by a receiver using the SACK option. The data ACKed without using SACK mechanisms is not recorded by this statistic.
Send Delay (CWND) (sec), *Send Delay (Nagle's) (sec),* *Send Delay (RCV-WND) (sec)*	These statistics record the delay in seconds encountered in sending buffered data. It is available separately for three different causes of the delay: CWND, Nagle's, and RCV-WND. The delay value is measured from the time TCP started refraining from sending data due to the specific cause until the time that the data was sent out. The three causes are CWND due to the congestion window being too small, Nagle's when delay is caused by Nagle's algorithm, and RCV-WND due to the receive window being too small.
Sent Segment ACK Number, *Sent Segment Sequence Number*	These statistics record the sent ACK and the segment numbers for a connection respectively.
Node Statistics **UDP**	
Traffic Received (bytes/sec), *Traffic Received (packets/sec)*	These statistics record the amount or rate of data received by UDP from the lower layer to be forwarded to the higher layer.
Traffic Sent (bytes/sec), *Traffic Sent (packets/sec)*	These statistics record the amount or rate of data received by UDP from the higher layer for transmission.

9 Network Layer
Introduction to the IP Protocol

9.1 INTRODUCTION

There are several different Layer 3 or Network Layer protocols available in today's networks. However, **IP** is the most widely used network layer protocol. IP is a fairly complex protocol and it is often considered to be the glue that holds today's Internet together. Even though IP has a wide variety of responsibilities, its primary task is to deliver data between the end systems in the Internet. Such data delivery may span multiple links and networks and requires that each node is uniquely identified through an IP address. An **IP address** identifies a particular interface that connects a node to a network. In end systems or hosts, which typically contain a single interface, an IP address uniquely identifies the node's interface or the node itself and therefore is often referred to as the host's IP address. Routers, on the other hand, are typically connected to multiple networks and therefore contain multiple interfaces. Each active interface in a router has a different IP address allocated to it. In this book, we will refer to an IP address as the address that identifies a particular interface in a node. In addition to end-to-end communication, the network layer handles data encapsulation, fragmentation and reassembly, data forwarding, routing, compression, multicasting, load balancing, preferential data treatment to support various Quality of Service (QoS) requirements, and other functions. OPNET software models the majority of these features and allows for custom configuration of various aspects of IP.

OPNET Technologies Inc. is constantly improving its software GUI by adding new options for quicker and easier configuration of various features of simulated systems. As a result of such growth, you may often achieve the same system configuration by performing a different set of configuration steps. This is particularly true for IP, which has been supported by OPNET software since the earliest releases. In particular, OPNET software allows specification of various IP features through such configuration tools as

- Model attributes and configuration parameters within individual nodes.
- Global attributes and configuration parameters pertaining to the whole simulation.
- **Protocols** pull-down menu options in the **Project Editor**.
- Utility nodes (objects that are used to define various scenario-wide features), which allow the defined profiles or mechanisms to be accessed by any object within the scenario.

In Chapters 9 and 10, we describe the configuration steps for various IP features supported in OPNET. We describe the most common and the easiest ways for configuring IP features, while occasionally omitting certain configuration alternatives that are too cumbersome and error-prone. It is not our intention to provide a comprehensive description of all available IP features in OPNET and all the possible ways to configure them. However, these two chapters provide a detailed description of the main IP features supported by OPNET. Specifically, the current chapter begins by introducing basic IP configuration attributes available in the end and core node models in Section 9.2. It continues by describing IP addressing and the steps for specifying IP addresses and subnet masks on the node interfaces in Section 9.3. In Section 9.4, we provide an overview of configuration steps for such OPNET-supported IP features as compression schemes, routing, interfaces, and load balancing. After that, Section 9.5 provides a description of the Internet Control Message Protocol (ICMP) and its attributes supported in OPNET. This chapter concludes in Section 9.6 with a description of common IP statistics and other features for performance evaluation of IP networks. Advanced IP features such as Network Address Translation (NAT), IP Multicast, IP Version 6 (IPv6), and QoS are discussed in Chapter 10.

9.2   BASIC IP CONFIGURATION ATTRIBUTES

In OPNET, each node model that supports Layer 3 functionality (e.g., routers, gateways, servers, workstations, cloud, LAN, and other objects) contains an **IP** module that implements various features of the IP protocol. Notice that node models of hubs and Layer 2 switches do not contain the IP module and therefore are not discussed in this chapter. Typically, model attributes for configuring IP are aggregated together under a single compound attribute called **IP**. However, the number and the type of subattributes for configuring IP vary depending on the type of the node model. For example, end nodes (such as workstations and servers) usually do not require all the IP functionality; thus, they contain fewer IP configuration attributes available to the user for configuration as compared to core node models (such as routers, gateways, level-3 switches, and IP cloud), which typically include a wider array of IP attributes. However, the number of attributes also varies based on the specific responsibilities of a node. Since the configuration of the IP protocol could become fairly complex, IP attributes that are less likely to be needed are hidden to avoid clutter and to make IP configuration more manageable. This section describes the main IP configuration attributes available in the end and core node models. Description of some of the more advanced attributes, which are available only within specific node models, is omitted.

9.2.1   BASIC IP CONFIGURATION ATTRIBUTES OF AN END NODE MODEL

To configure IP attributes in an end node model, you need to open the **Attributes** window by right-clicking on a node and selecting the **Edit Attributes** option. All the attributes for specifying IP properties, with the exception of the attributes for configuring IP multicasting using Internet Group Management Protocol (IGMP), are arranged under the compound attribute called **IP**, which consists of three compound subattributes:

- **IP Host Parameters** specifies IP addressing and forwarding features on this node.
- **IP Processing Information** defines such hardware characteristics as the number of processors or slots, switching and forwarding rates, and the amount of main memory available for packet forwarding.
- **IP QoS Parameters** defines the QoS mechanisms deployed at the node. We will discuss QoS configuration in Chapter 10.

IP Host Parameters consists of the compound attribute **Interface Information** and several other attributes for configuring various routing properties at the node. Attribute **Interface Information** specifies characteristics of the node's interface and consists of the following subattributes:

- **Name**—an alphanumeric name for the interface. In OPNET, each interface in a node has a name that starts with the letters IF (to signify interface) followed by the interface number. OPNET numbers the node's interfaces in ascending order starting from 0. Since end nodes have only one interface, the value of this attribute in an end node is IF0, as shown in Figure 9.1.
- **Address**—specifies the IP address of the interface.

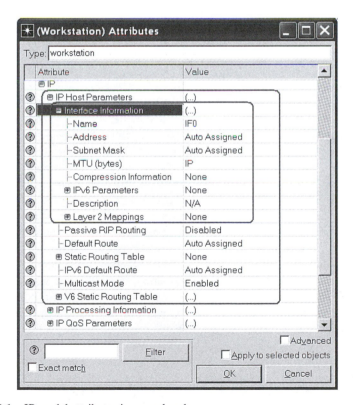

FIGURE 9.1   IP model attributes in an end node.

- **Subnet Mask**—specifies the subnet mask of the interface. We discuss attributes **Address** and **Subnet Mask** in greater detail in Section 9.3.
- **MTU (bytes)**—the value of the Maximum Transmission Unit (MTU), the largest possible packet size accepted by the underlying network. MTU is one of the parameters that determine how IP fragments incoming data before forwarding it into the network. OPNET provides several preset MTU values: `Ethernet`, which is 1500 bytes; `FDDI`, which is 4470 bytes; `ATM`, which is 4470 bytes; `IP`, which is 1500 bytes; `Token Ring`, which is 4464 bytes; `Frame Relay`, which is 1500 bytes; `WLAN`, which is 2304 bytes; or `TDMA`, which is also 2304 bytes. The smallest allowed value of MTU is 20 bytes, which corresponds to the size of the IP header.
- **Compression Information**—the attribute that specifies the name of the compression scheme used on this interface. The actual scheme is defined through the utility object **IP Attribute Config**, which will be discussed in Section 9.4.1.
- **IPv6 Parameters**—specifies address and other configuration parameters for IPv6.
- **Description**—a textual description of this interface. This attribute has no effect on the simulation.
- **Layer 2 Mappings**—specifies the mapping between this interface and Layer-2 permanent virtual circuits (PVC) such as ATM, Frame Relay (FR), or Virtual LAN (VLAN) connections.

The remaining subattributes of **IP Host Parameters** are described as follows:

- **Passive RIP Routing**—specifies if the current end node forwards data based on the routing information obtained from the Routing Information Protocol (RIP), which operates in passive mode.
- **Default Route**—specifies the IP address of the default gateway. The node forwards packets to the default gateway if IP routing is disabled or if the IP Forwarding Table yields no route to the destination. The default setting for attribute **Default Route** is `Auto Assigned`, which means that the default gateway address will be determined automatically during the simulation.
- **Static Routing Table**—defines the IP Version 4 (IPv4) static routing table. Each entry in the static routing table consists of such attributes as the destination's IP address and mask, the IP address of the next hop on the path to the destination, and administrative weight that specifies the route priority. If there are multiple routes to a destination, then the route with the lowest administrative weight is selected. The node uses the static routing table if there is no routing protocol enabled or if a route in the static routing table has a lower administrative weight than a route provided by a routing protocol such as RIP.
- **IPv6 Default Route**—specifies the IPv6 address of a default gateway.
- **Multicast Mode**—determines if the current node supports multicast applications or not. This attribute accepts only two possible values: `Enabled` and `Disabled`.

- **V6 Static Routing Table**—specifies the IPv6 static routing table. Each entry in this table consists of the destination's IPv6 address, prefix length (the number of leftmost bits in the address that define the network portion of the address), IPv6 address of the next hop, and administrative weight.

9.2.2   BASIC IP CONFIGURATION ATTRIBUTES OF A CORE NODE

As shown in Figure 9.2, the core node models also have IP configuration parameters grouped under the compound attribute **IP**, which contains the following subattributes:

- **APS Parameters**—configures the Automatic Protection Switching (APS) mechanism for supporting fault tolerance in the network. APS configures backup interfaces in the event of the primary interface failure. This is an advanced feature available in various Cisco routers and is not discussed further in this book.
- **IP Processing Information**—specifies such hardware characteristics of the node as the number of processors or slots, switching and forwarding rates, and the amount of main memory available for packet forwarding.
- **IP QoS Parameters**—specifies QoS settings on the router's interfaces.
- **IP Routing Parameters**—specifies IP addressing and routing protocol settings for each interface in this node.
- **IP Slot Information**—configures individual processors or slots if the current node supports a multiprocessor environment (i.e., when the value of the **IP Processing Information…Processing Scheme** attribute is set to Slot Based Processing). This attribute allows specifying such

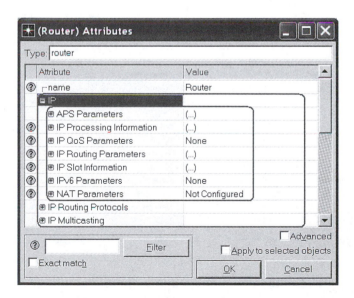

FIGURE 9.2   IP model attributes in a core node.

information as the service rate and the capacity of input and output buffers for each of the slots.

- **IPv6 Parameters**—configures various IPv6 addressing and routing features on this node.
- **NAT Parameters**—specifies Network Address Translation (NAT) configuration on this node.

The **IP Routing Parameters** compound attribute serves the same purpose as the **IP Host Parameters** attribute of the end node models. However, as shown in Figure 9.3, this attribute allows configuration of a wide array of additional features through numerous additional subattributes that are not available in end nodes. We provide a summary of the most commonly used **IP Routing Parameters** subattributes in the following:

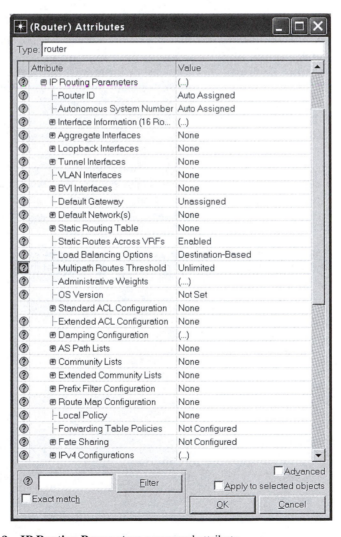

FIGURE 9.3   IP Routing Parameters compound attribute.

- **Router ID**—specifies a unique router identifier used by the Open Shortest Path First (OSPF) routing protocol.
- **Autonomous System Number**—is a 16-bit number represented as an integer value between 1 and 65535 that uniquely identifies an autonomous system (AS) that this router belongs to. The AS number is typically employed by the Border Gateway Protocol (BGP).
- **Interface Information**—is a compound attribute that specifies addressing and routing information for each physical interface of this node.
- **Aggregate Interfaces**—is a compound attribute for configuring Link Aggregation Control Protocol (LACP) by mapping multiple physical interfaces into a single logical or aggregate interface. This attribute defines characteristics of the aggregate interface. Mapping a physical interface to this aggregate interface requires you to set the value of the attribute **IP... IP Routing Parameters...Interface Information...<interface name>... Aggregation Parameters...Aggregate Interface** to the name of this aggregate interface. We discuss the steps for configuring aggregate interfaces in Section 9.4.3.
- **Loopback Interfaces**—is a compound attribute for configuring loopback interfaces on this node. By default, all router models include a single loopback interface. Additional loopback interfaces can be added by configuring the **Loopback Interfaces** attribute.
- **Tunnel Interfaces**—is a compound attribute for defining tunnel interfaces on this node.
- **VLAN Interfaces**—is a compound attribute for configuring VLAN interfaces. You can only specify this attribute in the switch node models that include a Layer 3 card (e.g., *ethernet16_switch_with_layer3*). In other node models, this appears as a simple attribute with the value None.
- **BVI Interfaces**—is a compound attribute for specifying configuration parameters of Bridged VLAN interfaces.
- **Default Gateway**—specifies the IP address of the default gateway. However, this attribute has been deprecated and has no effect on the simulation.
- **Default Network(s)**—is a compound attribute that allows specifying the default gateway.
- **Static Routing Table**—is a compound attribute for defining a static routing table.
- **Static Routes Across VRFs**—indicates whether static routes can be added on outgoing interfaces that belong to different Virtual Routing and Forwarding (VRF) instances. This attribute is applicable to Cisco routers only and is not discussed further in this book.
- **Load Balancing Options**—specifies the type of load balancing mechanism. We discuss load balancing configuration in Section 9.4.4.
- **Multipath Routes Threshold**—is an attribute that specifies the maximum number of routes to the same destination that will be considered during the routing process. Note that all the available routes to the destination will be recorded, but only the number specified by this attribute will be considered for routing.

- **Administrative Weights**—is a compound attribute that specifies priorities of various routing protocols. A lower value of administrative weight designates higher priority. If the IP Forwarding Table contains multiple routes that have been provided by various protocols, then the route that was computed by the routing protocol with the smallest administrative weight value will be considered first.
- **OS Version**—is the attribute that specifies the version of the operating system employed at this node. However, this attribute has no effect on the simulation.

The remaining attributes configure various advanced features such as access control lists, route maps, prefix filters, and route dampening, which are not discussed in this book.

As mentioned above, the attribute **Interface Information** contains parameters for configuring IP addressing and routing characteristics on all physical interfaces available in this core node model. Figure 9.4 illustrates the subattributes of **Interface Information** for a single physical interface (e.g., interface IF0). Notice that attributes **Name**, **Address**, **Subnet Mask**, **MTU (bytes)**, **Compression Information**, **Description**, and **Layer 2 Mapping** are available for interface configuration in both core and end node models. In both cases, these attributes have the same meaning, so we do not describe them in this section again. The only difference is that in end node models, these attributes are stored under the **IP…IP Host Parameters** attribute, while in the core node models, they are located under **IP…IP Routing Parameters…Interface Information…<interface name>**. The following list provides a short description of several other commonly used attributes for configuring IP properties on the physical interfaces of the core node models, while description of attributes that configure various advanced features (e.g., **Packet Filter**, **Policy Routing**, **Routing Instance**, and **VRF Sitemap**) are omitted.

- **Status**—this attribute specifies the current administrative status of the interface. This attribute accepts two values: Active, which means that the interface is configured to be operational, and Shutdown, which means that the interface is disabled and will not forward or receive any data.
- **Operational Status**—this attribute specifies if the current interface is physically operational or not. The value of this attribute is only relevant when attribute **Status** is set to Active. This attribute accepts values Up (e.g., operational), Down (e.g., not operational), and Infer (e.g., status is currently unknown and will be determined during simulation).
- **Secondary Address Information**—allows specifying additional addresses, to support multiple logical subnets over a single physical network.
- **Subinterface Information**—specifies subinterface configuration parameters.
- **Routing Protocol(s)**—specifies dynamic routing protocols enabled on this interface. More than one protocol can be enabled on the same interface. Also, different interfaces of the same node may have different routing protocols enabled.

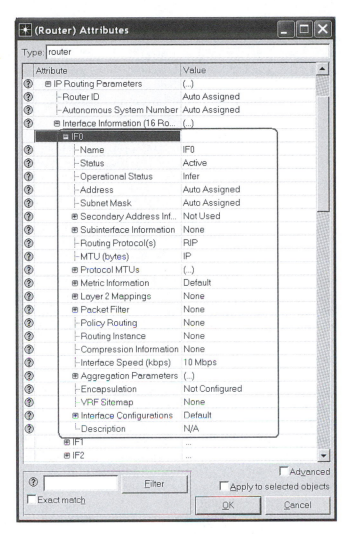

FIGURE 9.4 Subattributes of **Interface Informations** for a single physical interface (IF0).

- **Protocol MTUs**—specifies MTU sizes for various protocols (e.g., IP, MPLS, and IPv6).
- **Metric Information**—specifies various metric values (e.g., bandwidth, delay, reliability, and load) used by the routing protocols to determine the interface cost.
- **Aggregation Parameters**—specifies interface aggregation parameters for when the current physical interface is configured to be part of the logical aggregate interface. You can only configure this attribute if an aggregate interface has been defined through the attribute **IP...IP Routing Parameters...Aggregate Interfaces**.

- **Interface Configurations**—configures various additional interface characteristics such as specifying if directed broadcast is enabled on an interface, defining a helper address used for forwarding UDP broadcast messages such as BOOTP generated by the Dynamic Host Configuration Protocol (DHCP), listing ICMP features enabled on the interface (e.g., ICMP redirect, ICMP mask request, and ICMP unreachable messages), and so on.

Please refer to OPNET documentation and attribute description for a complete account of IP configuration attributes available in various node models. You can access OPNET's IP documentation through the **Protocols > IP > Model User Guide** pull-down menu option.

9.3   MANAGING IP ADDRESSES

9.3.1   IP Addresses and Masks

Currently, IPv4 is the predominant version of IP. Each IPv4 address, or simply an IP address, is a 32-bit value. Typically, an IP address is displayed in a dotted decimal notation where each byte of the address value is written as a decimal number separated by a period. Since 1 byte corresponds to 8 bits, each dot-separated byte of IP address represents an integer value between 0 and 255. For example, using the dotted decimal notation, the binary value of IP address 10101000 00000000 11101010 11111111 is written as 168.0.234.255.

An IP address is often thought to consist of two parts: network prefix and host address. The network prefix or the network address is **N** consecutive leftmost bits of the IP address. The network prefix uniquely identifies the network to which the node is connected. The host address consists of the remaining (**32 – N**) rightmost bits and it identifies the node's interface within a particular network. An IP address mask, or simply a mask, allows isolating the network prefix portion of an IP address. The mask is also a 32-bit value with the bits that correspond to the network prefix set to 1 and the remaining host address bits set to 0. Typically, the mask is represented using dotted decimal notation or a Classless Inter-Domain Routing (CIDR) notation. The CIDR notation represents a mask as a forward slash ("/") followed by the length of the network prefix. In CIDR notation, the mask is appended to the end of an IP address. For example, consider the IP address 133.234.7.56 and its mask 255.192.0.0 written in dotted decimal notation. The corresponding value of this IP address–mask pair in CIDR notation is 133.234.7.56/10. You can easily compute the network prefix by performing a bit-wise **AND** operation between the IP address and its mask. The mask is an essential element of many routing protocols because the routing information is typically stored and forwarding decisions are usually made based on the network prefix portion of an IP address.

To configure the IP address and the mask in an end node, you need to change the values of the attributes called **Address** and **Subnet Mask**, both of which are located under the compound attribute **IP...IP Host Parameters...Interface Information**. In core node models, the **Address** and **Subnet Mask** attributes are located under the compound attribute **IP...IP Routing Parameters...Interface Information...<interface number>**, where **<interface number>** identifies one of the node's interfaces. Recall that OPNET numbers the node interfaces starting from 0 and that each interface name is preceded by the letters IF. Therefore, if a core node contains **n** interfaces, then the interface numbers will range from IF0 to IF<n-1>.

By default, the IP address is set to Auto Assigned, which means that the value of the **Address** attribute will be automatically assigned at the start of the simulation. The attribute **Address** also accepts the value No IP Address, which means that this node will not be part of any IPv4 routing/forwarding activities during the simulation (i.e., a node could be configured to run IPv6). Double-clicking on the attribute's **Value** field or selecting the option Edit... allows you to specify a desired IP address for the node's interface. The IP address value should be specified using the dotted decimal notation. Specifying an invalid IP address value may result in the simulation failing to execute properly.

Similarly, the **Subnet Mask** value should be specified using the dotted decimal notation as well. By default, **Subnet Mask** is also set to Auto Assigned. In addition, the **Subnet Mask** attribute also accepts the following predefined values:

- Class A (natural), which corresponds to the 255.0.0.0 mask.
- Class B (natural), which corresponds to the 255.255.0.0 mask.
- Class C (natural), which corresponds to the 255.255.255.0 mask.

Notice that in core nodes, each physical interface has its own set of **Address** and **Subnet Mask** attributes. Furthermore, core nodes do not necessarily have all their interfaces active during a simulation. Therefore, you often may need to identify (i.e., discover the names of) the interfaces you will configure. Typically, you would want to configure only some of those few active interfaces that correspond to the links that connect the current node with other nodes in a network. Notice that this is not a problem in the edge node models that usually contain a single interface named IF0.

9.3.2   IDENTIFYING NAMES OF INTERFACES ATTACHED TO A LINK

To identify the name of an interface attached to a link connecting the current node with another object in a simulated network, you need to perform one of the following actions:

- "Hover" the mouse pointer over the link until a yellow hint box appears
- Examine the link's attributes
- Examine the link's port values

As shown in Figure 9.5a, the hint box provides basic information about the link, including the interface names of the nodes connected through that link. OPNET relies on the following interface naming convention: *<node name>.<interface type>_<interface number>_<channel number>*. To identify the interface name, you are only interested in the number that follows the *<interface type>*. For example,

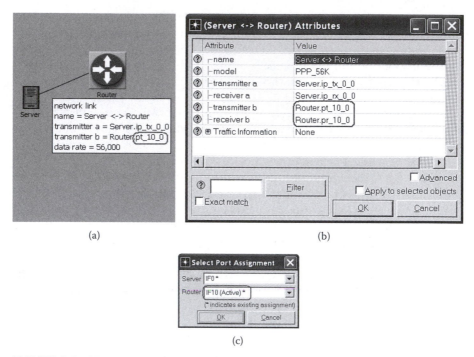

(a)   (b)

(c)

FIGURE 9.5   Three methods for identifying the interface numbers for the ends of a link: (a) Yellow hint box that appears after "hovering" the mouse pointer over the link; (b) link attributes that help identify interface number; and (c) identifying interface name through the **Edit Ports** option.

as shown in Figure 9.5a, the link is connected to interface number 10 in the node named Router and to interface number 0 in the node named Server. The same information can be obtained by examining the transmitter and/or receiver attributes of the connecting link, as shown in Figure 9.5b. Finally, you can also identify the interface names by right-clicking on the link and selecting the **Edit Ports** option that opens a **Select Port Assignment** window. As shown in Figure 9.5c, the **Select Port Assignment** window lists the names of the node interfaces attached to this link. Therefore, if you would like to configure the interface used by Router to connect to Server, then you will need to change the attribute values of the interface named IF10. Notice that, as expected, Server, which is an end node, has a single interface named IF0.

9.3.3   COMMON MISTAKES IN IP ADDRESS CONFIGURATION

One of the common novice mistakes in configuring IP addresses is specifying different *network* addresses (e.g., through an IP address or mask attribute) on the interfaces attached by a link between two nodes. Correct configuration requires that the IP address and the mask of the interfaces attached to the same link resolve to the same network address.

TABLE 9.1

Examples of Incorrect Interface Configuration

	Interface IF0 in Node Server		Interface IF10 in Node Router	
	IP Address	Mask	IP Address	Mask
Scenario #1	**192**.168.10.1	255.255.255.0	**193**.168.10.2	255.255.255.0
Scenario #2	192.168.10.1	255.255.255.**0**	192.168.10.2	255.255.255.**192**

Consider the example shown in Figure 9.5 and the two possible scenarios for incorrect interface configuration given in Table 9.1. In scenario #1, the connecting link has different network addresses specified on the corresponding node interface due to an error in the IP address assignment. `Server's` IP address starts with `192` while `Router's` IP address starts with `193`, which indicates that the nodes belong to different networks. In scenario #2, the mask values specified on the node interfaces result in different network addresses: `192.168.10.0` and `192.168.10.2`. (notice that the `Server's` mask value ends with `0` and `Router's` mask value ends with `192` which maps the corresponding addresses to different networks). In both cases, the simulation will likely execute without an abnormal termination, but the DES Log will contain warning messages. In scenario #1, the error log message will state that incompatible subnet (i.e., network) addresses have been detected and that no traffic will be sent between those nodes, while in scenario #2, the error log message will report a configuration error stating that incompatible subnet masks have been used. Notice that in both scenarios, the DES Log clearly identifies the nodes and the interfaces where the misconfiguration has occurred, which is extremely helpful when debugging your simulation. Recall that you can access the DES Log by selecting the option **DES > Open DES Log** from the pull-down menu in the **Project Editor** (see Section 4.8).

9.3.4 AUTO-ASSIGNMENT OF IP ADDRESSES

In many situations, assigning IP addresses by hand is a tedious and error-prone task. It is often sufficient to leave the IP address and the mask values set to `Auto Assign` and let OPNET handle IP addressing during the execution of the simulation. However, in some cases, it may be desirable to have IP addresses assigned before simulation execution begins. For that purpose, OPNET provides a set of options in the **Protocols** pull-down menu in the **Project Editor**. In this section, instead of stating a complete sequence of pull-down menu options for IP addressing (e.g., **Protocols > IP > Addressing > Auto-Assign IPv4 Addresses**), we will only specify the last name in the menu option (e.g., **Auto-Assign IPv4 Addresses**).

As shown in Figure 9.6, which illustrates all IP addressing options available through the **Protocols** menu, to auto-assign IPv4 addresses on all attached interfaces in the simulated network, you need to select the option **Auto-Assign IPv4 Addresses**. If this operation is successful, then the bottom panel of the **Project Editor** will display a message that states how many addresses have been assigned. If the operation was unsuccessful, then the message will state "`No IP addresses assigned.`"

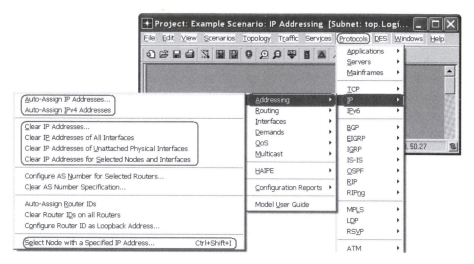

FIGURE 9.6   IP addressing options available through the **Protocol** menu.

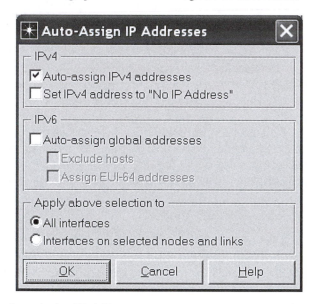

FIGURE 9.7   **Auto-Assign IP Addresses** window.

You can also auto-assign IP addresses using the option **Auto-Assign IP Addresses…** . This option opens the **Auto-Assign IP Addresses** window, as shown in Figure 9.7, in which you can provide additional details for the auto-addressing operation. Specifically, this window allows you to select among the following configuration choices:

a. *Auto-assign IPv4 addresses*—auto-assigns IPv4 addresses.
b. *Set IPv4 addresses to "No IP Addresses"*—disables IPv4 address assignment so that the nodes in the network communicate using IPv6 only. Options a and b are mutually exclusive, that is, you can select only one of those options at a time.

c. *Auto-assign global addresses (IPv6)*—auto-assigns IPv6 addresses.
  i. *Exclude hosts*—IPv6 addresses are only assigned to routers, while the end nodes discover their IPv6 addresses through router advertisement. This is called *stateless address auto-configuration*.
  ii. *Assign EUI-64 addresses*—assigns IPv6 addresses using IEEE's Extended Unique Identifier-64 (EUI-64) format. The EUI-64 format requires that the 64 bits of an IPv6 address (i.e., the host part) be based on the corresponding hardware address. Since hardware addresses are usually 48 bits, the EUI-64 format fills the remaining 16 bits with FF-FF or FF-FE, depending on whether the hardware address is in MAC-48 format (i.e., network hardware) or in EUI-48 format (other devices and software). If this checkbox is not selected, then OPNET assigns non-EUI-64 IPv6 addresses, that is, the whole 128-bit number is created without consideration for hardware addresses.
d. *All interfaces*—applies the auto-address configuration to all node interfaces in the network.
e. *Interfaces on selected nodes and links*—applies the configuration only to the selected objects in the network. Options d and e are mutually exclusive. To use option e, you should first select some network objects (nodes and/or links) before you try to auto-assign addresses.

9.3.5 CLEARING IP ADDRESS ASSIGNMENT

The **Protocols** menu also contains the following options for automatically clearing an IP address assignment in a simulated network:

- **Clear IP Addresses...** opens the **Clear IP Addresses** window, as shown in Figure 9.8, which allows you to configure details of the clear address assignment operation. Through this window, you can specify whether you

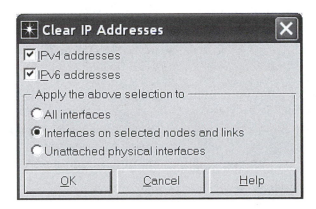

FIGURE 9.8   Clear IP Addresses window.

need to clear IPv4, IPv6, or both types of addresses and if you want to clear address assignment on all interfaces, on the interfaces of selected nodes and links only, or on the unattached physical interfaces.

- **Clear IP Addresses of All Interfaces** clears both IPv4 and IPv6 addresses on all (attached or unattached) interfaces in the network.
- **Clear IP Address of Unattached Physical Interfaces** clears both IPv4 and IPv6 addresses on all unattached interfaces in the network.
- **Clear IP Addresses for Selected Nodes and Interfaces clears** both IPv4 and IPv6 addresses on all interfaces of selected nodes and links.

Notice that the auto-assign operation assigns addresses to attached interfaces only. Therefore, it is prudent to perform IP address assignment only after completing the process of creating the network topology of a simulated system. If the network topology has been changed, then it is a good idea to first clear the current IP address assignment and then to perform the auto-assign operation again.

9.3.6 IDENTIFYING INTERFACE WITH A SPECIFIED IP ADDRESS

OPNET also contains a **Protocols** menu option for identifying the node that contains an interface with a specified IP address. You can open an **IP Address-based Node Selection** window, as shown in Figure 9.9, by choosing **Select Node with a Specified IP Address...** from the **Protocols** menu or by pressing Ctrl+Shift+I. This utility seems to work only with IPv4 addresses. You need to specify an IPv4 address in the corresponding textbox and click the **Find** button. You may need to click the **Reload** button to refresh the system. A warning message will appear if the simulated system does not contain any interfaces with the specified IP address. Otherwise, a window with a list of node entries will appear. Each node entry consists of two values: the name of the node and the name of its interface set to the specified IP address value. This feature is extremely useful when you are working on a simulation with a large number of nodes and need to quickly locate a node with a specific IP address.

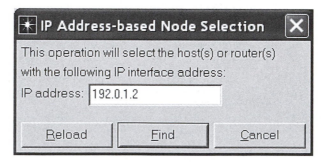

FIGURE 9.9   IP Address-based Node Selection window.

9.3.7   EXPORTING IP ADDRESS ALLOCATION

Finally, OPNET also allows you to export IP address allocation in the simulated system into a comma separated text file. You can configure the simulation to export the IP address allocation through the global attribute **IP...IP Interface Addressing Mode** in the **Configure/Run DES** window (see Section 4.3). This attribute accepts the following four values:

- `Auto Addressed`—all IP addresses on active interfaces will be automatically assigned during simulation, unless they have been manually assigned already. OPNET only assigns classful addresses (i.e., addresses in the classes A, B, or C). The IP address class assigned to an interface depends on the number of hosts in the corresponding network. The subnet mask for the interface is the default mask of the assigned address class.
- `Auto Addressed/Export`—in addition to automatically assigning IP addresses, OPNET also exports them into a file. The file is saved in the default model directory and is named according to the following convention: `<project-scenario-name>-ip_addresses.gdf`.
- `Manually Addressed`—the simulation relies on manually assigned IP addresses. Notice that for the simulation to run properly, all active or usable interfaces and the gateway of last resort (i.e., attributes **Default Route** and **Default Network(s)** in the end and core nodes respectively) must have a valid IP address assigned to them. However, if you would like to verify that your manual address assignment is correct, then you should set this attribute to `Auto Addressed` or `Auto Addressed/Export` (even though the addresses are manually configured).
- `Manually Addressed/Export`—has the same meaning as `Manually Addressed` attribute except that the assigned IP addresses on all active interfaces are exported to a file. The export file is named and saved the same way as when this attribute is set to `Auto Addressed/Export`.

9.4   CONFIGURING OTHER IP FEATURES

In this section, we describe the configuration of other IP features modeled in OPNET. Some of the features described in this section are configurable through node attributes while others are configurable through IP utility objects. OPNET contains three node models of utility objects for specifying scenario-wide IP characteristics:

- **IP Attribute Config**, which allows configuring scenario-wide details of such IP features as compression and ping probing. You can also configure your simulation to export routing tables into external files via this utility node. The IP features supported by the **IP Attribute Config** object are described later in this section.

- **QoS Attribute Config** allows specifying configuration for various QoS mechanisms. We discuss configuration and deployment of QoS support through the **QoS Attribute Config** node in Chapter 10.
- **IP VPN Config** specifies virtual private network (VPN) tunnels between nodes in the simulated network system. We do not discuss IP tunneling in this book.

9.4.1 IP Compression

IP compression decreases the packet size by compressing certain portions of the datagram. This feature is supported in OPNET. You can define desired compression schemes through the **IP Compression Information** attribute in the **IP Attribute Config** object and then deploy them on the node interfaces of your choice. By default, the **IP Compression Information** attribute contains a set of preconfigured compression schemes including TCP/IP header compression, per-interface compression, per-virtual circuit compression, image compression, and Telnet application compression. However, you can define new compression schemes, if needed, by performing the following steps:

- Add the **IP Attribute Config** object into the project workspace. You should omit this step if the **IP Attribute Config** object has been already added into the current scenario.
- Edit attributes of the **IP Attribute Config** object by right-clicking on it and selecting the **Edit Attributes** option.
- Expand the attribute **IP Compression Information**.
- Change the value of the attribute **Number of Rows** to a number greater than 6 (since there are 6 predefined compression schemes).
- Expand one of the newly created IP compression rows and define your new compression scheme by specifying the values of the following attributes:
  - **Name**—a unique name of a new compression scheme. This value will identify this IP compression scheme in the current scenario.
  - **Compression Method**—specifies the type of compression method. This attribute accepts only one of the following four values:
    - None—no compression,
    - TCP/IP Header Compression—only TCP/IP header will be compressed,
    - Per-Interface Compression—the entire packet will be compressed, and
    - Per-Virtual Circuit Compression—only the packet's payload (i.e., not the header information) will be compressed.
  - **Compression Ratio** and **Compression Ratio PDF** define the arguments for the probability distribution function (PDF) and the name of the PDF itself. **Compression Ratio PDF** computes the ratio between the compressed and uncompressed packet (typically a value between 0 and 1). If the PDF requires multiple arguments, then they are separated by a space. The value(s) specified in the **Compression Ratio** attribute

are fed into the function defined in the **Compression Ratio PDF** attribute, which in turn computes the actual ratio between the compressed and uncompressed packet. OPNET simulation uses this value to compute the size of the compressed packet.

- **Compression Delay (sec/bits)** and **Decompression Delay (sec/bits)** specify the time it takes to compress and decompress a single bit in a packet respectively. The actual compression or decompression delay is computed by multiplying the packet size in bits by the corresponding (de)compression delay value.
- You may repeat this process for as many compression schemes as necessary.
- Once all the desired compression schemes have been defined, press the **OK** button to save the changes.

After you have defined your compression schemes, you can deploy them on desired interfaces by setting the value of the attribute **Compression Information**. This attribute can be found under the **IP...IP Host Parameters...Interface Information** attribute of the end node models and under the **IP...IP Routing Parameters...Interface Information...<interface number>** attribute of the core node models. OPNET software comes with a set of example projects that illustrate various networking technologies. These projects are located under the example_ networks directory. In particular, the project **IP** contains four scenarios that illustrate the influence of different compression schemes on application performance.

9.4.2 BASIC CONFIGURATION OF ROUTING PROTOCOLS

OPNET contains models of the following routing protocols: Routing Information Protocol (RIP), Open Shortest Path First (OSPF), Interior Gateway Routing Protocol (IGRP), Enhanced Interior Gateway Routing Protocol (EIGRP), Intermediate System to Intermediate System (IS-IS), Border Gateway Protocol (BGP), and several Mobile Ad Hoc Networks (MANET) protocols. You can specify routing protocols deployed in the network using several different methods, which we briefly discuss in this section. Typically, each network interface has a routing protocol associated with it. You can specify the routing protocol deployed at a core node's interface by setting the value of the **Routing Protocol(s)** attribute. This attribute is located under **IP...IP Routing Parameters...Interface Information...<interface number>**. This option allows configuring one interface at a time, which could be very time-consuming when the routing protocol has to be deployed on multiple interfaces.

The **Routing Protocol Configuration** window, as shown in Figure 9.10, provides a method for deploying routing protocols on multiple interfaces in the network at once. You can open this window through the **Protocols > IP > Routing > Configure Routing Protocols** option. The **Routing Protocol Configuration** window allows you to specify such information as the name of the routing protocol(s) to be deployed (i.e., RIP, IGRP, OSPF, IS-IS, EIGRP, or None), whether the selected routing protocol(s) are to be also deployed on the subinterfaces, whether the routing protocols are to be deployed on all interfaces or only on those of selected links, and finally, whether the routing domains are to be visualized. By clicking the **OK** button,

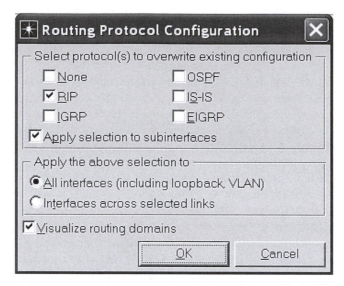

FIGURE 9.10   Routing Protocol Configuration window.

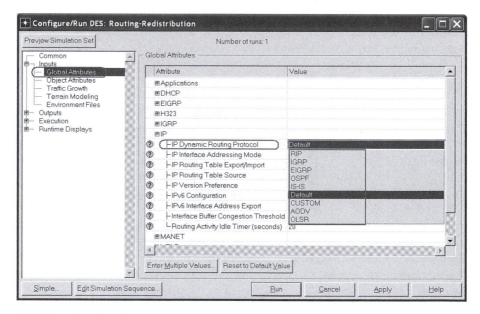

FIGURE 9.11   Specifying the routing protocol through the **Configure/Run DES** menu.

you save the routing protocol configuration and overwrite the values of the **Routing Protocol(s)** attributes on the specified interfaces.

Finally, you can also specify the name of the routing protocol to be deployed in the simulated system through the **Configure/Run DES** menu by setting the value of the global attribute **IP…IP Dynamic Routing Protocol**, as shown in Figure 9.11. The simulation uses routing protocols configured locally on individual interfaces when this attribute is set to `Default`. In all other cases, the specified routing protocol takes precedence over the protocols defined on each interface locally. The only exception is when an interface runs IPv6. In such a case, a locally defined non-MANET routing protocol will be used instead of the global value specified through the **IP...IP Dynamic Routing Protocol** attribute. You can also specify configuration details of the routing protocols by changing the values of the attribute **IP Routing Protocols** at various core node models. Chapter 11 discusses routing protocol configuration in greater detail.

9.4.3 CONFIGURING DIFFERENT TYPES OF IP INTERFACES

OPNET node models support the following interface types: physical, loopback, tunnel, aggregate, and subinterface. Physical interfaces are typically configured through attributes **IP...IP Routing Parameters...Interface Information...** and **IP...IP Host Parameters...Interface Information** in core and end node models, respectively, as described in Sections 9.2.1 and 9.2.2.

Loopback interfaces are available in core node models only. By default, there is one loopback interface defined for each core node model. You can view loopback interface configuration by examining the **IP...IP Routing Parameters...Loopback Interfaces** attribute. You can create additional loopback interfaces by expanding the attribute **Loopback Interfaces**, setting the value of its subattribute **Number of Rows** to the total number of loopback interfaces desired, and then configuring each individual newly added loopback interface. You can also create loopback interfaces through the **Protocols > IP > Interfaces > Create Loopback Interface...** option.

Similarly, you can create tunnel interfaces by setting values of the **IP...IP Routing Parameters...Tunnel Interfaces** attribute. Tunnel interfaces are also only supported in the core node models. However, the core node models do not contain any preconfigured default tunnel interfaces. You can add desired number of tunnel interfaces by expanding the attribute **Tunnel Interfaces**, setting the value of its subattribute **Number of Rows** to the total number of tunnel interfaces desired, and then configuring each individual newly added tunnel interface.

OPNET also supports interface aggregation, a mechanism that maps multiple physical interfaces into a single logical interface. The process of specifying interface aggregation consists of two distinct steps: (1) define aggregate interfaces, and (2) map physical interfaces to defined aggregate interfaces. The aggregate interfaces are defined through the attribute **IP...IP Routing Parameters...Aggregate Interfaces**. Once again, you will first need to set the value of the subattribute **Number of Rows** to the total number of aggregate interfaces to be configured and then configure each individual newly created aggregate interface. Each aggregate interface contains a set of configuration attributes that are identical to those used to configure physical interfaces. The attribute **Name** uniquely identifies each aggregate interface and is used to map physical interfaces to aggregate interfaces, which is the next step in the process. To specify that a physical interface belongs to a particular aggregate interface, you need to set the value of the attribute **IP...IP Routing Parameters...Interface Information... <interface number>...Aggregation Parameters...Aggregate Interface** to the name of the aggregate interface, as illustrated in Figure 9.12.

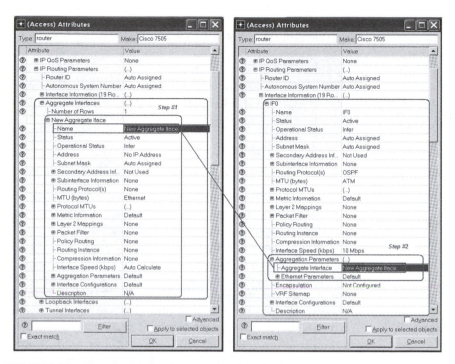

FIGURE 9.12   Steps for configuring aggregate interfaces.

Finally, core node models also support subinterfacing, a mechanism that allows a single physical interface to be partitioned into several logical subinterfaces. OPNET implements subinterfacing for the following protocols: ATM, FR, Serial, and VLAN. You can configure subinterfaces by setting values of the attribute **IP…IP Routing Parameters…Interface Information…<interface number>…Subinterface Information**. Again, you will need to set the value of the attribute **Number of Rows** and then configure each newly created subinterface. Typically, subinterfaces are named according to the following conventions: the name of their physical interface followed by a period and the subinterface number. For example, if there are three subinterfaces defined for interface IF12, then these subinterfaces should be named IF12.1, IF12.2, and IF12.3. In addition, OPNET documentation warns that you cannot set the IP address of a subinterface to Auto Assigned and instead must enter the IP address value explicitly.

Once basic subinterface information has been specified, you need to specify a Permanent Virtual Circuit (PVC) name in case of ATM/FR or VLAN/ELAN identifier attached to this subinterface. You can specify PVC name or identifier number through subattribute **Layer 2 Mapping** as shown in Figure 9.13. Notice that PVCs for ATM and FR are defined through utility objects **ATM_SPVx_Config** and **FR PVC Config** respectively. In addition, you can specify subinterfaces through the **Protocols > IP > Interfaces > Create sub-interfaces for selected ATM/FR PVXs…** option. The example project **IP** in the example_networks directory

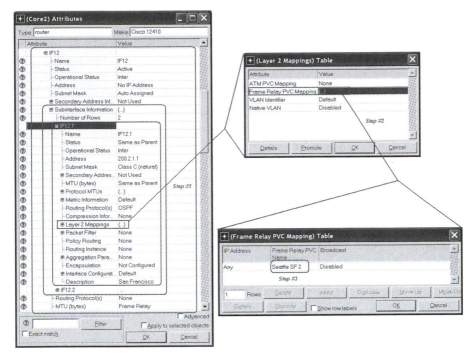

FIGURE 9.13 Steps for configuring subinterfaces.

contains a scenario called `Subinterfaces`, which provides an example of subinterface configuration.

9.4.4 CONFIGURING IP LOAD BALANCING

In the presence of multiple routes to a destination, you can configure a router to implement load balancing, a technique that attempts to distribute traffic over various paths to optimize network utilization. You can configure load balancing through the attribute **IP…IP Routing Parameters…Load Balancing Options**. This attribute accepts the following two values, which represent the types of load balancing supported by OPNET:

- `Destination-Based`, which distributes traffic based on source–destination IP address pair. In this approach, all packets that belong to the same traffic flow identified by source–destination IP address pair are guaranteed to follow the same route on their way to the destination. This is the default setting for the load balancing technique. The packets that belong to different flows might be routed over different paths, even though the flows may be going to the same destination.
- `Packet-Based`, which distributes traffic on a per-packet basis—successive packets are forwarded on alternative paths to the destination, where the paths are selected in a round-robin fashion. This approach provides

a fine-grained load distribution over multiple paths, but it might result in packet reordering for packets belonging to the same flow, which might not be acceptable for certain applications.

9.5   INTERNET CONTROL MESSAGE PROTOCOL

The Internet Control Message Protocol (ICMP) is often discussed in conjunction with IP because it addresses such shortcomings of IP as error-reporting and query management. There are two versions of ICMP available in today's networks: ICMPv4 and ICMPv6. OPNET Modeler and IT Guru software products only support ICMPv4. However, ICMPv6 is supported by OPNET's System-in-the-Loop (SITL) product, which is an additional module for OPNET Modeler Version 15.0 or later. We do not discuss SITL and ICMPv6 in this book due to space limitations, which is why in this chapter, we will refer to ICMPv4 as simply ICMP.

The only type of ICMP traffic supported by OPNET is ICMP echo-request and echo-reply messages that are modeled through IP ping demand objects. However, most of the router node models contain attributes that allow enabling or disabling certain ICMP messages. These attributes are defined under the **Interface Configurations** compound attribute and are typically used for verifying correctness of router configuration, which is beyond the scope of this book. The rest of this section discusses steps for deploying IP ping demands and examines various ping statistics.

9.5.1   SPECIFYING PING PATTERNS

Each IP ping demand object contains an attribute called **Ping Pattern**, which specifies various characteristics of ping traffic. You can define a ping pattern through the **IP Ping Parameters** attribute of the **IP Attribute Config** utility object. OPNET already provides a set of preconfigured ping patterns, but you can define your own by executing the following steps:

- Add an **IP Attribute Config** object into the project workspace if it is not already present in the current simulation scenario.
- Open the **Attributes** window of this object.
- Expand the attribute **IP Ping Parameters**, which is shown in Figure 9.14.
- Change the value of the attribute **Number of Rows** to a number greater than 4 (by default, there are 4 predefined ping patterns).
- Expand a newly created row and define your new ping pattern by specifying the values of the following attributes:
  - **Pattern**—a unique scenario-wide name for your ping pattern.
  - **Details**—a compound attribute that specifies traffic details for the ping pattern. It consists of the following subattributes:
    - **IP Version**—specifies the version of IP protocol used by the packets that carry ICMP messages. This attribute supports only two possible values: IPv4 and IPv6.
    - **Interval (sec)**—specifies the duration in seconds between successive ping messages.

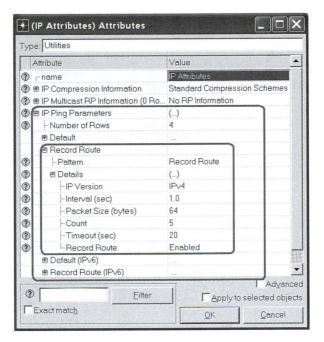

FIGURE 9.14   IP Ping Parameters attribute.

- **Packet Size (bytes)**—specifies the size of the ping packet. This value does not include the 8-byte ICMP header, which is added to each ping message before it is encapsulated within the IP packet.
- **Count**—specifies the number of ICMP echo-request packets generated by this pattern.
- **Timeout (sec)**—specifies the duration in seconds that the node will wait before it considers the ICMP message lost if there is no response from the pinged node.
- **Record Route**—enables or disables the option for recording the route taken by the ping messages. If ping demands are configured with a ping pattern that has the **Record Route** attribute enabled, then the path taken by ICMP echo-request and echo-reply messages is recorded. You can view the recorded path upon simulation completion by examining the **DES Run Tables** available through the **View Results** option. The recorded route is reported in the form of a table with the following columns: *IP address, Hop Delay, Node Name, MPLS Label,* and *MPLS EXP.*

- You may repeat this process for as many ping patterns as necessary.
- Once all desired patterns have been added, click the **OK** button to save the changes.

Once the ping patterns have been defined, you can deploy IP ping demands in the network. You can specify IP ping demands either by dragging and dropping *ip_ping_traffic* objects into the project workspace or by using the **Protocols > IP > Demands > Configure Ping Traffic on Selected Nodes** option. We examine each of these approaches in Sections 9.5.2 and 9.5.3.

9.5.2   DEPLOYING IP PING DEMANDS WITH THE *IP_PING_TRAFFIC* OBJECT

The first step in the process of deploying an IP ping demand by this technique is to add the necessary demand objects into the project workspace. To add an IP ping demand, you need to perform the following steps:

- Open the **Object Palette Tree** and find the *ip_ping_traffic* object (located in the *Demand Models*, *demands*, and *internet_toolbox* folders).
- Select the *ip_ping_traffic* object and then click on its icon in the **Object Palette Tree**.
- Click on the node where you would like the ping traffic to be generated and then click on the node which you would like to be "pinged." The result will be a dotted line joining these two nodes representing the IP traffic demand object.
- You can create many such objects between any pairs of nodes. When you are finished, right-click in the project workspace and select **Abort Demand Definition** from the pop-up menu.

If you wish to remove a demand object, then select the object and press the DELETE key on your keyboard.

Once all desired IP ping demands have been added, you may want to modify their configuration. The following list provides a brief description of all IP ping demand attributes, which are shown in Figure 9.15:

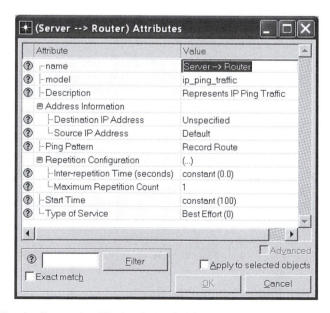

FIGURE 9.15   Attributes of an IP ping demand object.

- **name**—specifies the name of the IP ping demand object.
- **model**—specifies the name of the object model.
- **Description**—provides a brief hint regarding the purpose of this demand object. This attribute has no effect on simulation execution.
- **Address Information**—specifies IP addresses of the demand's source and destination nodes through subattributes **Destination IP Address** and **Source IP Address**.
- **Ping Pattern**—specifies the name of the ping pattern that was defined in the **IP Attribute Config** object.
- **Repetition Configuration**—specifies the repetition pattern for this demand. This compound attribute has the following preset values: Once at Start Time, Every 10 minutes, Every 30 minutes, and Every 1 hour. If desired, you can define your own repetition pattern by setting the values for subattributes **Inter-repetition Time (seconds)** and **Maximum Repetition Count**, which specify the time between successive ping pattern repetitions and the maximum number of ping pattern repetitions (including the first one) respectively.
- **Start Time**—specifies the time when the ping demand will begin sending ICMP messages.
- **Type of Service**—specifies the ToS value of the IP packets that will carry the ICMP messages.

The process of deploying IP ping demands is completed once you have configured all the desired demand objects. After that, you only need to select the ping statistics to be collected during the simulation, which we discuss in Section 9.5.4.

9.5.3 DEPLOYING IP PING DEMANDS USING THE PROTOCOLS MENU

Before deploying IP ping demands through the **Protocols** menu, you need to select at least two nodes that will participate in the IP ping traffic exchange. Once the desired nodes have been selected, choose the **Protocols > IP > Demands > Configure Ping Traffic on Selected Nodes** option, which opens the **Configure ICMP Ping Messages** window, as shown in Figure 9.16.

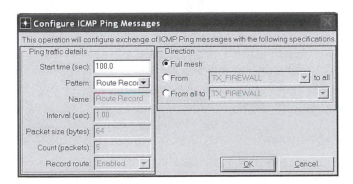

FIGURE 9.16   **Configure ICMP Ping Messages** window.

The **Configure ICMP Ping Messages** window consists of two panels called **Ping Traffic Details** and **Direction**. By default, the **Ping Traffic Details** panel has only the **Start time (sec)** and **Pattern** attributes enabled. The pull-down textbox of the **Pattern** attribute contains the list of all ping patterns defined through the **IP Attribute Config** object and an option New..., which when selected will create a new ping pattern and will enable the rest of the attributes. These attributes are identical to those that define ping patterns in the **IP Attribute Config** object and therefore their description is omitted.

Notice that the **Configure ICMP Ping Messages** window does not allow configuring ping demand repeatability, ToS, and address information. All the ping demands deployed through the **Protocols** menu will execute only once at the time specified in the **Start time (sec)** attribute, and the IP packets that carry ICMP messages will be set with the default ToS value of 0.

The **Direction** panel of the **Configure ICMP Ping Messages** window is used to configure a ping demand's source and destination address information. As shown in Figure 9.16, the **Direction** panel has the following three options, which specify how the ping demands will be deployed among the selected nodes:

- **Full mesh**—each selected node will send ping demand traffic to all the other selected nodes and will receive ping demand traffic from all the other selected nodes.
- **From <pull-down textbox> to all**—this option allows you to generate ping traffic from a single node to all the other selected nodes. The pull-down textbox contains the list of node names that have been selected. The node chosen in the pull-down textbox will serve as a source for the ping traffic.
- **From all to <pull-down textbox>**—this option allows you to generate ping traffic to a single node from all the other selected nodes. The pull-down textbox contains the list of node names that have been selected. The node chosen in the pull-down textbox will serve as a destination for the ping traffic.

Once you have completed configuration of ping demands, you need to press the **OK** button to save the changes and deploy the specified demands. If you did not use one of the ping patterns defined in **IP Attribute Config** and instead created a new one, then the new ping pattern is added as a new row under the **IP Ping Parameters** attribute in the **IP Attribute Config** object. Notice that even though the **Configure ICMP Ping Messages** window does not include attributes for configuring repetition pattern and ToS value, you can still specify this information by first creating ping demands through the **Protocols** menu and then changing their corresponding attribute values in the **IP Attribute Config** object.

9.5.4   PING STATISTICS

Before executing the simulation, you may want to select ping statistics to be collected during the simulation run. Ping-specific statistics are found in the **IP** category of Node Statistics:

- **Ping Replies Received (packets)**—records the number of ICMP echo-reply messages received at a node in response to echo-request messages sent to a particular destination.
- **Ping Requests Sent (packets)**—records the number of ICMP echo-request messages sent by a node to a specific destination.
- **Ping Response Time (sec)**—records the time passed since an ICMP echo-request was sent until the corresponding ICMP echo-reply was received.

Finally, recall that for any IP ping demand configured to have the **Record Route** attribute of its ping pattern set to `Enabled`, the path taken by its ICMP echo-request and echo-reply messages will be recorded. You can view the recorded route by clicking on the **DES Run Table** tab in the **Results Browser**. Once again, OPNET provides an excellent illustration of this feature in the `icmp_route_print` scenario of project **IP** in the `example_networks` directory, which comes standard with all distributions of IT Guru and Modeler software.

9.6   COMMON IP STATISTICS, TABLES, AND REPORTS

9.6.1   IP STATISTICS

OPNET provides a set of statistics for examining IP performance. Specifically, IP statistics are available in the following Node Statistics categories: **IGMP Host**, **IGMP Router**, **IP**, **IP Interface**, **IP Processor**, **IP Tunnel**, **IP VPN tunnel**, **IPv6**, **PIM-SM**, **Route Table**, **Router Convergence**, as well as in several other categories for routing protocols. Similarly, IP Global Statistics categories include **IP**, **IPv6**, **PIM-SM**, and routing protocols categories. In this section, we only provide a summary of statistics for general IP features. Statistics pertaining to specific IP modules such as IP multicasting (i.e., **IGMP Host**, **IGMP Router**, some statistics from **IP** and **IPv6** categories, and **PIM-SM**), QoS (i.e., **IP Interface**), IPv6 (i.e., **IPv6**), ICMP (i.e., certain statistics from the **IP** category), IP Processor (i.e., **IP Processor**), VPN (i.e., **IP VPN tunnel**), and various routing protocols are either omitted from discussion or described in the corresponding portions of this book. Table 9.2 provides a summary of the statistics for general IP features.

9.6.2   VISUALIZATION AND CONFIGURATION REPORTS

OPNET provides several other useful options that may help you better understand the configuration and performance of a simulated IP network. Specifically, OPNET allows visualizing various features of the IP layer and other protocols' configurations through **View > Visualize Protocol Configuration**, **View > Visualize Network Configuration**, and **View > Visualize Link Usage** pull-down menu options in the **Project Editor**. These options are fairly easy to use, can provide a snapshot view of certain network configuration aspects, and can be quite useful during the debugging process.

You can also generate various network configuration and performance reports. Specifically, you can generate configuration reports through the **Protocols > IP > Configuration Reports > Select/Generate...** option, which opens a **Generate**

TABLE 9.2

Statistics for General IP Features

Category	Name	Description
Global Statistics		
IP	Background Traffic Delay (sec)	Records the end-to-end delay experienced by the background traffic that traverses the network.
	Network Convergence Activity	Records convergence activity in the network (i.e., changes in the Forwarding Tables of all routers). Time periods when there is convergence activity are denoted as 1 and time periods when there is no convergence activity are denoted as 0.
	Network Convergence Duration (sec)	Records the time it took IP Forwarding Tables of all routers in the network to converge, that is, experience no change to the Forwarding Table.
	Number of Hops	Records the average number of hops traversed by IP packets to reach their corresponding destinations.
	Traffic Dropped (packets/sec)	Records the total number of IP packets discarded by all nodes in the network on all their interfaces.
Node Statistics		
IP	Background Traffic Delay (sec) ←, Background Traffic Delay (sec) →	Records the end-to-end delay experienced by the background traffic flows (i.e., the time it takes a unit of background traffic data to travel from the flow source to the flow destination). These statistics are recorded for all background flows either arriving at this node (i.e., this node is the flow's destination) or departing from this node (i.e., this node is the flow's source) respectively.
	Background Traffic Flow Delay (sec)	Records the end-to-end delay experienced by the background traffic flows. This statistic is recorded on a per-flow basis (i.e., a separate statistics vector is created for each flow) for the flows originating at this node.
	Broadcast Traffic Received (packets/sec), Broadcast Traffic Sent (packets/sec)	Records the number of broadcast packets received or sent by this node on all its interfaces. These statistics are recorded in units of packets/second.
	End-to-end Delay (sec)	Records the end-to-end delay experienced by a packet. The end-to-end delay is computed as the difference between the time the packet arrives at the destination and the time the packet was created. This statistic is recorded for unicast traffic only. A separate statistic vector is computed for each source–destination pair.

End-to-end Delay Variation (sec)

Records the end-to-end delay variation or jitter experienced by the packet as it reaches the destination node. This statistic is recorded for unicast traffic only. A separate statistic vector is computed for each source–destination pair.

Number of Hops ←,
Number of Hops →

Records the number of hops traversed by the packets, either arriving at this node (i.e., originating from multiple different sources) or departing from this node (i.e., traveling to multiple destinations). In the latter case, the statistic is recorded for this node by the destination nodes as they receive packets from this node.

Processing Delay (sec)

Records IP layer processing delay experienced by a packet. The delay value is computed from the time the packet arrives at the IP layer until the time the packet leaves the IP layer.

Traffic Dropped (packets/sec),
Traffic Received (packets/sec),
Traffic Sent (packets/sec)

These statistics record the total amount of IP traffic (i.e., unicast, multicast, and broadcast) dropped, received, and sent by this node on all its interfaces.

Node Statistics
IP Tunnel

Delay Variation (sec)

Records delay variation or jitter experienced by the packets traveling through the tunnel.

ETE Delay (sec)

Records the end-to-end delay experienced by the packet that travels through the tunnel. Recorded value includes the encapsulation delay, the time it takes the packet to travel through the tunnel, and the decapsulation delay.

Traffic Dropped (bits/sec),
Traffic Dropped (packets/sec)

Records the traffic dropping rate on the IP tunnel interface in units of bits/second and packets/second. The name of the resulting statistic has the name of a tunnel appended to it.

Traffic Received (bits/sec),
Traffic Received (packets/sec)

Records the traffic arriving rate at the IP tunnel interface in units of bits/second and packets/second. The name of the resulting statistic has the name of a tunnel appended to it.

Traffic Sent (bits/sec),
Traffic Sent (packets/sec)

Records the traffic sending rate on the IP tunnel interface in units of bits/second and packets/second. The name of the resulting statistic has the name of a tunnel appended to it.

(Continued)

TABLE 9.2 (Continued)
Statistics for General IP Features

Category	Name	Description
Node Statistics IP VPN Tunnel	Packets Received (packets), Packets Received (packets/sec)	Records the total and the average number of packets received from this VPN tunnel in units of packets and packets/second respectively.
	Packets Sent (packets), Packets Sent (packets/sec)	Records the total and the average number of packets sent over this VPN tunnel in units of packets and packets/second respectively.
	Tunnel Delay (sec)	Records the average end-to-end delay experienced by the packets traveling through this VPN tunnel.
Node Statistics Route Table	Number of Next Hop Updates	Records the number of times the next hop field value was changed in the IP Forwarding Table for a particular route.
	Number of Route Additions	Records the number of times a route has been added to the IP Forwarding Table.
	Number of Route Deletions	Records the number of times a route has been deleted from the IP Forwarding Table.
	Size (number of entries)	Records the size of the IP Forwarding Table. By default, this statistic is collected in a bucket mode using summary function; therefore, min, max, and average sizes of the IP Forwarding Table are recorded for the duration of each bucket period.
	Time Between Updates (sec)	Records the average length of the time between two successive updates of the IP Forwarding Table.
	Total Number of Updates	Records the number of times the IP Forwarding Table at this node was updated.
Node Statistics Router Convergence	Convergence Activity	Records convergence activity at the router (i.e., changes in the Forwarding Table). Time periods when there is convergence activity are denoted as 1 and time periods when there is no convergence activity are denoted as 0.
	Convergence Duration (sec)	Records the time it takes the IP Forwarding Table at this router to converge to a steady state (i.e., until there are no more changes to the Forwarding Table). Each recorded value corresponds to the time it took the IP Forwarding Table to reach a convergence state.

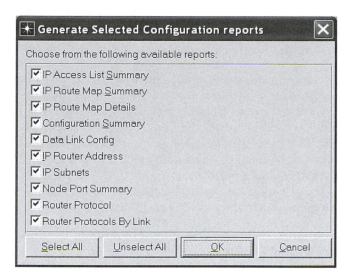

FIGURE 9.17   **Generate Selected Configuration reports** window.

Selected Configuration reports window, as shown in Figure 9.17. By selecting the checkboxes within that window, you can choose which configuration reports are to be generated. Alternatively, you can select the **Protocols > IP > Configuration Reports > Generate All** option, which will generate all the available reports. Generated reports can be viewed through the **Configuration Reports Tables** tab in the **Results Browser** window.

9.6.3 Viewing Forwarding and Routing Tables

Often it is important to view the IP Forwarding Table and routing tables that were created at various nodes during a simulation. Each routing protocol deployed at a node's interface produces a routing table that stores information about the routes known to that routing protocol at this node. The IP protocol pulls together all these routing tables at the node into a common IP Forwarding Table (also sometimes called the IP Common Route Table) for that node. The IP Forwarding Table is used by IP to make forwarding decisions for each packet that arrives at the node. OPNET documentation is inconsistent in its usage of the term "forwarding table" as this term is used in many places to include the routing tables as well, while at other places, the term "routing table" is used to refer to the IP Forwarding Table. In this book, we use the term IP "Forwarding Table" to only refer to IP's Forwarding Table, and we refer to the tables used by the routing protocols as routing tables.

You can configure the nodes to export various forwarding/routing tables through the **Reports** attribute. This attribute is available in most core and edge node models and allows you to configure the node to export various routing tables, including the IP Forwarding Table. Let us examine how to export the IP Forwarding Table generated in a node. The steps for exporting routing tables created by specific routing protocols are very similar and we omit their description to avoid redundancy.

The simplest way to export the IP Forwarding Table is to change the value of the **Reports…IP Forwarding Table** attribute at the nodes whose IP Forwarding Table you would like to examine. This attribute has the following predefined values:

- `Do Not Export` indicates that the IP Forwarding Table will not be exported.
- `Export at End of Simulation` indicates that the IP Forwarding Table created at this node will be exported upon simulation completion.
- `Edit…` allows you to specify the times during the simulation when the IP Forwarding Table is to be exported. Figure 9.18 shows an example of this attribute configuration, which results in the IP Forwarding Table being exported at the end of simulation and at time 10 seconds. Notice that to have the IP Forwarding Table exported, you must set the value of the **Reports… IP Forwarding Table…Status** attribute to `Enabled`.

You can also export the IP Forwarding Table by selecting the **Protocols > IP > Routing > Export Routing Tables…** option. This option opens a window in which you can specify if you would like to export the IP Forwarding Table of all the nodes in the network or only of the selected nodes (if you had pre-selected some nodes). When you click the **OK** button to apply your selection, the value of the **Reports…IP Forwarding Table** attributes will be changed to `Export at End of Simulation` on the chosen nodes. In addition, a window will

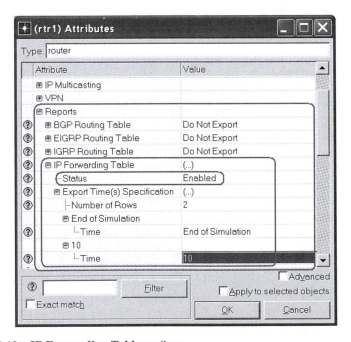

FIGURE 9.18   IP Forwarding Table attribute.

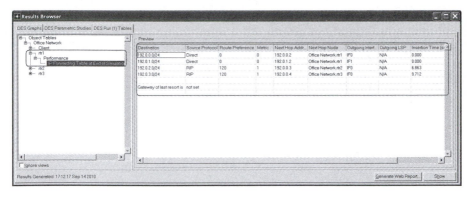

FIGURE 9.19   Viewing exported forwarding tables.

appear, which will notify you how many **IP Forwarding Table** attributes have been set. Finally, you can configure the simulation to export the IP Forwarding Table at all nodes through the **Configure/Run DES** window. Setting the value of the global attribute **IP...IP Routing Table Export/Import** to Export will configure the simulation to export the IP Forwarding Table for all nodes at the end of simulation.

Notice that the last two methods are only applicable to exporting the IP Forwarding Table. If you would like to export the routing table generated by a specific routing protocol, then you would need to set the **Reports...** attribute to export the routing table of the desired routing protocol at the nodes of your choice. For example, to export the routing table generated by the BGP protocol, you would need to configure the attribute **Reports...BGP Routing Table** at the desired nodes.

All these methods export the forwarding/routing tables of selected nodes into a standard output file with extension .ot. You can view the exported tables upon simulation completion through the **DES Run Tables** tab in the **Results Browser** window. As shown in Figure 9.19, you can generate a web report or view the table in an editor by clicking on the **Generate Web Report** or **Show** buttons, respectively. Notice that the default OPNET editor allows you to export the displayed forwarding/ routing table into a spreadsheet, XML, or comma-separated text file.

You can also configure your simulation to export the IP Forwarding Table of all nodes into a text file by setting the value of the **IP Route Table Export** attribute in the **IP Attribute Config** utility node. The **IP Route Table Export** attribute consists of the same subattributes as the attribute **Reports...IP Forwarding Table** and there- fore is configured in a similar fashion. Note that while the attribute is called **IP Route Table Export**, it actually exports only the IP Forwarding Table. The **IP Route Table Export** attribute also allows configuring the simulation to export the IP Forwarding Tables not only at the end of simulation but at the user-defined times during simulation. The exported information is saved as a text file with extension .gdf in your default directory. The exported file is typically named <scenario name>_<DES>-<run #>_ip_route_tables_<file #>.gdf. You can open .gdf files with any text editor.

10 Advanced IP Protocol Features

This chapter describes various advanced IP features supported in OPNET. We start the chapter by describing available OPNET options for configuring Network Address Translation (NAT), followed by IP multicasting and IPv6. We conclude the chapter by providing a detailed description of QoS mechanisms modeled in OPNET.

10.1 NETWORK ADDRESS TRANSLATION (NAT)

10.1.1 OVERVIEW OF NAT

Network Address Translation (NAT) is a mechanism for mapping private IP addresses (e.g., 10.0.0.0/8, 169.254.0.0/16, 172.16.0.0/11, and 192.168.0.0/16) into globally unique addresses. With the help of NAT, all hosts in a domain with a limited number of global IP addresses can gain access to the Internet. This technique was developed to tackle the issue of a dwindling number of global IP addresses (the Internet finally ran out of IP addresses in February 2011). To simplify the discussion of NAT, we will use the following definitions in the rest of this section:

- A *private* or *local address* is an unregistered IP address that cannot be officially used in the Internet.
- A *global address* is a globally unique registered IP address that can be used for data delivery in the Internet.
- An *inside network* is a network that contains devices with private IP addresses. These devices require NAT to access the Internet.
- An *outside network* is a network such as the Internet that contains devices with global IP addresses.
- A *gateway* is a device that resides at the boundary between an inside network and an outside network. An interface attached to a link that connects a gateway to an inside or outside network is called an *inside* or *outside interface*, respectively.

NAT is typically configured in a gateway or router node between inside and outside networks, as shown in Figure 10.1. NAT is responsible for maintaining a mapping between local and global IP addresses. There are several types of NATs including *static NAT*, which provides one-to-one local to global IP address translation; *dynamic NAT*, which dynamically selects a translation match for local IP addresses from a pool of available global addresses; *port address translation* (PAT), where

FIGURE 10.1   Example network topology with a NAT-enabled router.

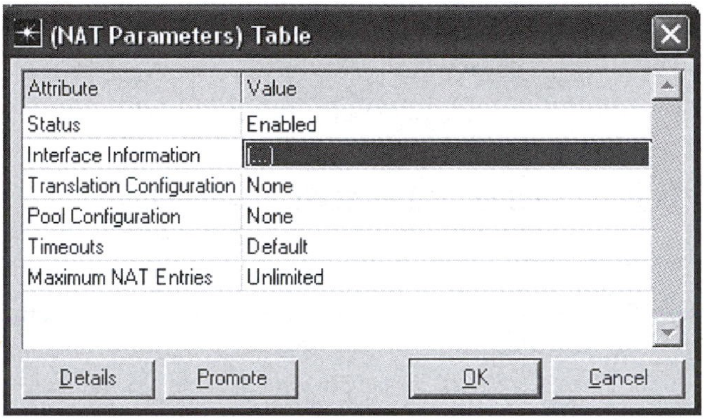

Attribute	Value
Status	Enabled
Interface Information	[...]
Translation Configuration	None
Pool Configuration	None
Timeouts	Default
Maximum NAT Entries	Unlimited

Details   Promote   OK   Cancel

FIGURE 10.2   Attribute **NAT Parameters**.

multiple local addresses are translated into a single global address with different port numbers, and others.

Typically, NAT operates as follows: An inside network node generates a request to be sent to a destination node in the outside network. Before forwarding this request to the outside network, the gateway uses NAT to overwrite the request packet's source IP address and possibly port number IP header fields with a corresponding global address and possibly a port number. NAT records the mapping between local and global IP addresses and port numbers in its translation table. When the response arrives from the external network at the gateway, NAT consults its translation table and then changes the response packet's destination IP address and possibly port number to the corresponding local IP address and port number. The updated response packet is then forwarded into the inside network.

In OPNET, you can define a NAT configuration in the models of core nodes with layer-3 functionality (e.g., routers, gateways, firewalls, etc.), whereas edge node models usually do not support NAT functionality. You can configure NAT functionality on a node by setting the values of its **IP...NAT Parameters** attribute. As Figure 10.2 illustrates, the attribute **NAT Parameters** consists of the following subattributes:

- **Status**—specifies if the NAT mechanism is enabled on this router.
- **Interface Information**—identifies and configures interfaces that support NAT functionality. Typically, the process of configuring NAT functionality consists of specifying translation rules deployed on the NAT-enabled interfaces.

- **Translation Configuration**—defines NAT translation rules.
- **Pool Configuration**—defines a set of global addresses and possibly port numbers. This information may be used for translation of local addresses.
- **Timeouts**—specifies NAT timeouts (i.e., the duration of time NAT waits before removing translation information for an idle connection) for various types of traffic.
- **Maximum NAT Entries**—specifies the maximum size of the NAT translation table.

10.1.2 CONFIGURING NAT

Typically, the process of configuring NAT requires the following steps:

1. Identify the names of the inside and outside interfaces in a gateway router, which will perform the NAT.
2. If needed, specify address pools for translation rules.
3. Define translation rules.
4. Deploy translation rules on the interfaces identified in step 1.

The first step of identifying the inside and outside interface names is fairly straightforward. It follows the same process described in Section 9.3.2 to identify the name of any interface attached to a link (i.e., "hover" the mouse pointer over the link, examine the transmitter and receiver link attributes, or open the **Select Port Assignment** window using the **Edit Ports** option). The remaining three steps are explained in detail in Sections 10.1.3 through 10.1.5.

10.1.3 SPECIFYING ADDRESS POOLS

This step may be omitted if the translation rules do not rely on address pools. To define an address pool, you need to configure the **Pool Configuration** attribute, as shown in Figure 10.3.

First, expand the **Pool Configuration** attribute and set the value of the attribute **Number of Rows**, which specifies the total number of address pool definitions to be configured. Next, perform the following steps for each newly created row:

- Specify the name of the address pool by setting the value of the attribute **Pool Name**.
- Set the value of the attribute **Number of Rows**, which corresponds to the number of different address ranges to be defined within the current address pool.
- Configure each address range by specifying globally routable addresses and possibly port numbers. The list below provides a description of the key attributes for defining address ranges.
  - **Interface Name** specifies the names of the outgoing interfaces for which this address range can be used. You can set this attribute value to All, which will make this address range applicable to all outgoing interfaces.

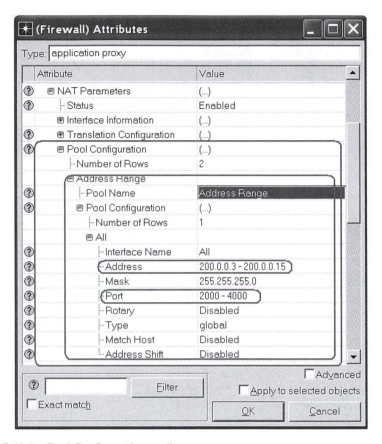

FIGURE 10.3 **Pool Configuration** attribute.

- **Address** specifies an actual IP address or a range of IP addresses for the translation rules. The address ranges are specified using the following notation: `<start address>` – `<end address>` (e.g., `134.13.4.0` – `134.13.4.255`).
- **Mask** specifies the subnet mask to be used together with the **Address** attribute. This attribute accepts the value `Not Assigned`, which will cause the simulation to ignore the mask value.
- **Port** specifies a port number or a range of port numbers for the current address range. The port ranges are specified using the following notation: `<start port number>` – `<end port number>` (e.g., `2000 – 3000`). The attribute **Port** also accepts the following two values: (1) `Not Assigned`, which causes the simulation to ignore the port numbers during translation, and (2) `Automatic`, which causes the simulation to select port numbers automatically.
- **Type** specifies if the address range defines a unique global address or local addresses for remote VPN clients, which is not the same as private addresses.

10.1.4 SPECIFYING TRANSLATION RULES

The third step for configuring NAT is to define the actual translation rules. The translation rules are defined via the **Translation Configuration** attribute, as shown in Figure 10.4, by performing the following steps:

1. Expand the **Translation Configuration** attribute and set the value of the attribute **Number of Rows**, which corresponds to the total number of NAT rules to be defined on this node.
2. Configure each of the NAT rules by specifying:
   a. Classification rules for identifying the original packets to be translated by NAT.
   b. Description of the translated packet, that is, how the original packet will look like after NAT translation.
   c. Various translation characteristics.

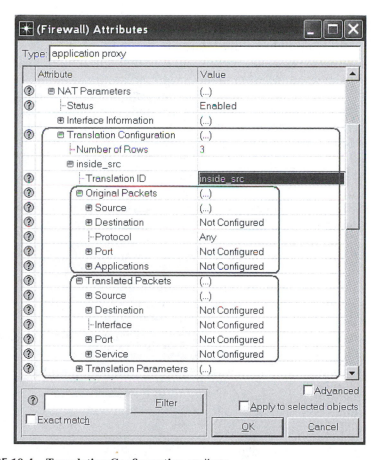

FIGURE 10.4 **Translation Configuration** attribute.

The classification rules for identifying original packets are specified via the **Original Packets** attribute, which allows defining source, destination, protocol, port, and application information to be matched against the corresponding information of arriving packets. The source and destination information can be specified via such attributes as **IP address** and **Subnet Mask, Access Control List, Route Map,** and **Network Object Group.** The attribute **Protocol** accepts only values Any, TCP, or UDP; the attribute **Port** allows specifying the actual port numbers; whereas the attribute **Applications** requires you to specify the name of the application to which the matching packet belongs.

The definition of translated packet characteristics is specified via the attribute **Translated Packets**, which consists of the subattributes **Source, Destination, Interface, Port,** and **Service.** You do not need to specify values for all of these subattributes. In most situations, you will only need to set either source or destination information. Both the **Source** and **Destination** attributes have the same structure and define source and destination characteristics via only one of the following attributes:

- **IP Address**—the translated packets will have their source or destination IP address field set to the specified IP address.
- **Interface**—the translated packets will have their source or destination IP address set to the address of the outgoing interface.
- **Pool**—this attribute accepts only the names of the address pools that have already been defined. The source or destination IP address of translated packets will be selected from the address pool specified in this field.
- **Network Object Group**—the source or destination information will be set based on the Network Object Group value, which is configurable only on a firewall node. We do not discuss this feature in this book.

The attribute **Translation Parameters**, as shown in Figure 10.5, specifies the following information:

- The type of translation mode (i.e., Dynamic—translation address is determined from a pool of address, Static—one-to-one address translation, or Overload—port address translation (PAT))
- The direction of translation (i.e., Inbound or Outbound, typically Outbound)
- Whether certain NAT characteristics (i.e., **Extendable Static Translation, Generate Translation Alias, Payload Translation,** and **Add Route**) are enabled or disabled. In most cases, these attributes remain set to their default values.

10.1.5 DEPLOYING TRANSLATION RULES ON GATEWAY INTERFACES

Once the translation rules have been defined, you can deploy them on the corresponding interfaces that you have identified in the first step of Section 10.1.2. To deploy the translation rules, expand the attribute **Interface Information**, as

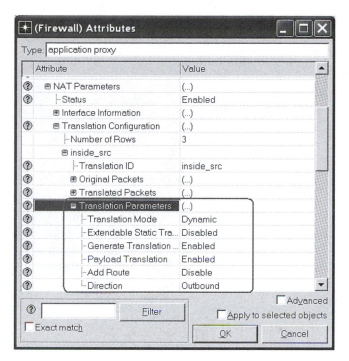

FIGURE 10.5   Translation Parameters attribute.

shown in Figure 10.6, and then set the value of the attribute **Number of Rows,** which represents the total number of NAT-enabled interfaces. Each newly created row represents the NAT configuration for a single interface. You need to configure each individual row by specifying the following information:

- **Name**—the name of the interface to be configured for the NAT (e.g., IF0, IF5, etc.). You should have identified such an interface in the first step of the process.
- **Status**—indicates if the current interface connects the router to an inside or outside network. This attribute is applicable to Cisco IOS NAT only and is ignored in Cisco PIX firewall devices.
- **Translations**—a compound attribute that specifies the translation rules deployed on this interface. You can configure an interface to support multiple rules defined in the **Translation Configuration** attribute.
- **Dual Translations**—a compound attribute that allows configuring dual-translation rules on the interface. Dual-translation rules are specified via the **Dual NAT Configuration** attribute, which is available only in certain node models. In most cases, this attribute remains set to Not Configured.
- **Subinterface Information**—allows specifying NAT on the node's subinterfaces. Note that you cannot configure this attribute if the node has no defined subinterfaces.

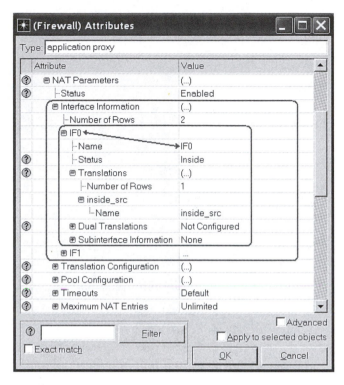

FIGURE 10.6   Interface Information attribute.

Project **IP**, which is a part of the standard OPNET software distribution, contains a scenario called *NAT*, which provides an excellent example of NAT configuration in a network.

10.2   IP MULTICAST

10.2.1   IP MULTICAST FEATURES SUPPORTED IN OPNET

IP multicast is a technique that allows one-to-many communication, that is, a single source transmits data to multiple destinations. OPNET Modeler and IT Guru support IP multicast and allow development of network models that include the Internet Group Management Protocol (IGMP) and Protocol Independent Multicast-Sparse Mode (PIM-SM). In addition, OPNET's IP multicast models allow specifying a Rendezvous Point (RP) using the following methods:

- *Static-RP*—all multicast routers are explicitly configured with RP information.
- *Auto-RP*—only certain multicast routers are configured with RP information. The rest of the routers discover RP information dynamically during simulation. The auto-RP method relies on mapping agents to distribute RP information in the simulated network.

- *Bootstrap*—also does not need all routers to have RP information configured statically. Instead, a Bootstrap Router (BSR) is responsible for distributing RP information among other routers in the network.

Currently, you can define IP multicast traffic using certain standard applications running over UDP, unidirectional custom applications running over UDP, and IP multicast demands (i.e., the demand object *ip_mcast_traffic_flow*). In particular, you can specify multicast traffic for Voice, Video, Database, and FTP standard applications only. For voice and video applications, only workstation nodes can serve as a destination for multicast traffic. For the remaining supported standard applications, only server nodes can receive multicast traffic.

10.2.2 Overview of Steps for Deploying IP Multicast Traffic

Typically, the process of deploying multicast traffic in a network consists of the following steps:

1. Define multicast traffic (i.e., applications or demand).
2. Configure source nodes.
   a. Deploy defined multicast traffic (applicable only if using standard applications).
   b. Enable multicast on the interfaces.
3. Configure destination nodes.
   a. Enable support for multicast application services (applicable only if using standard applications).
   b. Specify details for joining/leaving a multicast destination group.
   c. Enable multicast on the interfaces.
4. Configure multicast routers.
   a. Enable multicast on the interfaces that will carry multicast traffic.
   b. Specify RP using static RP, Auto-RP, or bootstrap mechanism.

We will discuss these steps in detail in Sections 10.2.3 through 10.2.7. However, IP multicast is a fairly complex topic and we describe only the key configuration steps for deploying IP multicast. For additional information on IP multicast, refer to the RFCs 1112, 2236, 2362, and 4601 or other external sources. Chapter STM-13 in the OPNET documentation is a great source of information regarding OPNET support for IP multicast. In addition, Modeler and IT Guru contain a preconfigured sample project named **IP_Multicast**, which contains several scenarios with an excellent illustration of IP multicast deployment and configuration in OPNET.

10.2.3 Defining Multicast Traffic

You can define user profiles and standard applications for IP multicast the same way as any other standard application. Since the destination for this application's traffic is a multicast group, we suggest that you set the value of the **Symbolic Destination Name** attribute to a name that indicates this fact. For example, in the **Multicast** scenario of the

IP_Multicast project, the value of the **Symbolic Destination Name** attribute for the multicast voice application is set to `Multicast Receiver`. Moreover, you should ensure that the multicast applications run over UDP. Recall that in advanced workstation and server models, you can control the type of transport protocol used by an application via the attribute **Applications…Application: Transport Protocol Specification**.

The process of adding an IP multicast demand object is slightly more complex and consists of the following steps:

- Double-click on the *ip_mcast_traffic_flow* object in the **Object Palette Tree**. You can find this object in the *demands* category.
- Click on the node that will serve as the source of the multicast traffic. However, do not connect the multicast demand object to any of the destination nodes.
- Double-click anywhere in the empty space of the project workspace, which will add a multicast demand object to the selected node.
- You can repeat this process and add other demand objects or you can stop adding demand objects by right-clicking anywhere in the project workspace and selecting the **Abort Demand Definition** option from the menu that appears.

Once all multicast demand objects have been added, you need to configure each of them by specifying at least the destination's IP address and the traffic rate. Figure 10.7 shows the configuration attributes of an IP multicast demand object. You can specify the multicast group address by setting the value of the attribute **Destination IP Address**. The provided address value must correspond to a valid multicast address

FIGURE 10.7 Attributes of an IP multicast demand object.

(e.g., any address in the range 224.0.0.0/4). The multicast demand traffic rate is specified via the attributes **Traffic (bits/second)** and **Traffic (packets/second)**. You may also want to set the value of the attribute **Traffic Start Time**, which controls the time when the demand starts generating traffic. By default, this attribute is set to `Same As Global Setting` (i.e., the value defined via the global attribute **Traffic... Background Traffic Start Delay (seconds)** in the **Config/Run DES** window). Notice that if the destination nodes join the multicast group after the demand starts traffic transmission, then all the packets sent prior to the destinations joining the multicast group will be discarded. Typically, you can easily identify such a problem because the DES Log records a message in the event of a multicast demand packet being discarded due to the absence of destinations in the multicast group.

10.2.4 Configuring Source Nodes

When configuring source nodes, the first step is to enable IP multicast on the nodes. You can enable IP multicast on the desired source nodes by either of the following:

- Setting the value of the **IP...IP Host Parameters...Multicast Mode** attribute to `Enabled`.
- Selecting all nodes that you would like to function as multicast sources and then choosing the option **Protocols > IP > Multicast > Enable Multicasting on Selected Hosts** from the **Project Editor**'s pull-down menu, which in turn will change the value of the **IP...IP Host Parameters... Multicast Mode** attribute to `Enabled` on the selected nodes.

If you have deployed all multicast traffic via demand objects, then this step completes the source node configuration. However, if IP multicast traffic is defined via custom or standard applications, then you need to deploy profiles that will run the defined multicast applications on the desired source nodes. OPNET does not support automated deployment of custom applications, which means that you cannot deploy multicast applications using the **Protocols > Applications > Deploy Defined Applications...** option. Therefore, to deploy multicast applications, you need to manually configure both source and destination nodes.

To define a node as a source for a multicast application, perform the following two steps illustrated in Figure 10.8:

1. Configure the node to support a multicast application profile by setting the attribute **Applications...Application: Supported Profiles**.
2. Define destination preferences for the multicast application by mapping the application's symbolic server name to the multicast group address. Recall that you can configure such a mapping by setting the value of the attribute **Applications...Application: Destination Preferences**. Note that the actual server address (i.e., the attribute **Name**) must be set to a valid multicast address.

As shown in Figure 10.8a, the node `Admin_Sender` is configured to support a profile called `Video`, which includes a multicast application with the symbolic

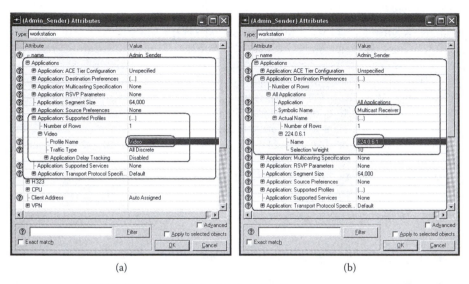

(a)   (b)

FIGURE 10.8 Steps for deploying a multicast application at a source node: (a) Example of deploying a user profile with a multicast application in the node `Admin_Sender`. (b) Example of mapping symbolic name `Multicast Receiver` to a specific multicast group address.

server name set to `Multicast Receiver`. Figure 10.8b shows that the node `Admin_Sender` maps the symbolic server name `Multicast Receiver` to the multicast address `224.0.6.1`, which in effect defines the multicast group address.

10.2.5 Configuring Destination Nodes

Before you start configuring multicast destnation nodes, you need to ensure that the destination nodes are represented via intermediate or advanced node models (i.e., the model name has *_adv* or *_int* modifier). Otherwise, certain features necessary for configuring IP multicast will not be available. Typically, the process of configuring destination nodes consists of the following steps:

1. Specify multicast group on the destination nodes.
2. Enable multicast on the destination nodes.
3. Configure the destination nodes to support multicast application services. This step is only applicable when the multicast traffic is specified using standard or custom applications.

You can specify that a destination node is a part of a multicast group by one of the following approaches:

1. For each destination node that belongs to the multicast group, configure the attribute **Applications...Application: Multicasting Specification** by setting the values for at least the following attributes:
   a. **Application Name**—the name of the multicast application.

b. **Membership Addresses**—the multicast group address, that is, the IP address value mapped to the multicast application's symbolic server name or set in the **Destination IP Address** attribute of the IP multicast demand object.

c. **Joining Time (seconds)**—the time when the node joins the multicast group.

d. **Leave Time (seconds)**—the time when the node leaves the multicast group.

After that, you will also need to enable multicast on the destination nodes by setting the value of the attribute **IP…IP Host Parameters…Multicast Mode** to Enabled. Figure 10.9 illustrates the **Application: Multicasting Specification** attribute. Notice that this attribute is available only in advanced end node models.

2. Alternatively, you can achieve the same result by selecting the multicast destination nodes and then using the **Protocols > IP > Multicast > Set Multicast Group on Selected Destination Nodes** option, which opens the **Destination nodes** window, as shown in Figure 10.10. Notice that all the fields in the **Destination nodes** window will be disabled, unless all selected nodes use advanced end node models. Once again, you need to specify values for the **Application**, **IP Multicast Group Address**, **Join**

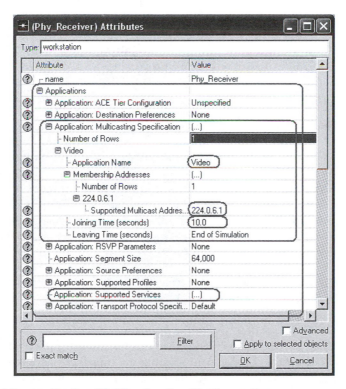

FIGURE 10.9 **Application: Multicasting Specification** attribute.

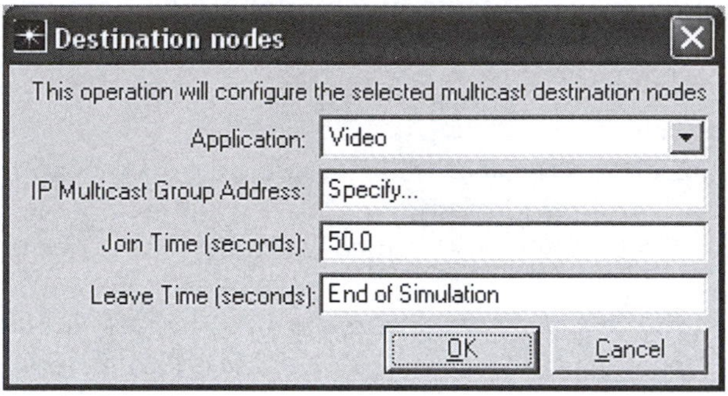

FIGURE 10.10   Destination nodes window.

Time (seconds), and **Leave Time (seconds)** attributes. You can save the introduced changes by clicking the **OK** button. Notice that the resulting system configuration will be the same as if you have performed the steps described in the first approach, that is, all selected nodes will have the attribute **Multicast Mode** set to Enabled and the attribute **Applications... Application: Multicasting Specification** will be configured according to the setting specified in the **Destination nodes** window.

Finally, if multicast traffic is specified via standard or custom applications, then you need to specify that each destination node supports the multicast application's services by configuring the node's attribute **Applications...Application: Supported Services**, circled at the bottom of Figure 10.9.

10.2.6 CONFIGURING ROUTER NODES

Configuring the core routers to forward multicast traffic is a two-step process:

1. Enable multicast on the interfaces that will carry multicast traffic.
2. Specify the Rendezvous Point (RP). RP is often defined as a point in the network where the multicast data from sources and receivers meet. In effect, it is a point in the network where the multicast data that originates at a source starts being distributed on different paths of a shared distribution tree on its way to various destinations. OPNET supports static-RP, auto-RP, and bootstrap mechanisms for specifying the RP.

To enable multicast on the router interfaces that will carry multicast traffic, you need to perform the following actions:

1. Select all the router links that will carry multicast traffic.
2. Enable multicasting across the selected links by choosing the **Protocols > IP > Multicast > Enable Multicasting on Interfaces across Selected Links** option. As a result of this operation, IP multicast will be enabled on all router interfaces attached to the selected links.

The first step in the process of specifying the RP is to identify the router that will serve as RP in the distribution tree of the multicast network. Specifically, you need to find the IP address of an active multicast-enabled interface on the RP router. We will refer to such a value as the *RP address*. If you have not configured IP addresses in your network, then you can either manually input IP addresses on the corresponding router interfaces or, as we would recommend, use the **Protocols > IP > Addressing > Auto-Assign IPv4 Addresses** option to specify IP addresses in all the nodes in the network. Once IP addresses have been assigned, you need to examine the **IP...IP Routing Parameters... Interface Information** attribute to identify the RP address. The IP address of any active multicast-enabled interface in the RP router can serve as the RP address.

Now you are ready to specify RP in the simulated system. The configuration steps differ depending on the type of RP mechanism selected. We describe configuration steps for each of the mechanisms in turn. Notice that you can configure multicast routers by manually changing attribute values at individual nodes. However, manually enabling multicast on the router interfaces and then manually defining RP is a tedious and error-prone task, which is why we strongly encourage you to rely on the **Protocols** menu option as described in this section.

10.2.6.1   Static-RP Mechanism

Execute the following steps to define RP using the static-RP mechanism:

- Select the routers that forward multicast traffic. You can omit this step if you want to specify RP on all routers in the network.
- Select the **Protocols > IP > Multicast > Configure Rendezvous Point Using Static RP Configuration...** option, which opens the window shown in Figure 10.11.
- Set the value of the *IP Multicast Group Address/Mask* field to the multicast group address. Notice that the inputted value must be a valid multicast address written in CIDR notation.

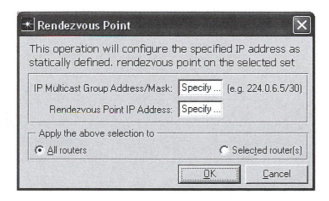

FIGURE 10.11   Configuring RP using the static-RP mechanism.

- Set the value of the *Rendezvous Point IP Address* field to the RP address that you have determined earlier.
- Choose whether your configuration applies to all routers in the network or to selected routers only.
- Click the **OK** button to save your changes and define RP using the static-RP mechanism.

10.2.6.2   Auto-RP Mechanism

Unlike the static-RP mechanism, where each router has RP statically specified prior to simulation, the auto-RP mechanism dynamically distributes RP information during simulation. To configure the auto-RP mechanism, which is a proprietary mechanism of Cisco, you need to execute the following steps:

- On any router that carries multicast traffic, specify the RP candidate for the auto-RP mechanism by performing the following actions. These steps are illustrated in Figure 10.12:

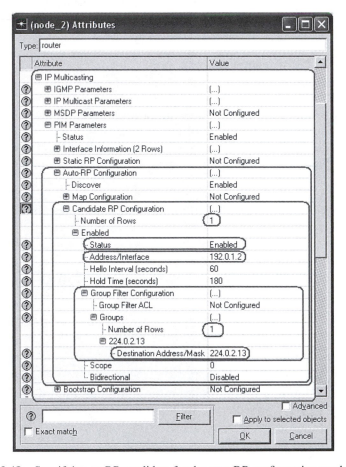

FIGURE 10.12   Specifying an RP candidate for the auto-RP configuration mechanism.

- Expand the attribute **IP Multicasting...PIM Parameters...Auto-RP Configuration.**
- Set the attribute **Discover** to Enabled.
- Expand the attribute **Candidate RP Configuration** and set the value of the attribute **Number of Rows** to 1.
- Expand the newly created row and set the attribute **Status** to Enabled and the attribute **Address/Interface** to the RP address that you have determined earlier.
- Expand the attribute **Group Filter Configuration...Groups**.
- Set the value of the attribute **Number of Rows** (i.e., subattribute of **Groups**) to 1.
- Expand the newly created row and set the value of the **Destination Address/Mask** attribute to the multicast group address written in CIDR notation.
- Click the **OK** button to save the changes.

- Enable the auto-RP mechanism on the multicast routers in the network by first selecting all the multicast routers and then clicking on the **Protocols > IP > Multicast > Enable Auto-RP on Selected Routers** option.
- Enable the mapping agent on at least one of the multicast routers by setting the value of the attribute **IP Multicasting...PIM Parameters...Auto-RP Configuration...Map Configuration...Status** to Enabled, as shown in Figure 10.13.

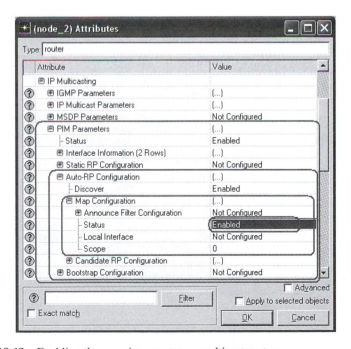

FIGURE 10.13   Enabling the mapping agent on a multicast router.

10.2.6.3 Bootstrap Mechanism

The bootstrap or bootstrap router (BSR) mechanism also dynamically distributes RP information and overall is very similar to the auto-RP mechanism with the exception of several implementation details, which are beyond the scope of this book. BSR is an industry standard mechanism. To configure the bootstrap mechanism, you need to execute the following steps:

1. On any router that carries multicast traffic, specify the RP candidate for bootstrap configuration by performing the following actions, as shown in Figure 10.14:
   a. Expand the attribute **IP Multicasting...PIM Parameters...Bootstrap Configuration...Candidate RP Configuration** and set the value of the attribute **Number of Rows** to 1.
   b. Expand the newly created row and set the attribute **Status** to Enabled and the attribute **Address/Interface** to the RP address that you have determined earlier.
   c. Expand the attribute **Group Filter Configuration...Groups**.

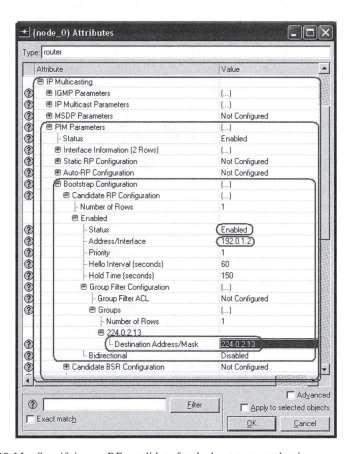

FIGURE 10.14   Specifying an RP candidate for the bootstrap mechanism.

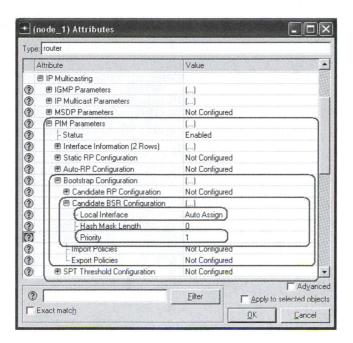

FIGURE 10.15   Specifying a candidate bootstrap router for the bootstrap mechanism.

d. Set the value of the attribute **Number of Rows** (i.e., subattribute of **Groups**) to `1`.

e. Expand the newly created row and set the value of the **Destination Address/Mask** attribute to the multicast group address written in CIDR notation.

f. Click the **OK** button to save the changes.

2. Select one of the multicast routers to serve as a candidate BSR by performing the following tasks, illustrated in Figure 10.15:

a. Expand the attribute **IP Multicasting...PIM Parameters...Bootstrap Configuration...Candidate BSR Configuration**.

b. Set the value of the attribute **Local Interface** to either `Auto Assign` or to an IP address of this router's interface that will be advertised as a candidate BSR.

c. Set the attribute **Priority** to a value greater than `0`. This attribute represents the bootstrap candidacy priority. The router with the highest priority value is elected as BSR. If multiple routers have the highest priority value, then the router with the highest IP address value is chosen as BSR.

10.2.7   OTHER MULTICAST CONFIGURATION PARAMETERS

In Sections 10.2.2 through 10.2.6, we described the basic procedure for deploying multicast traffic in a simulated network and briefly described some of the available attributes. OPNET software contains numerous other attributes for configuring IP multicast and related features, including IGMP versions 1–3, source-specific multicast, multicast VPNs,

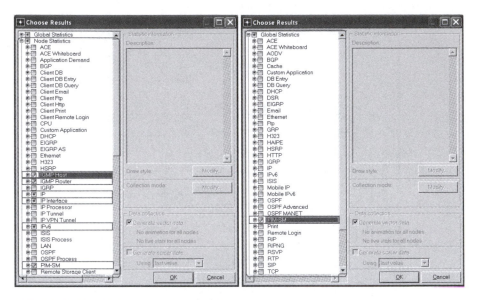

FIGURE 10.16   Multicast-specific statistics categories.

and others. For more information about the supported multicast features, please refer to the product documentation accessible via **Protocols > IP > Multicast > Model User Guide**.

10.2.8   MULTICAST STATISTICS AND REPORTS

OPNET provides a set of statistics for evaluating the performance of multicast traffic and related protocols. As shown in Figure 10.16, multicast-specific Node Statistics are located in the **IGMP Host**, **IGMP Router**, **IP**, **IP Interface**, **IPv6**, and **PIM-SM** statistics categories, while only the **PIM-SM** category contains multicast-specific Global Statistics. Notice that the **IP**, **IP Interface**, and **IPv6** categories contain numerous IP-related statistics, only a few of which are related to multicast. A summary of multicast-specific statistics is provided in Table 10.1.

In addition to the statistics, you may find it useful to examine multicast routing tables. Specifically, OPNET allows exporting group-to-RP mappings and PIM-SM routing tables via the **Reports...IP Multicast Group-to-RP Table** and **Reports... PIM_SM Routing Table** attributes, respectively. The exported tables can be viewed via the **DES Run Tables** tab in the **Results Browser** window.

10.3   IPV6

10.3.1   OVERVIEW OF SUPPORTED IPv6 FEATURES

OPNET provides support for IPv6, the Internet Protocol Version 6. However, you need to have **IPv6 for R&D** module licenses in order to use IPv6 models in a simulation. OPNET supports such IPv6 features as assigning IPv6 addresses on interfaces, setting up dual-stack node configuration (i.e., the nodes can be configured to support both IPv4 and IPv6 protocols), specifying IPv6 static routing tables, creating IPv6 tunnels,

TABLE 10.1
Summary of IP Multicast Statistics

Category	Name	Description
Global Statistics **PIM-SIM**	*Network Convergence Activity*	The time periods when the signs of convergence were detected are recorded as 1. The time periods when there were no signs of convergence activity are recorded as 0. The resulting graph consists of periods of 1s and 0s, which correspond to periods of convergence and no convergence activities in the whole network.
	Network Convergence Duration (sec)	This statistic records the duration of the convergence cycles for the PIM-SM routing tables across the whole network. A separate statistic vector is recorded for each multicast group.
	PIM-SM Control Traffic Received (bits/sec), PIM-SM Control Traffic Received (packets/sec), PIM-SM Control Traffic Sent (bits/sec), PIM-SM Control Traffic Sent (packets/sec)	These statistics record the total amount of PIM-SM control traffic received or sent by all nodes in the network. These statistics are available in units of bits/second and packets/second.
	PIM-SM Register Traffic Received (bits/sec), PIM-SM Register Traffic Received (packets/sec), PIM-SM Register Traffic Sent (bits/sec), PIM-SM Register Traffic Sent (packets/sec)	These statistics record the total amount of PIM-SM Register packets received or sent by all nodes in the network. These statistics are available in units of bits/second and packets/second.
Node Statistics **IGMP Host** **IGMP Router**	*IGMP Traffic Received (bits/sec), IGMP Traffic Received (packets/sec), IGMP Traffic Sent (bits/sec), IGMP Traffic Sent (packets/sec)*	These statistics record the total number of IGMP messages received or sent by this node (i.e., host or router) across all of its IP interfaces. These statistics are available in units of bits/second and packets/second.

(Continued)

TABLE 10.1 (*Continued*)
Summary of IP Multicast Statistics

Category	Name	Description
Node Statistics IP	*Multicast Traffic Dropped (bits/sec), Multicast Traffic Dropped (packets/sec)*	These statistics record the total number of IP multicast packets discarded by this node on all of its IP interfaces. These statistics are available in units of bits/second and packets/second.
	Multicast Traffic Received (packets/sec), Multicast Traffic Sent (packets/sec)	These statistics record the total number of IP multicast packets received or sent by this node on all of its IP interfaces.
Node Statistics IP Interface	*Multicast Traffic Received (bits/sec), Multicast Traffic Received (packets/sec), Multicast Traffic Sent (bits/sec), Multicast Traffic Sent (packets/sec)*	These statistics record the total number of IP multicast packets received or sent by this node on a specific interface. These statistics record packets with multicast destination and Register messages that carry data. These statistics are available in units of bits/second and packets/second.
Node Statistics IPV6	*Multicast Traffic Received (packets/sec), Multicast Traffic Sent (packets/sec)*	These statistics record the total number of IPv6 multicast packets received or sent by this node on all of its IP interfaces.
Node Statistics PIM-SM	*PIM-SM Control Traffic Received (bits/sec), PIM-SM Control Traffic Received (packets/sec), PIM-SM Control Traffic Sent (bits/sec), PIM-SM Control Traffic Sent (packets/sec)*	These statistics record the amount of PIM-SM control traffic received or sent by this node on all of its interfaces. These statistics are available in units of bits/second and packets/second.
	PIM-SM Register Traffic Received (bits/sec), PIM-SM Register Traffic Received (packets/sec), PIM-SM Register Traffic Sent (bits/sec), PIM-SM Register Traffic Sent (packets/sec)	These statistics record the amount of PIM-SM Register messages received or sent by this node on all of its interfaces. These statistics are available in units of bits/second and packets/second.

TABLE 10.1 (*Continued*)
Summary of IP Multicast Statistics

Category	Name	Description
	Network Convergence Activity	This statistic records the time convergence periods at this node. The periods of convergence activity are designated as 1 while the periods of no convergence activity are designated as 0. The resulting graph consists of 1s and 0s, which correspond to periods of convergence and no convergence activities at this node.
	Network Convergence Duration	This statistic records the amount of time required by this router to have its PIM routing table converge for a particular multicast group. This statistic is recorded as the amount of time elapsed from the time the router exhibited signs of convergence activity until the period of no convergence activity.

deploying certain IPv6 routing protocols, creating traffic demand and standard application sources within an IPv6 network, generating ICMPv6 ping messages, deploying QoS support, connecting mobile nodes via Mobile IPv6, and others. However, OPNET does not support path MTU discovery in IPv6 networks, and therefore all IPv6-enabled interfaces are configured to support an MTU size of 1500 bytes only. In this section, we describe only the basic IPv6 configuration parameters because many of the supported IPv6 features are configured in a manner similar to that of IPv4. For a more comprehensive description of IPv6 features, please refer to the IPv6 User Guide available via the **Protocols > IPv6 > Model User Guide** option. OPNET also includes a preconfigured project called IPv6, which provides an excellent illustration of various IPv6 features.

10.3.2   IPv6 Addressing

In OPNET, all node models that support layer-3 functionality are IPv6-capable. You can specify IPv6 addresses in the network by using the **Protocols > IPv6 > Auto-Assign IPv6 Addresses** option. Specifically, if you have selected nodes and/or links prior to this operation, then OPNET will enable IPv6 (i.e., it will set the link-local IPv6 address to the Default EUI-64 value) and will specify global IPv6 addresses on all corresponding interfaces of selected nodes and links that do not have IPv6 addresses assigned yet. If you do not select any nodes prior to selecting the **Auto-Assign IPv6 Addresses** option, then OPNET will enable IPv6 and will specify global IPv6 addresses on all connected interfaces in the network. You can clear assigned IPv6 addresses by using the **Protocols > IPv6 > Clear IPv6 Addresses** option. This operation clears all assigned global IPv6 addresses and sets all link-local IPv6 addresses to the value Not Active, which indicates that the interface is not IPv6 enabled.

You may recall from Section 9.3 that an IPv4 address on a node's interface is specified via the attribute **Address**. In end node models, this attribute is located under **IP...IP Host Parameters...Interface Information**, and in core node models, the **Address** attribute resides under **IP...IP Routing Parameters...Interface Information...<interface name>**. IPv6 addresses are 128 bits long and are specified via attributes called **Link-Local Address** and **Global Address(es)...Address**, which define link-local and global IPv6 addresses, respectively. In end node models, these attributes are located under **IP... IP Host Parameters...Interface Information...IPv6 Parameters**, whereas in core node models, they reside under **IP...IPv6 Parameters...Interface Information... <interface name>**, as shown in Figure 10.17. To specify a link-local IPv6 address, you need to set the attribute **Link-Local Address**, which supports the following values:

- `Default EUI-64`—indicates that the link-local IPv6 address on this interface will be determined automatically upon simulation execution. This value also indicates that the interface is IPv6 enabled.
- `Not Active`—indicates that this interface is not IPv6 enabled.
- `Edit...`—allows you to manually specify the IPv6 value. However, the first 8 bytes of a manually specified IPv6 link-local address must correspond to the value `FE80::` as stated in RFC 2373. Notice that all link-local addresses have a prefix length of 64 bits.

To specify a global IPv6 address on an interface, you need to perform the following steps:

1. Expand the subattribute **Global Address(es)**.
2. Set the value of the attribute **Number of Rows** to an integer number, which corresponds to the total number of global IPv6 addresses to be configured on that interface. Notice that each interface may be configured with multiple global IPv6 addresses.
3. For each newly created row, you need to set the value of the attribute **Address** to a global IPv6 address. The global IPv6 address value cannot be auto-assigned and must be configured manually. However, if the node is configured to support DHCP, then you can set the value of the attribute **Address** to `From DHCP` or `From DHCP Prefix Delegation`, which will obtain the IPv6 address automatically. Refer to the *OPNET IPv6 Model User Guide* for the IPv6 DHCP configuration instructions. Further, if the value of the **Address Type** attribute is set to `EUI-64` (Extended Unique Identifier format), then you only need to specify the first 64 bits of the IPv6 address. The remaining 64 bits are automatically computed based on the 48-bit MAC address value of the interface as suggested by RFC 2373.

Typically, each layer-3 node model can support both IPv4 and IPv6 protocols. Such node models are often referred to as dual-stack capable. Each dual-stack capable node can be configured to support IPv4, IPv6, or both IPv4 and IPv6. You can enable or disable IPv6 functionality on the node interfaces using the following **Protocols** menu options:

- **Protocols > IPv6 > Configure Interface Status...**—opens a window, which allows you to enable or disable IPv6 functionality on selected nodes and links or on all nodes.

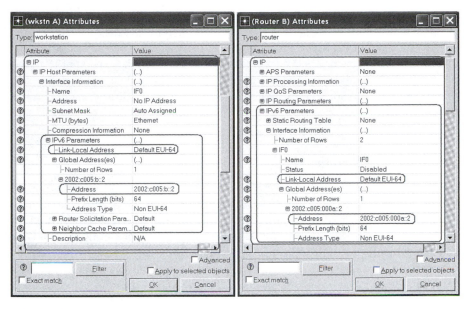

FIGURE 10.17   Specifying IPv6 link-local and global addresses.

- **Protocols > IPv6 > Enable IPv6 on All Interfaces**—enables IPv6 functionality on all node interfaces in the network.
- **Protocols > IPv6 > Disable IPv6 on All Interfaces**—disables IPv6 functionality on all node interfaces in the network.

You can also manually enable IPv6 functionality on an interface by setting the value of the IPv4 **Address** attribute to `No IP Address` and having the IPv6 address set to a valid IPv6 value. Similarly, you can manually disable IPv6 functionality on the interface by setting the value of the **Link-Local Address** attribute to `Not Active` and having the value of the IPv4 **Address** attribute set to a valid IPv4 value. Finally, you can also configure the node's interface to support dual-stack setting by specifying both IPv4 and IPv6 addresses on the interface. You can assign IPv4 and IPv6 addresses either by auto-assignment using the **Protocols > IP > Addressing > Auto-Assign IP Address** option (which opens the **Auto-Assign IP Addresses** window described in Chapter 9), or by manually specifying IPv4 and IPv6 addresses as described earlier.

10.3.3 Configuring Traffic for IPv6 Networks

OPNET allows deploying any standard application and traffic demands in an IPv6-enabled network without any additional configuration steps. However, the version of IP used for data transfer depends on the configuration of the source and destination nodes. Specifically, if all source and destination nodes are dual-stack enabled, then the value of the attribute **IP...IP Version Preference** located in the **Inputs...Global**

Attributes category of the **Configure/Run DES** window (Section 4.3) determines the version of the IP protocol used to transfer the application data. By default, the value of **IP Version Preference** is set to IPv6. In situations when at least one of the source and destination nodes is not dual-stack enabled, then the version of IP that is common to all the nodes is used to transfer the application data. If the source and destination nodes do not support a common version of IP, then traffic cannot be sent between these nodes.

10.3.4   OTHER IPv6 OPTIONS

OPNET allows configuring IPv6 interfaces to run the following routing protocols: RIPng, OSPFv3, IS-IS, BGP, AODV, DSR, and OLSR. Similarly to IPv4, you can configure these routing protocols via the **Protocols > IPv6 > Configure IPv6 Routing Protocols...** option, which opens the **IPv6 Routing Protocol Configuration** window shown in Figure 10.18. Through this window you can specify which routing protocol(s) (i.e., could be more than one) you want to deploy and where you want these routing protocols to be deployed (i.e., all interfaces including loopback and tunnels, physical interfaces only, or physical interfaces on selected links only). Please refer to the *OPNET IPv6 Model User Guide* for instructions on configuring IPv6 routing protocols manually.

IPv6 parameters are specified slightly differently in the end and core node models. End node models contain the following IPv6 parameters:

- **IP...IP Host Parameters...Interface Information...IPv6 Parameters** attribute allows specifying such information as IPv6 addresses, router solicitation, and neighbor cache parameters.

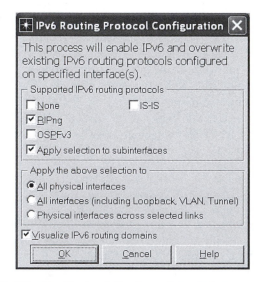

FIGURE 10.18   IPv6 Routing Protocol Configuration window.

- **IP...IP Host Parameters...IPv6 Default Route** attribute defines the default gateway for this end node. The specified value must represent a valid IPv6 address of a directly connected IPv6-enabled router.
- **IP...IP Host Parameters...V6 Static Routing Table** attribute allows defining an IPv6 static routing table at the end node. This attribute is primarily used in situations when the end node operates as a router (e.g., in MANET environments and/or when passive RIP configuration is enabled). The routing table entry is defined via such attributes as **Destination Address**, **Prefix Length**, **Next Hop**, and **Administrative Weight**.

The core node models support a wider array of IPv6 features, all of which are accumulated under the **IP...IPv6 Parameters** attribute. Specifically, core node models allow specifying the following IPv6 configuration parameters:

- **IP...IPv6 Parameters...Static Routing Table** attribute specifies the IPv6 static routing table deployed at this router. Each routing table entry of the core node models contains additional attributes for defining an IPv6 static route, including such information as whether the entry is permanent or not, the name of the VPN routing/forwarding (VRF) instance table, multicast reverse path forwarding (RPF) weight, and others.
- **IP...IPv6 Parameters...Interface Information** attribute allows specifying and configuring IPv6-enabled interfaces of the router. Configuration of an IPv6-enabled interface requires you to specify such information as the name of the interface, IPv6 status of the interface, link-local and global IPv6 addresses, supported IPv6 routing protocols, route advertisement and neighbor cache parameters, available subinterfaces, packet filters for the packets arriving to and departing from this interface, and others.
- **IP...IPv6 Parameters...Loopback Interfaces** attribute specifies the IPv6 configuration for the loopback interfaces defined on this node. Notice that you need to define loopback interfaces via the **IP...IP Routing Parameters...Loopback Interfaces** attribute prior to specifying their IPv6 configuration.
- **IP...IPv6 Parameters...Tunnel Interfaces** attribute specifies the IPv6 configuration for the tunnel interfaces defined on this node. You need to define the tunnel interfaces via the **IP...IP Routing Parameters...Tunnel Interfaces** attribute prior to specifying their IPv6 configuration.
- **ACL Configuration**, **ACL Parameters**, and **Prefix Filter Configuration** attributes allow specifying various IPv6 access control and prefix filter lists.
- **Maximum Static Routes** attribute specifies the maximum number of entries allowed in the IPv6 static routing table.
- **General Prefixes** attribute specifies various short IPv6 address prefixes used to define longer more specific IPv6 prefixes.
- **Local Policy** attribute specifies a route map policy for this interface. However, this attribute is currently not supported and has no effect on the discrete event simulation.

In addition to the IPv6 node attributes and **Protocols > IPv6** options, OPNET also contains global IPv6 attributes that influence the IPv6 configuration in the whole scenario. These attributes are accessible via the **Inputs...Global Attributes** category in the **Configure/Run DES** window:

- **IP...IPv6 Configuration** attribute allows enabling or disabling IPv6 configuration in the whole scenario. This attribute accepts two values:
  - Consider—forces the simulation to model the configured IPv6 behavior (i.e., IPv6 is enabled in the simulation).
  - Ignore—forces the simulation to ignore the IPv6 configuration (i.e., IPv6 is disabled in the simulation).
- **IP...IPv6 Interface Address Export** attribute allows exporting into a file IPv6 addresses specified on all interfaces in the simulated network. This attribute accepts only two values: Enabled and Disabled, which enables or disables IPv6 address exporting, respectively. The exported IPv6 addresses are saved into a comma-separated plain text file that can be opened with any text editor or spreadsheet application. The exported file is named <project>-<scenario>-ipv6_addresses.gdf, where <project> and <scenario> are the names of this project and scenario. The exported file is saved in your default directory.
- **Simulation Efficiency...IPv6 ND Simulation Efficiency** attribute specifies how the DES performs translation of IPv6 addresses into MAC addresses. This attribute accepts the following two values:
  - Enabled—forces the simulation to use a global table and may result in the simulation executing faster.
  - Disabled—forces the simulation to rely on explicit neighbor solicitation/ advertisement messages to discover IPv6 to MAC address mapping.

10.3.5 IPv6 Statistics and Other Performance Evaluation Options

OPNET provides a set of statistics for evaluating the performance of IPv6 networks. Specifically, the simulation can be configured to record the following IPv6 statistics:

- **Global Statistics...IPv6...Traffic Dropped (packets/second)** statistic records the total number of IPv6 packets dropped per second on all IPv6-enabled interfaces in all the nodes in the simulated network.
- **Node Statistics...IPv6...Traffic Dropped (packets/second)** statistic records the total number of IPv6 packets per second discarded on all IPv6-enabled interfaces in the current node.
- **Node Statistics...IPv6...Traffic Received (packets/second)** statistic records the total number of IPv6 packets per second received from the network on all IPv6-enabled interfaces in the current node.
- **Node Statistics...IPv6...Traffic Sent (packets/second)** statistic records the total number of IPv6 packets per second sent into the network on all IPv6-enabled interfaces in the current node. Note that this statistic records all forwarded packets including those that may later get discarded by the QoS mechanisms deployed on the node's interfaces.

OPNET also provides Global Statistics for evaluating the performance of Mobile IPv6 networks and Node Statistics that record the amount of IPv6 Multicast traffic sent and received by a node. OPNET contains an IPv6 visualization option, which provides a snapshot view of the IPv6 routing protocols deployed within the network. You can enable this feature by selecting the **View > Visualize Protocol Configuration > IP Routing Protocols > IPv6 Routing Protocols** option, which places icons that represent various IPv6 routing protocols over the links where these protocols are deployed.

Finally, you can also generate user-defined IPv6 reports by performing the following actions:

1. Select the **Scenarios > User-Defined Reports > Generate Report from Template...** option from the pull-down menu in the **Project Editor**, which will open the **Generate User-Defined Report** window.
2. In the **Generate User-Defined Report** window, as shown in Figure 10.19, choose the IPv6 reports that you would like to generate.
3. Once the desired reports are selected, click the **Generate** button.
4. You can access the generated reports via the **User Tables** tab in the **Results Browser** window (accessed from the pull-down menu **DES > Results > View Results**) after executing the simulation.

10.4   QUALITY OF SERVICE

Originally, the Internet was designed primarily for the delivery of user data between network nodes. However, as the number and the type of Internet applications grew, requirements as to how the data has to be delivered through the network began to appear. The Internet became responsible not only for data delivery, but also for performance criteria or levels of service experienced by the applications. Typically, these service levels are defined in terms of achieved bandwidth and/or experienced delay, jitter, and/or loss. Furthermore, since applications usually have diverse requirements, the Internet also needs to differentiate between traffic flows and their requirements

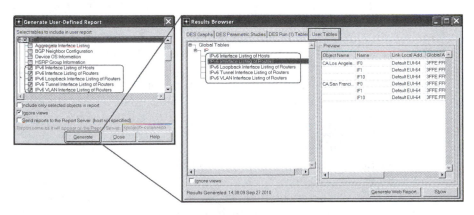

FIGURE 10.19   IPv6 User-Defined Reports.

so that each flow or group of flows receives its desired level of service. In general, Quality of Service (QoS) can be defined as the capability of the network to guarantee certain levels of service to various traffic flows. OPNET supports the following mechanisms for providing QoS guarantees:

- **Traffic Classifier**—classifies traffic flows based on certain characteristics.
- **Traffic Marker**—sets certain packet header fields (e.g., ToS byte) to values that can later be used to identify the packet's traffic class.
- **Traffic Policer**—limits the transmission rate of the traffic arriving at or leaving from an interface. Traffic policing rules may cause nonconforming packets to be discarded, marked as nonconforming, or simply left as is. Traffic policing is typically implemented at the network edges to prevent congestion by limiting the amount of traffic that enters the network.
- **Congestion Avoidance mechanism**—probabilistically discards arriving packets when a core node's buffer occupancy reaches a certain threshold. Discarding arriving packets forces the corresponding TCP sources to reduce their congestion windows, which effectively slows down their transmission rates and reduces the probability of congestion. OPNET supports the Random Early Detection (RED) and Weighted RED (WRED) mechanisms.
- **Traffic Scheduler**—determines how the packets buffered in the logical queues are scheduled for departure. OPNET supports the following scheduling mechanisms: Priority Queuing (PQ), Low Latency Queues (LLQ), Custom Queuing (CQ), Weighted Round Robin (WRR), Weighted Fair Queuing (WFQ), and others.

Currently, OPNET (i.e., IT Guru and Modeler Version 16.0) does not support such QoS mechanisms as flow-based WFQ, flow-based WRED, traffic shaping (i.e., delaying packets of the traffic until they conform to the rate-limit specification), rate-limit access control lists, table maps, and dropping profiles.

The process of deploying QoS in a simulated network consists of two main steps: (1) define and configure QoS mechanisms or profiles and (2) deploy defined QoS profiles on desired interfaces in the network.

You can deploy QoS profiles on any physical interface, subinterface, aggregated interface, or VLAN interface. However, you are not allowed to specify QoS profiles on loopback or tunnel interfaces. There are two types of QoS profiles: **local** and **global**. You can define local QoS profiles by changing attribute values at a node. A local QoS profile can only be accessed by the node where this profile has been defined, that is, the nodes in the network only have access to their own local QoS profiles. On the other hand, global QoS profiles are scenario-wide, that is, all nodes within the scenario have access to global QoS profiles.

In this section, we discuss the steps for configuring and deploying QoS profiles. Specifically, the rest of this section is organized as follows. First, we briefly introduce various QoS mechanisms and discuss how each of those mechanisms can be defined as a global QoS profile. Next, we will discuss the steps for configuring local QoS profiles, followed by the instructions for deploying the defined QoS profiles in the network. We conclude this section with an overview of the available statistics for evaluating QoS performance. It appears that OPNET closely follows Cisco standards for modeling the supported QoS features. You might find it useful to consult Cisco documentation regarding the meaning of the various QoS features and certain QoS configuration attributes.

10.4.1 SPECIFYING GLOBAL QoS PROFILES

To specify global QoS profiles, you need to add a utility object called *QoS Attribute Config* to your project workspace. The *QoS Attribute Config* object can be found in the *internet_toolbox*, *QoS*, and *utilities* object palettes of the **Object Palette Tree**. All global QoS profiles are defined via the *QoS Attribute Config* object and you need only one such object per scenario. As shown in Figure 10.20, the *QoS Attribute Config* object consists of the following compound attributes:

- **CAR Profiles**—defines global Committed Access Rate (CAR) profiles for traffic policing.
- **Custom Queuing Profiles**, **FIFO Profiles**, **MWRR/MDRR/DWRR Profiles**, **Priority Queuing Profiles**, **WFQ Profiles**—specifies global QoS profiles for various packet scheduling mechanisms.
- **RSVP Flow Specification** and **RSVP Profiles**—specifies global RSVP configuration. We do not discuss Integrated Services and the RSVP protocol in this book.

In Sections 10.4.1.1 through 10.4.1.7, we discuss each of the compound attributes that define the global policer and scheduler profiles.

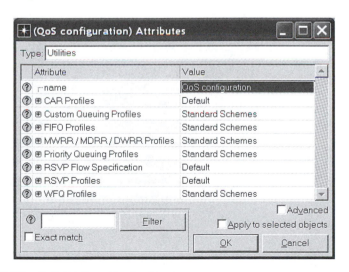

FIGURE 10.20 *QoS Attribute Config* object.

10.4.1.1   Committed Access Rate Profiles

You can specify global CAR profiles for traffic policing using the **CAR Profiles** attribute shown in Figure 10.21. By default, OPNET provides two preconfigured CAR profiles: one that limits the transmission rate of all the traffic and another for throttling HTTP traffic only. In addition, the **CAR Profiles** attribute contains an empty CAR profile, ready to be configured. If desired, you can specify additional CAR profiles by changing the value of the attribute **Number of Rows**. Each CAR profile definition consists of the following attributes:

- **Profile Name**—provides a descriptive name of this profile. The value of this attribute identifies this QoS profile and it is used when the profile is being deployed in the network.
- **Details**—contains an actual configuration of the current CAR profile.

Specifically, under the attribute **Details** you can specify multiple Class of Service (COS) profiles (i.e., the number of COS profiles is controlled via the attribute

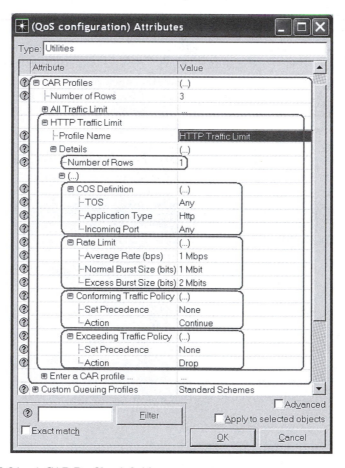

FIGURE 10.21   A **CAR Profiles** definition.

Number of Rows). Each COS profile consists of a traffic classifier and a traffic policer configured via the following attributes:

- **COS Definition**—configures the traffic classifier by specifying rules for determining if arriving packets belong to the current COS. COS traffic can be configured to classify arriving packets based on the ToS byte (also referred to as the DSCP value or precedence value), application type, and/ or incoming port number.
- **Rate Limit**—specifies the rate limits for the current COS. The rate limits are defined via such criteria as average rate and normal and excess burst size. Traffic that arrives at a rate below the average rate or below the normal burst size is considered conforming. Traffic bursts that arrive at a rate higher than the normal burst size but smaller than the excess burst size are considered nonconforming with a probability that increases as the traffic burst size increases. All traffic that arrives at a rate larger than the excess burst size is considered nonconforming with a probability of 1.
- **Conforming Traffic Policy**—configures the rules for dealing with conforming traffic. Specifically, the policy consists of two steps:
  - **Marking**: you need to specify how, if at all, conforming packets will have their precedence value (i.e., ToS byte value) changed.
  - **Forwarding**: you need to specify if conforming packets will be transmitted right away (i.e., value `Transmit`) or forwarded to the next COS definition within the current CAR profile (i.e., value `Continue`).
- **Exceeding Traffic Policy**—configures the rules for dealing with nonconforming traffic. Once again, the exceeding traffic policy rule consists of two steps:
  - **Marking**: specify how, if at all, nonconforming packets will have their precedence value changed.
  - **Forwarding**: specify if nonconforming packets will be discarded (i.e., value `Drop`), transmitted right away (i.e., value `Transmit`), or forwarded to the next COS definition within the current CAR profile (i.e., value `Continue`).

10.4.1.2   Custom Queuing Scheduler

You can configure the CQ scheduler via the attribute **Custom Queuing Profiles** shown in Figure 10.22. The CQ scheduler processes logical queues in a round-robin fashion. During a queue's turn, the CQ scheduler processes the queue's share of traffic. In OPNET, the traffic share of a logical queue is defined via **byte count**, which specifies the number of bytes to be transmitted from the queue upon its turn.

By default, OPNET provides three preset CQ profile configurations and an empty CQ profile ready to be configured. Each CQ profile definition consists of the following:

- **Profile Name** is a basic attribute that specifies a descriptive profile name used to identify this profile within the scenario.
- **Details** is a compound attribute, which contains the actual definition of this CQ profile.

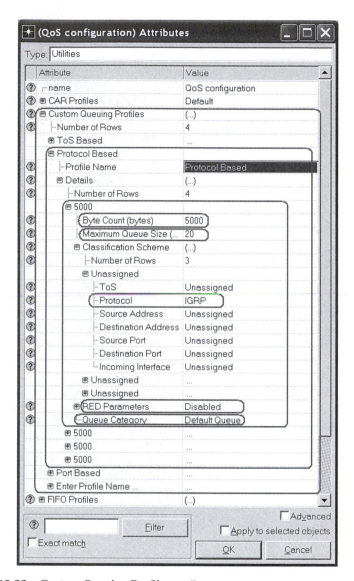

FIGURE 10.22   Custom Queuing Profiles attribute.

Through the attribute **Details**, you can configure each individual logical queue within the CQ scheduler. The total number of logical queues in the CQ scheduler is controlled via the **Number of Rows** attribute, and each logical queue is configured via the following attributes:

- **Byte Count (bytes)**—specifies the amount of traffic to be transmitted from this queue upon its turn. When servicing a queue, the packets are transmitted until the number of bytes sent becomes larger than the value of the attribute **Byte Count (bytes)**, or until the queue becomes empty.

- **Maximum Queue Size (pkts)**—specifies the total number of packets allowed in this queue during congestion (i.e., during congestion, if the number of packets in this logical queue exceeds the value of this attribute, then all packets arriving into this queue are discarded).
- **Classification Scheme**—specifies classification rules for identifying the packets to be placed into this logical queue. The packets can be identified based on their ToS value, the type of protocol, source IP address, destination IP address, source port number, destination port number, and/or incoming interface. This attribute allows you to specify multiple classification rules. The total number of rules is controlled via the attribute **Number of Rows** and each newly created row represents a separate classification rule.
- **RED Parameters**—allows configuring the RED or WRED congestion avoidance mechanism deployed on this queue. We discuss the configuration of the RED and WRED mechanisms in Section 10.4.1.3.
- **Queue Category**—specifies the queue type. This attribute accepts the following values:
  - `Default Queue`—all the packets that do not match any of the classification rules are placed into the default queue. For each physical queue, there can be only one default queue (i.e., you cannot configure multiple queues within the scheduler to operate as default queues).
  - `Low Latency Queue`—packets in the LLQ are served ahead of all other traffic, and only when the LLQ is empty the other queues will be processed according to the CQ scheduling mechanism. LLQs provide preferential treatment to delay-sensitive traffic such as voice or video. For each physical queue, there can be only one LLQ (i.e., you cannot configure multiple queues within the scheduler to operate as LLQs).
  - `Default and Low Latency Queue`—the queue is both default and low latency.
  - `None`—the queue is neither default nor low latency.

With the exception of **Byte Count (bytes)**, all attributes for configuring logical queues in the CQ scheduler are also used for configuring logical queues in PQ (i.e., **Priority Queuing Profiles**), WRR (i.e., **MWRR/MDRR/DWRR Profiles**), and WFQ (i.e., **WFQ Profiles**) global profiles. That is why we do not provide a description of these attributes again.

10.4.1.3 RED and WRED Configuration

As mentioned in Section 10.4.1.2, logical queue configuration attributes for any global scheduling mechanism profile include the attribute **RED Parameters**, which allows configuring the RED/WRED congestion avoidance mechanisms on a corresponding logical queue. The idea of the RED mechanism is fairly simple: when the queue occupancy reaches a certain level, notify the sources about possible congestion by marking or discarding arriving packets. Specifically, RED maintains an estimate of the average queue occupancy, which we call *avg*. When the average queue occupancy reaches the minimum threshold, min_{th}, RED starts marking packets with a certain probability called the packet-marking probability. When the average queue occupancy exceeds the maximum threshold, max_{th}, then RED marks all arriving packets. Marked packets typically have their ECN field (i.e., two right-most bits of the ToS byte in the IP header) set to the Congestion Encountered (CE) value, which notifies network edges about congestion in the network. Alternative RED implementations discard packets instead of marking them, which also effectively serves as a notification to TCP sources about congestion in the network. RED estimates the average queue occupancy using the exponentially weighted moving average (EWMA) algorithm. The packet-marking probability, p_b, is computed as shown in Equation 10.1, where p_{max} is the RED configuration parameter that specifies the maximum value of the packet-marking probability when the queue occupancy reaches max_{th}.

$$p_b = p_{max} \frac{avg - min_{th}}{max_{th} - min_{th}} \qquad (10.1)$$

WRED is a variation of RED, which uses a different set of configuration parameter values (i.e., min_{th}, max_{th}, and p_{max}) for packets with different ToS values. In OPNET, the WRED configuration for a logical queue of global scheduler profiles only allows varying the value of p_{max}, whereas the min_{th} and max_{th} values remain the same for all the packets. As shown in Figure 10.23, the RED/WRED mechanism is configured via the following attributes of **RED Parameters**:

- **RED Status**—specifies which mechanism is enabled: RED, WRED, or neither.
- **Exponential Weight Factor**—a value used for estimating the average queue occupancy.
- **Maximum Threshold**—maximum queue occupancy threshold.

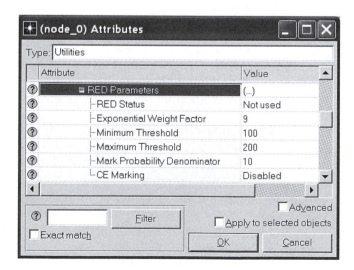

FIGURE 10.23   Attribute **RED Parameters**.

- **Minimum Threshold**—minimum queue occupancy threshold.
- **Mark Probability Denominator**—denominator for a maximum packet-marking probability. The actual maximum marking probability, p_{max}, is computed as 1/ (value of the **Mark Probability Denominator** attribute). When configuring WRED, you can input multiple values of **Mark Probability Denominator** separated by a comma.
- **CE Marking**—when this attribute is set to `Enabled`, the packets that were determined to be marked have their ECN field in the IP header set to the CE value. When this attribute is set to `Disabled`, such packets are discarded.

Notice that OPNET allows configuring local WRED profiles, which support the full WRED configuration (i.e., you can specify multiple sets of min_{th}, max_{th}, and p_{max} parameters).

10.4.1.4 FIFO Profiles

OPNET allows specifying First-In First-Out (FIFO) scheduling profiles. The FIFO profile definition is structured the same way as all the other schedulers and consists of two attributes: **Profile Name** and **Details**. Since FIFO queues are not divided into logical subqueues, the attribute **Details** configures the FIFO scheduler via attributes **Maximum Queue Size (pkts)** and **RED Parameters**. These attributes have the same meaning as the corresponding attributes of the CQ scheduler described in Section 10.4.1.2.

10.4.1.5 MWRR/MDRR/DWRR Profiles

OPNET also supports three variations of the WRR scheduler: Modified Weighted Round Robin (MWRR), Modified Deficit Round Robin (MDRR), and Deficit Weighted Round Robin (DWRR). All these mechanisms rely on a deficit counter, which specifies the amount of data in bytes that can be serviced during each cycle of round-robin processing. MWRR and MDRR service the packets from a nonempty queue as long as the queue's deficit counter is greater than zero. Notice that during subsequent rounds, the queue may service fewer packets to compensate for the excess traffic processed during previous rounds. DWRR, on the other hand, services packets as long as the deficit counter value is greater than the size of the packet to be processed, which means that during subsequent rounds the queue may service more data to compensate for the service deficit in previous rounds. After each turn of round-robin processing, the deficit counter is incremented by a value proportional to the queue's weight. Both MWRR and MDRR maintain the deficit counter in units of bytes, whereas DWRR maintains the deficit counter in units of 53 bytes, which is the size of an ATM cell.

To configure a global MWRR/MDRR/DWRR profile, as shown in Figure 10.24, you need to perform the following steps (notice that the structure of the attributes **MWRR/MDRR/DWRR Profiles** and **Custom Queuing Profile** is the same):

1. Specify the number of profiles you would like to create. By default, the attribute **MWRR/MDRR/DWRR Profiles** contains four predefined profiles and one empty profile ready to be configured.

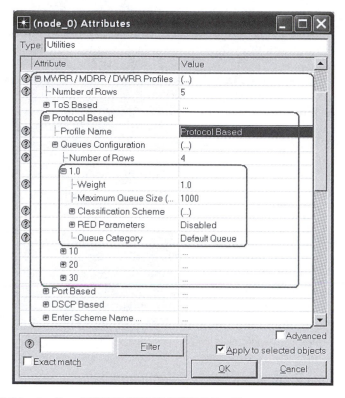

FIGURE 10.24   Attribute **MWRR/MDRR/DWRR Profiles**.

2. For each profile, specify the name (i.e., attribute **Profile Name**) and the details of the queue configuration (i.e., attribute **Queues Configuration**).
3. The queue configuration requires specifying the number of logical queues in the WRR profiles (i.e., attribute **Number of Rows**) and configuring each logical queue separately.
4. All attributes for configuring the logical queue for the MWRR/MDRR/ DWRR profiles are the same as in the CQ profiles. The only exception is the attribute **Weight**, which specifies the amount of bandwidth allocated for servicing from the current logical queue upon its turn. In the case of MWRR, **Weight** represents the percentage of link bandwidth, whereas for MDRR/DWRR, the value of the attribute **Weight** is an integral multiple of the MTU on the corresponding interface.

Notice that there is no attribute for specifying the type of round-robin mechanism for a given profile definition (i.e., MWRR, MDRR, or DWRR). In fact, global profiles for the MWRR, MDRR, and DWRR schedulers are configured through the same set of attributes. However, when deploying QoS profiles in the network, OPNET requires you to first specify the type of QoS profile (i.e., Custom Queuing, MWRR, FIFO, DWRR, etc.) and only then its name. We discuss these configuration steps in Section 10.4.3.

10.4.1.6   Priority Queuing Profiles

You can configure the PQ scheduling mechanism by modifying the **Priority Queuing Profiles** attribute. The PQ mechanism services queues in order of their priority: the lower priority queue is serviced only when all the higher priority queues are empty. OPNET structures the **Priority Queuing Profiles** attribute the same way as attributes for configuring the global CQ and MWRR/MDRR/DWRR profiles. Configuration of individual queues in the CQ and WRR profiles relies on the attributes **Byte Count (bytes)** and **Weight**, respectively, which determine the share of resources dedicated to each logical queue. Similarly, the PQ mechanism contains the attribute **Priority Label**, which specifies the priority of a logical queue. This attribute is the only configuration difference between the PQ, the CQ, and the WRR schedulers.

10.4.1.7   WFQ Profiles

WFQ provides fair bandwidth distribution among the logical queues based on the weight values associated with each logical queue. Once again, the OPNET structure of the global WFQ profiles is similar to that of the global CQ, WRR, and PQ profiles. As before, each WFQ profile configuration consists of attributes for specifying the profile name and configuration of the logical queues. In addition, each WFQ profile contains the attribute **Buffer Capacity**, which specifies the buffer size in packets on the interface where the corresponding WFQ profile will be deployed.

The attributes for configuring logical queues in the global WFQ scheduler profile are identical to those of the global MWRR/MDRR/DWRR profiles shown in Figure 10.24. The only differences are the meanings of the attributes **Maximum Queue Size (pkts)** and **Weight**. In the WFQ scheduler, the attribute **Maximum Queue Size (pkts)** determines the maximum number of packets that can be accumulated in the logical queue when the number of packets in the physical queue reaches the value of the attribute **Buffer Capacity**. Specifically, as long as the total physical buffer occupancy does not exceed the **Buffer Capacity** value, an individual logical queue can accumulate as many packets as needed. As soon as the buffer occupancy reaches **Buffer Capacity**, all incoming packets are discarded until either the logical queue occupancy decreases below **Maximum Queue Size (pkts)** or the buffer occupancy drops below the **Buffer Capacity** value. Notice that OPNET does not discard OSPF and IGRP packets even when **Buffer Capacity** is exceeded. In the global WFQ profiles, the attribute **Weight** specifies the share of the allocated bandwidth for the corresponding queue, that is, larger weight values correspond to a larger share of the allocated bandwidth.

10.4.2   Specifying Local QoS Profiles

You can configure local QoS profiles and deploy the defined QoS profiles (i.e., both local and global) via the **IP...IP QoS Parameters** attribute shown in Figure 10.25. Both the end and the core node models contain the same set of attributes for configuring QoS profiles. However, some of the features are not supported yet even

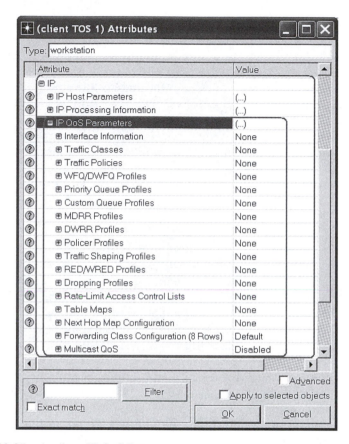

FIGURE 10.25   Attribute **IP QoS Parameters**.

though the node models may contain the attribute for configuring these unsupported features. Specifically, the QoS features defined via the following attributes have not been implemented yet: **Traffic Shaping Profiles**, **Dropping Profiles**, **Rate-Limit Access Control Lists**, and **Table Maps**. These attributes are ignored during the simulation. In this section, we provide a brief overview of the steps for configuring various local QoS profiles.

10.4.2.1   Traffic Classes and Traffic Policies

You can define traffic classes and specify QoS policies applicable to the defined classes via the **Traffic Classes** and **Traffic Policies** attributes shown in Figures 10.26 and 10.27, respectively. To define new traffic classes, you need to perform the following two steps:

1. Specify the number of traffic classes by setting the value of **Number of Rows** located under the **Traffic Classes** attribute.
2. Configure each newly created traffic class.

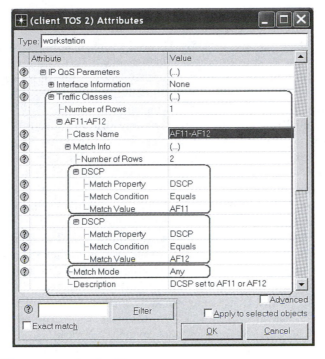

FIGURE 10.26   Attribute **Traffic Classes**.

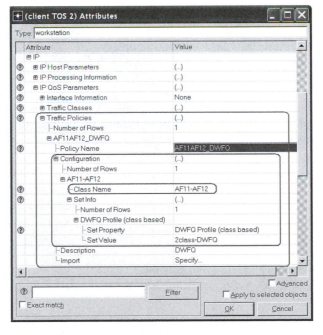

FIGURE 10.27   Attribute **Traffic Policies**.

In the second step, a traffic class is configured by setting the following attributes:

- **Class Name**—name of the class. This is the name that will be used to identify the current class of traffic within this node.
- **Match Info**—a compound attribute that specifies the classification rules for the current traffic class. You can define multiple rules if needed. For each rule, you need to set the following attributes:
    - **Match Property**—the traffic property that will identify the current class (e.g., DSCP, Incoming Interface, Packet Size, etc.).
    - **Match Condition**—a condition (e.g., Equals, Greater Than, Less Than, Not Equals) used to compare **Match Property** against **Match Value**.
    - **Match Value**—the value of match property used to identify this traffic class.
- **Match Mode**—specifies whether the incoming packet must satisfy all classification rules (value All), or at least one of them (value Any), in order to be categorized as being part of this traffic class. You can think of setting this attribute value to All as creating a single classification condition in which all individual rules are joined by logical AND, while setting this attribute to the value Any as joining individual classification rules by logical OR. You must be careful to avoid creating classification rules that result in empty traffic classes. For example, consider a situation where **Match Mode** is set to All and the **Match Info** attribute configures the following two rules: (rule #1) packets with DSCP value equal to AF11 and (rule #2) packets with DSCP value equal to AF12. Since a packet's DSCP value cannot be equal to AF11 and AF12 simultaneously, there will be no packets matched into this traffic class.
- **Description**—provides a brief description of this traffic class and does not influence the simulation execution.

Traffic policies specify treatment for various traffic classes in the current node. Traffic treatment may include various actions to be performed on the local QoS profiles to be applied to the packets of the traffic class. Note that you need to define the necessary traffic classes and the corresponding local QoS profiles prior to specifying the new traffic policies. Furthermore, you can specify the traffic policies using only certain local QoS profiles. Specifically, traffic policies cannot be defined using any global QoS profiles and local PQ and CQ QoS profiles. These profiles are deployed directly on the node's interface, which we will describe in Section 10.4.3.

To create new traffic policies, you need to perform the following actions:

- Expand the attribute **Traffic Policies** and set the value of the attribute **Number of Rows**, which corresponds to the total number of traffic policies to be defined.
- For each newly created policy row, specify the following attribute values:
    - **Policy Name**—name used to identify this traffic policy within the current node.

- **Configuration**—the actual policy configuration.
- **Description**—textual description of this traffic policy.

You can specify a single traffic policy (i.e., set values of the attribute **Configuration**) for multiple traffic classes. The number of traffic classes covered by the policy is controlled by the attribute **Configuration...Number of Rows**. For each newly created row that represents a separate traffic class policy, you need to configure the following attributes:

- **Class Name**—specifies the name of the traffic class defined via the attribute **IP QoS Parameters...Traffic Classes**.
- **Set Info**—specifies the policies to be applied to this traffic class.
  - **Number of Rows**—controls the number of policies defined for this traffic class.
  - For each row you need to specify the following:
    - **Set Property**—specifies the property to be applied to the packets of this class. Currently, OPNET supports only the following set properties: `WFQ Profile`, `DWFQ Profile`, `DWRR Profile`, `MWRR Profile`, `Policer Profile`, `RED/WRED Profile`, `DSCP`, and `Precedence`.
    - **Set Value**—specifies the value or name of the set property to be applied to this traffic class.

10.4.2.2   WFQ/DWFQ Profiles

OPNET also supports local WFQ profiles. Specifically, you can define the Distributed WFQ (DWFQ) and class-based WFQ (CBWFQ) scheduling mechanism profiles. However, flow-based WFQ profiles are currently not supported in OPNET. Even though the QoS configuration attributes (i.e., **IP QoS Parameters**) in the node models include the compound subattribute **Flow-based WFQ/DWFQ Profiles**, its configuration values are ignored during the simulation.

DWFQ or VIP-DWFQ is a variation of the WFQ mechanism that runs over the Versatile Interface Processor (VIP). VIP-DWFQ is supported by certain Cisco routers and is not discussed in this book. CBWFQ is a variation of WFQ that operates on traffic classes. Each local CBWFQ profile represents the configuration of a single FIFO queue of a traffic class within the CBWFQ scheduler. The actual CBWFQ scheduler is defined via traffic policies (i.e., configured through the attribute **Traffic Policies**), which specify mappings between the traffic classes (i.e., configured through the attribute **Traffic Classes**) and the local CBWFQ profiles. To create the CBWFQ profiles, you need to perform the following steps:

1. Expand the attribute **IP QoS Parameters...Class-based WFQ Profiles**.
2. Set the value of the attribute **Number of Rows**, which represents the total number of CBWFQ profiles.
3. Configure each newly created row, which corresponds to a single FIFO queue of a traffic class within the CBWFQ scheduler, by specifying the following attribute values:

a. **Name**—the name of the CBWFQ profile.
b. **Bandwidth Type**—specifies how the minimum bandwidth guarantee for a traffic class during congestion is represented: `Relative`—percentage of the link's capacity, `Absolute`—the actual value in units of bits per second, or `Remaining Percentage`—the percentage of bandwidth left after all other queues have received their share of the bandwidth.
c. **Bandwidth Value**—specifies the actual amount of minimum bandwidth guaranteed to a traffic class. Depending on the **Bandwidth Type** value, this attribute accepts either the absolute amount of bandwidth in units of bits per second or a percentage of link capacity.
d. **Priority**—this attribute enables or disables priority queuing (i.e., LLQ) on this queue. If priority queuing is enabled, then, to prevent starvation of all other queues within the CBWFQ scheduler, the value of the attribute **Bandwidth Value** serves as a rate limit for this queue.
e. **Queue Limit (packets)**—specifies the maximum queue capacity in packets.
f. **Burst Size (bytes)** and **Default Class Dynamic Queues** attributes are currently not supported.

10.4.2.3   Priority Queue Profiles and Custom Queue Profiles

To specify a local PQ profile, you need to expand the **Priority Queue Profiles** attribute, set the value of the **Number of Rows** attribute, which represents the total number of profiles, and then configure each individual PQ profile by setting the following attributes:

- **List Number**—specifies the name of the current PQ profile.
- **Configuration**—specifies packet classification rules, which determine the queue into which a packet is placed, and the configuration detail of individual queues. You can specify up to four queues within a single PQ profile by setting the value of the **Number of Rows** attribute. Each logical queue consists of the following configuration attributes:
  - **Queue Priority**—specifies the priority of the current queue. This attribute accepts the values `Low`, `Normal`, `Medium`, and `High` only.
  - **Match Info**—specifies the rules for identifying the packets to be stored in this queue.
  - **Queue Limit (packets)**—specifies the maximum capacity of this queue in units of packets.
- **Default Queue**—specifies which of the queues within the profile (i.e., `Low`, `Normal`, `Medium`, or `High`) will serve as a default queue. The default queue accepts all packets that do not match any of the classification rules within the PQ profile.

The definitions of the local PQ and CQ profiles have a very similar structure. To specify a local CQ profile, you also need to expand the **Custom Queue Profiles** attribute, set the value of the **Number of Rows** attribute, which represents the total number of profiles, and then configure each individual CQ profile by specifying the following attribute values:

- **List Number**—specifies the name of the CQ profile.
- **Configuration**—specifies packet classification rules, which determine the queue into which a packet is placed and the configuration detail of individual queues. You can specify up to 16 queues within the CQ profile by setting the value of the **Number of Rows** attribute. Each queue consists of the following configuration attributes:
  - **Queue Number**—specifies the number of the current queue. This attribute accepts any integer value between 1 and 16.
  - **Match Info**—specifies the rules for identifying the packets to be stored in this queue.
  - **Number of Bytes Allowed (bytes)**—specifies the amount of traffic in bytes served from this queue during each round.
  - **Queue Limit (packets)**—specifies the maximum capacity of this queue in units of packets.
- **Default Queue**—identifies the default queue. This attribute accepts an integer number, which corresponds to the value of the attribute **Queue Number** of a queue within this CQ profile that will serve as the default queue.

10.4.2.4   MDRR Profiles and DWRR Profiles

The steps for configuring local MDRR and DWRR profiles are almost identical to those for CBWFQ profiles. Unlike local PQ and CQ profiles, which contain configuration details for multiple logical queues under a single profile, a single CBWFQ/MDRR/DWRR profiles specifies configuration details of a single logical queue within the corresponding scheduler. The actual scheduler is defined via traffic policies, which provide mapping between CBWFQ/MDRR/DWRR profiles and traffic classes.

To specify a local MDRR or DWRR profile, you need to expand the **MDRR Profiles** or **DWRR Profiles** attribute, respectively, and then set the value of the **Number of Rows** attribute, which represents the total number of logical queues to be created. Then, for each newly created row, you need to configure the attributes **Name**, **Bandwidth Type**, **Bandwidth Value**, **Priority**, and **Queue Limit**. In DWRR profiles, the attribute **Queue Limit** is called **Queue Limit Value** and it is located under the attribute **Queue Configuration**. The meaning of these attributes is identical to that of the corresponding attributes for configuring CBWFQ profiles. Other attributes available under DWRR Profiles are currently not supported by OPNET.

10.4.2.5   Policer Profiles

The local policer profiles can be deployed directly on node interfaces or can be used for defining traffic policies. To specify a local policer profile, you need to execute the following steps:

1. Expand the **Policer Profiles** attribute.
2. Set the value of the **Number of Rows** attribute, which corresponds to the total number of policer profiles.

3. Configure the individual profiles by setting the following attributes:
  - **Name**—specifies the name of the policer profile.
  - **Policer Details**—specifies policing rules within the current policer profile.

Notice that each policer profile may contain multiple policing rules (i.e., a single policing rule corresponds to a row under the **Policer Details** attribute). The total number of policer rules within the profile is controlled by the attribute **Number of Rows**. To configure each rule within the profile, you need to specify the following attributes:

- **Match Property**—specifies the property used to identify policed traffic (e.g., the name of the incoming interface, DHCP value, etc.). Please note that OPNET currently does not support traffic matching using the QoS Group and Rate-Limit Access Control List properties.
- **Match Value**—the value used to identify the policed traffic. The actual value of this attribute depends on the value of the **Match Property** attribute.
- **Bandwidth Type**—controls bandwidth specification. This attribute accepts only two values: Absolute and Relative, although currently, OPNET only supports the absolute bandwidth type.
- **Average Rate**—specifies the long-term average traffic rate. Traffic that arrives at a rate below this value is considered conforming.
- **Peak Rate Information**—this attribute is currently not supported.
- **Conform Burst Size (bits)**—specifies the maximum size of the traffic burst below which the traffic is considered conforming (i.e., normal burst size).
- **Excess Burst Size (bits)**—specifies the maximum size of the traffic burst above which the traffic is considered nonconforming (i.e., exceeding rate limit). A traffic burst between the conform and excess burst size is considered nonconforming with a certain probability, which increases as the size of the traffic burst increases.
- **Action Configuration**—is a compound attribute, which allows specifying actions to be performed on a packet that falls in the rate-limit categories Conform, Exceed (i.e., nonconforming), or Violate (i.e., exceed the sum of conform and excess burst size). As with global policers, you can perform two actions on the packet: (1) change packet property (OPNET currently only supports changing the DSCP or Precedence value) and (2) transmit the packet (i.e., value Transmit), discard it (i.e., value Drop), or send it to the next policer in the list (i.e., value Continue). If the attribute **Action** is set to Continue and there are no more policers defined for this traffic type, then the packet is transmitted.

10.4.2.6 RED/WRED Profiles

You can specify local RED and WRED profiles with the attribute **RED/WRED Profiles**. As usual, the total number of profiles is controlled via the attribute **Number of Rows**, where each row represents a new RED or WRED profile. To configure a single RED or WRED profile, you need to set the following attributes:

- **Name**—the name of the profile.
- **Match Property**—specifies the packet property (e.g., DSCP, Precedence, COS, etc.) that will be used to differentiate between various packet types in the WRED mechanism.
- **Threshold**—specifies configuration parameters for a RED or WRED mechanism. The RED mechanism relies on a single set of configuration attributes (i.e., min_{th}, max_{th}, and p_{max}) or consists of a single RED queue, whereas WRED applies a different set of configuration parameters to each traffic class (i.e., consists of multiple RED queues). The attribute **Number of Rows** controls the number of RED queues defined within this profile. If there is more than one RED queue, then the current profile operates as WRED, otherwise it works as RED. Each set of RED parameters is configured via the following attributes:
    - **Match Value**—identifies packets, which will be placed into the current RED queue. All packets within this queue are subject to the current RED configuration. The value of this attribute is compared against the value of the packet's property specified via the **Match Property** attribute to determine if the packet will be placed into this RED queue or not.
    - **Minimum Threshold (packets)**—minimum queue occupancy threshold.
    - **Maximum Threshold (packets)**—maximum queue occupancy threshold.
    - **Mark Probability Denominator**—denominator for a maximum packet-marking probability.
- **Exponential Weight Constant**—a constant for estimating the average queue occupancy.
- **Flow Based**—this feature is currently not supported by OPNET.

We do not describe the rest of the QoS profiles because they are either not supported by OPNET or configure features that are beyond the scope of this book.

10.4.3 Deploying Defined QoS Profiles on an Interface

The final step of deploying QoS in a simulated network is to configure the node interfaces to support the defined QoS profiles. Certain local profiles, such as MDRR, DWRR, and CBWFQ, cannot be directly deployed on the node interfaces. These profiles can be deployed on the interfaces via traffic policies only. To deploy QoS profiles on the node interfaces, you need to expand the node attribute **IP...IP QoS Parameters...Interface Information** and then set the attribute **Number of Rows** to a value that corresponds to the total number of interfaces in the node that will support the defined QoS profiles. Next, you need to configure each individual interface by setting the following attributes shown in Figure 10.28:

- **Name**—the name of the interface to be configured to support QoS. You can only set this attribute to the name of an active interface available at that node. The value field of this attribute contains a pull-down menu, which lists all the active interfaces in the node.

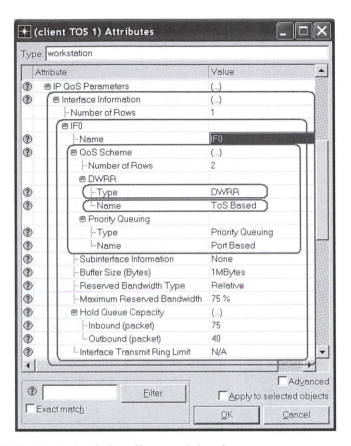

FIGURE 10.28   Deploying QoS profiles on node interfaces.

- **QoS Scheme**—configures the QoS profiles to be deployed on this interface. You can deploy multiple QoS profiles on a single interface. The number of QoS profiles deployed on the interface is controlled via the attribute **Number of Rows**. Each deployed QoS profile is configured via the following two attributes:
    - **Type**—specifies the type of QoS profile (e.g., Inbound Traffic Policy, Outbound Traffic Policy, Custom Queuing, DWFQ (Class Based), DWRR, FIFO, MDRR, MWRR, Priority Queuing, etc.)
    - **Name**—specifies the name of the deployed QoS profile. You can only set this attribute to QoS profiles of the type specified in the **Type** attribute. The value field of the attribute contains a pull-down menu, which provides a list of all the defined QoS profiles of the type specified via the **Type** attribute.
- **Subinterface Information**—configures QoS profiles on subinterfaces of this interface.
- **Buffer Size (bytes)**—specifies the size of the buffer on an interface. All traffic is discarded when the interface buffer is full. This attribute defines the

total amount of buffer space available to and shared by all the logical queues of the QoS profiles deployed on this interface. You must be careful to configure the individual logical queues so that their total maximum size in bytes does not exceed the value of this attribute. Since the maximum queue size of the logical queues is defined in units of packets, to determine the maximum queue occupancy in bytes, you need to sum up the maximum queue sizes and then multiply the resulting value by the MTU size on this interface.

- **Reserved Bandwidth**—specifies the reserved bandwidth type. This attribute accepts the following two values:
  - `Relative`—the value specified in the attribute **Maximum Reserved Bandwidth** represents a percentage of link bandwidth.
  - `Absolute`—the value specified in the attribute **Maximum Reserved Bandwidth** represents the actual value of the bandwidth in units of bits per second.
- **Maximum Reserved Bandwidth**—specifies the processing rate of the QoS scheduling mechanisms deployed on this interface (i.e., specifies how fast the scheduler can service outgoing traffic). This value provides the upper limit on the guaranteed bandwidth for outgoing traffic. When this attribute is set to the value `N/A`, then the entire link bandwidth is dedicated to the scheduler.
- **Hold Queue Capacity**—is a compound attribute that specifies the capacity of the inbound (currently not supported) and outbound queues. The outbound queue capacity represents the maximum capacity in packets of the interface buffer and is configured via the attribute **Hold Queue Capacity... Outbound (packets)**.
- **Interface Transmit Ring Limit**—is currently not supported.

10.4.4 CLOSING REMARKS

As you have probably realized, configuring and deploying QoS profiles in an OPNET simulation can be quite a challenge. In this section, we would like to reiterate some of the key points described earlier:

- Global QoS profiles are accessible by all nodes in the scenario, whereas local QoS profiles can be accessed only by the nodes that define them.
- OPNET supports LLQ configuration in global and local QoS profiles. Only the CQ, WFQ, and WRR scheduling mechanisms allow designating one of their logical queues as LLQ and there can be only one LLQ per profile.
- Global QoS profiles designate a logical queue as an LLQ by setting the attribute **Queue Category** to the `Low Latency Queue` value.
- Local QoS profiles designate a logical queue as an LLQ by setting the attribute **Priority** to `Enabled`.

You can set the ToS field value in generated packets by changing the attribute values of custom and standard applications, as well as traffic demands. Definitions of all custom and standard applications and traffic demands contain the attribute **Type of**

Service, which specifies the value of the ToS byte (i.e., precedence or DSCP value). The ToS value is typically used for traffic classification and it allows you to explicitly organize traffic flows into different classes.

10.4.5 QoS-Related Statistics

OPNET provides several statistics directly related to the performance evaluation of QoS mechanisms. However, in general, examining other statistics such as end-to-end delay, jitter, and traffic rates provides a fairly good indication about the performance of various QoS mechanisms. All QoS-specific statistics supported by OPNET are NODE STATISTICS located under the **IP Interface** category. These statistics are recorded for each interface separately. Furthermore, if an interface deploys a QoS mechanism that divides the physical buffer into several logical queues (e.g., WFQ, WRR, CQ, or PQ), then these statistics are recorded for each logical queue separately. Table 10.2 provides a summary of all QoS-related statistics.

TABLE 10.2
QoS-Related Node Statistics in the *IP Interface* Category

Name	Description
Buffer Usage (bytes), *Buffer Usage (packets)*	These statistics record the size of the queue on each interface in units of bytes and packets.
CAR Incoming Traffic Dropped (bits/sec), *CAR Incoming Traffic Dropped (packets/sec),* *CAR Outgoing Traffic Dropped (bits/sec),* *CAR Outgoing Traffic Dropped (packets/sec)*	These statistics record the total amount of traffic dropped by the CAR traffic policer on a single incoming or outgoing interface. Typically, the traffic policer discards nonconforming packets and packets that exceed the rate limit. These statistics are recorded in units of bits/second and packets/second.
Queue Delay Variation (sec)	This statistic records delay variation in seconds experienced by the packets in each queue of an interface.
Queuing Delay (sec)	This statistic records delay in seconds experienced by the packets in each queue of an interface.
RED Average Queue Size	This statistic records the average queue size for each queue as estimated by the RED mechanism deployed on an interface.
Traffic Dropped (bits/sec), *Traffic Dropped (packets/sec),* *Traffic Received (bits/sec),* *Traffic Received (packets/sec),* *Traffic Sent (bits/sec),* *Traffic Sent (packets/sec)*	These statistics record the total amount of traffic dropped/received/sent by each queue of an interface. These statistics are recorded in units of bits/second and packets/second.

11 Network Layer
Routing

11.1 INTRODUCTION

Routing is the process of computing possible paths between the nodes in a ne work.
Routing is a function of the network layer, yet many routing protocols in the Internet
depend on transport layer protocols for the communication of their control packets. Typically, routing is performed by core nodes in the network, called routers,
which store path information in their routing tables. To keep the routing process
scalable given the immense size of the Internet, the Internet is partitioned into
smaller groups of routers and networks. Each such partition is managed by a single
authority, which defines administrative rules and routing policies, and is called
an **Autonomous System (AS)**. A wide variety of routing protocols is deployed in
the Internet today as each AS can independently decide which routing protocol to
employ within its confines. Protocols used within an AS are known as intra-AS
routing protocols. On the other hand, inter-AS routing protocols are responsible for
maintaining routing information for data delivery between ASs. OPNET supports
many common intra-AS protocols including RIP, RIPng, OSPF, OSPFv3, EIGRP,
IGRP, and IS-IS. OPNET also implements BGP, an inter-AS routing protocol. We
only describe two intra-AS protocols, RIP and OSPF, in this chapter. The other
intra-AS protocols and BGP are beyond the scope of this book and are not discussed here.

Each intra-AS routing protocol constructs routes to be used within its domain and
stores them in the form of routing tables in each router in the domain. Some routers
called boundary routers may belong to multiple ASs, and these routers may then
run multiple routing protocols. Boundary routers typically contain multiple routing
tables, one for each domain they belong to. All these routing tables are merged into
a single *IP Forwarding Table*, which is maintained by IP within the network layer.
When a packet arrives at a router for transmission to a specific destination, IP looks
up the *IP Forwarding Table* for that destination to decide the outgoing interface to
which the packet will be assigned for transmission.

The rest of this chapter is organized as follows. We dedicate the remainder of
this section to describing how to deploy a specific routing protocol in a network. In
Sections 11.2 and 11.3, we provide a detailed account of how OPNET models two
of the most commonly used intra-AS routing protocols, RIP and OSPF. The first of
these, **Routing Information Protocol (RIP)**, is a distance-vector protocol that runs
over UDP. It only constructs one path to each destination and does not do any load
balancing. On the other hand, **Open Shortest Path First (OSPF)** is a link-state protocol that runs directly over IP and can do load balancing because it can construct
multiple paths to each destination. We conclude the chapter with an overview of

OPNET statistics available for evaluating the performance of routing protocols and steps for exporting routing information computed by the routers within the modeled network during a simulation.

11.1.1 Deploying Routing Protocols in a Simulated Network

Typically, the first step in configuring a network simulation to run desired routing protocols is to deploy those routing protocols within the network. Either RIP or OSPF can be configured to run on the entire network or on a portion of it. We define a *routing domain* to be an interconnected set of router interfaces that run the same routing protocol. You can deploy one or more routing protocols on various router interfaces in a network. The attribute **IP...IP Routing Parameters... Interface Information...<interface name>...Routing Protocol(s)** in the core node models controls which routing protocols (i.e., could be more than one) are deployed on the interface corresponding to **<interface name>**. Clicking on the value field of the **Routing Protocol(s)** attribute opens the **Select Dynamic Routing Protocol(s)** window shown in Figure 11.1, which allows you to specify which routing protocols should be enabled and which should be disabled on the corresponding interface.

However, when deploying routing protocols on multiple interfaces in a network, configuring one interface at a time is a tedious and time-consuming task. Instead, it is much easier to deploy the desired routing protocols in the network using the **Protocols** menu option. The steps to deploy one or more routing protocols on a subset of the interfaces in a network are as follows:

1. Select the links attached to the interfaces of interest. Multiple links can be selected by first clicking on one link and then shift clicking on all subsequent ones (i.e., hold the shift key and click on all subsequent links). You can omit this step if you wish to deploy the routing protocol on all router interfaces in the network. When a routing protocol is deployed on a selected link, then the interfaces attached on both sides of the link will be configured to run the corresponding routing protocol.

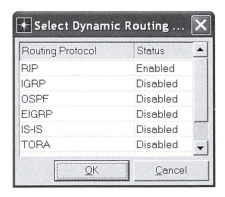

FIGURE 11.1   Select Dynamic Routing Protocol(s) window.

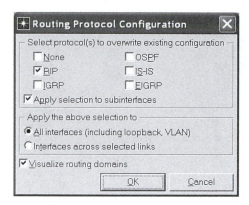

FIGURE 11.2 **Routing Protocol Configuration** window.

2. Select the **Protocols > IP > Routing > Configure Routing Protocols** option, which opens the **Routing Protocol Configuration** window shown in Figure 11.2.

3. Configure routing protocol deployment in the network by specifying the routing protocols to be deployed and the links they will be deployed on. Notice that the **Routing Protocol Configuration** window consists of three panels: (1) routing protocol selection, (2) interface selection, and (3) visualize routing domains.

   a. The routing protocol selection panel defines which routing protocols will be deployed in the network. You can select multiple protocols to be deployed on the interfaces by clicking on the corresponding checkboxes in the routing protocol selection panel. As a result, selecting OSPF will not de-select the default selection of RIP. You must explicitly uncheck the box for RIP if you want to deploy only OSPF on the selected interfaces.

   b. In the interface selection panel, you can specify whether the selected routing protocols will be deployed on selected interfaces only (i.e., interfaces attached to the selected links) by checking the radio button *Interfaces across selected links* or on all active interfaces in the network by checking the radio button *All interfaces (including loopback, VLAN)*.

   c. Finally, placing a checkmark in the *Visualize routing domains* checkbox will create a routing protocol deployment view that displays an icon with a letter identifying the routing protocol on all the links in the simulated network, as illustrated in Figure 11.3.

4. Click the **OK** button to deploy the selected routing protocols in the network, which results in the corresponding interfaces (i.e., interfaces chosen via the interface selection panel) configured to run the selected routing protocols.

Let us consider the example shown in Figure 11.3. Router 7 runs the OSPF routing protocol on the interfaces attached to the links that connect to nodes Router 5

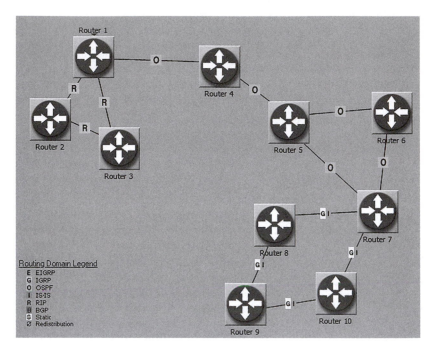

FIGURE 11.3   Example of routing protocol deployment.

and Router 6, and it runs the IGRP and IS-IS protocols on the interfaces connecting to Router 8 and Router 10.

You can examine the routing protocol deployment on all interfaces in a router by performing the following steps:

1. Right-click on the router node and select the **Edit Attributes** option.
2. In the **Attributes** window, expand the compound attribute **IP...IP Routing Parameters**.
3. Click on the value field of the **Interface Information** attribute to open a window, which displays the configuration attributes for all node interfaces in the form of a table.

Figure 11.4 shows the **Interface Information** table for the node Router 7. The figure shows a total of 12 interfaces, of which interfaces IF4 and IF5 are configured to run two routing protocols, IGRP and IS-IS; interfaces IF10 and IF11 run the OSPF protocol, whereas all the remaining interfaces are configured to run RIP, which is the default routing protocol.

You will observe later that OPNET contains several different subattributes named **Interface Information**. To avoid confusion, we will refer to each of them using different qualifiers. In the remainder of this chapter, we will refer to **IP...IP Routing Parameters...Interface Information** as the physical interface-specific attribute. There are other subattributes also called **Interface Information** for each of the RIP and OSPF routing protocols, and we will call them RIP/OSPF interface-specific attributes.

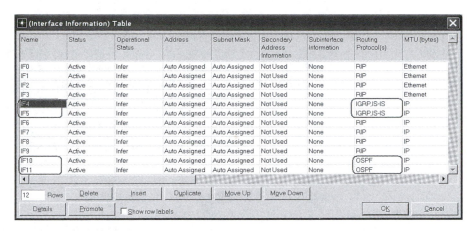

Name	Status	Operational Status	Address	Subnet Mask	Secondary Address Information	Subinterface Information	Routing Protocol(s)	MTU (bytes)
IF0	Active	Infer	Auto Assigned	Auto Assigned	Not Used	None	RIP	Ethernet
IF1	Active	Infer	Auto Assigned	Auto Assigned	Not Used	None	RIP	Ethernet
IF2	Active	Infer	Auto Assigned	Auto Assigned	Not Used	None	RIP	Ethernet
IF3	Active	Infer	Auto Assigned	Auto Assigned	Not Used	None	RIP	Ethernet
IF4	Active	Infer	Auto Assigned	Auto Assigned	Not Used	None	IGRP,IS-IS	IP
IF5	Active	Infer	Auto Assigned	Auto Assigned	Not Used	None	IGRP,IS-IS	IP
IF6	Active	Infer	Auto Assigned	Auto Assigned	Not Used	None	RIP	IP
IF7	Active	Infer	Auto Assigned	Auto Assigned	Not Used	None	RIP	IP
IF8	Active	Infer	Auto Assigned	Auto Assigned	Not Used	None	RIP	IP
IF9	Active	Infer	Auto Assigned	Auto Assigned	Not Used	None	RIP	IP
IF10	Active	Infer	Auto Assigned	Auto Assigned	Not Used	None	OSPF	IP
IF11	Active	Infer	Auto Assigned	Auto Assigned	Not Used	None	OSPF	IP

12 Rows | Delete | Insert | Duplicate | Move Up | Move Down

Details | Promote | ☐ Show row labels | OK | Cancel

FIGURE 11.4   Interface Information table.

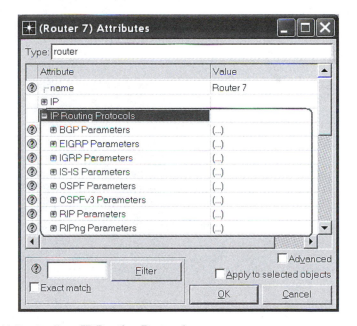

FIGURE 11.5   Attribute **IP Routing Protocols**.

11.1.2 CONFIGURING ROUTING PROTOCOL ATTRIBUTES

Often, when performing simulation studies of routing protocols, you may need to configure the routing protocols deployed in the network. Parameters for configuring individual routing protocols are located under the node attribute **IP Routing Protocols**. As shown in Figure 11.5, you can change configuration settings for such routing protocols as BGP, EIGRP, IGRP, IS-IS, OSPF, OSPFv3, RIP, and RIPng. To configure a particular routing protocol, you need to expand the corresponding compound attribute (e.g., **BGP Parameters, EIGRP Parameters**, etc.) and specify the desired values for the routing protocol configuration attributes. Notice that there is no mechanism for specifying routing protocol parameters simulation-wide. Therefore, you may need to configure routing protocols one node at a time. Recall that OPNET allows you to specify attribute values on multiple nodes simultaneously, which becomes handy when specifying the same routing protocol configuration on multiple nodes. Specifically, you need to perform the following steps when specifying a routing protocol configuration on multiple nodes:

1. Select all routers on which you wish to specify the routing protocol configuration.
2. Open the **Edit Attributes** window on any of the selected routers.
3. Expand the attribute **IP Routing Protocols** and then specify the configuration parameters for the desired routing protocol.
4. Check the *Apply to selected objects* checkbox, which will copy the introduced changes on all the selected routers and click the **OK** button to save your configuration.

11.2   ROUTING WITH RIP

11.2.1   INTRODUCTION TO RIP

RIP is a distance-vector protocol that uses the distributed Bellman–Ford algorithm to compute shortest paths. Each router uses RIP to exchange routing information with its neighboring routers. Each route has a metric or cost associated with it, which is the hop count to the destination. The hop count is limited to 15, and a hop count of 16 is used to represent infinity. OPNET implements both RIP version 1 (RFC 1058) and version 2 (RFC 1723), which is the version described here; OPNET also implements RIPng, which is the next generation of RIP version 2 for IPv6.

In RIP, each router sends routing update messages to its neighbors at regular intervals and upon changes in the network topology. A routing update message contains the metric (i.e., typically a hop count), which represents the cost to reach each destination network from this router. When a router receives a routing update, it increments each hop count in the update message by 1 and compares the result with its locally stored values (i.e., routing table entries) for each destination. If the new metric value for a destination (received value plus 1) is smaller than the locally stored value, then the router updates the routing table entry for that destination with the new value. If a destination's entry is updated, then the next hop address for that destination is set to the address of the node from which the routing update message arrived. A router considers a destination to be unreachable if the destination's metric value is equal to 16. In addition to the basic route update mechanism described above, RIP provides features such as stability during periods of rapidly changing network topology, and mechanisms for holddown, split horizon, and poisoned reverse to prevent incorrect routing information from being propagated.

All parameters related to the configuration of RIP are located under the compound attribute **IP Routing Protocols...RIP Parameters**. As shown in Figure 11.6, the attribute **RIP Parameters** consists of the following subattributes:

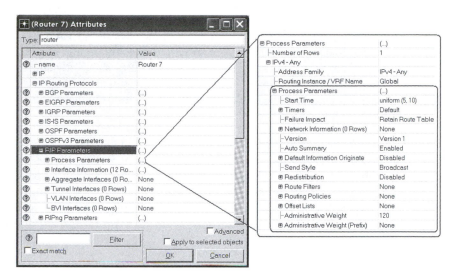

FIGURE 11.6   Attribute **RIP Parameters**.

- **Process Parameters**—contains local, router-wide RIP configuration parameters that are common to all interfaces within this router. We describe these attributes in Section 11.2.2.
- **Interface Information**—contains interface-specific RIP configuration parameters. This is a compound attribute in which the total number of rows equals the total number of physical interfaces in the router. Each row contains subattributes for configuring the interface corresponding to that row. We describe these attributes in Section 11.2.3. Also, as mentioned previously, we will call this subattribute RIP interface-specific attribute to distinguish it from the physical interface-specific attribute of the same name under **IP...IP Routing Parameters**.
- **Tunnel Interfaces**, **VLAN Interfaces**, **BVI Interfaces**—contains RIP configuration parameters for each of the tunnel interfaces, VLAN interfaces, and BVIs defined on the router. These topics are beyond the scope of this book and are not discussed here.

Typically, the process of configuring RIP is fairly straightforward and consists of the following steps:

1. Assign IP addresses to the router interfaces in the network (i.e., **Protocols > IP > Addressing > Auto-Assign IP Addresses**). You can omit this step but having IP addresses explicitly assigned on the router interface is helpful when configuring individual interfaces, debugging the simulation, and examining the results. For example, after you have assigned IP addresses in the network, you can easily identify the attached interfaces in a router. Only the attached interfaces will have their IP address and mask values explicitly specified in the **Interface Information** table, whereas all unattached interfaces will have these attributes set to `Auto Assign`.

2. Specify RIP as a routing protocol on the desired interfaces (by using pull-down option **Protocols > IP > Routing > Configure Routing Protocols**).
3. Specify local router-wide and interface-specific RIP configuration parameters. You may also want to configure such RIP features as advertisement mode, triggered extension mode, RIP start time, RIP timers, RIP simulation efficiency mode, and others. We describe these features in the next few subsections of this section.

For more information about the RIP features supported in OPNET, please refer to the product documentation available via the **Protocols > RIP > Model User Guide** menu option. In addition, OPNET software includes a preconfigured project called RIP, which provides an excellent illustration of various RIP features and also contains a brief description of the configuration attributes in the README scenario.

11.2.2 Local RIP Configuration Attributes

As stated earlier, local RIP parameters common to all interfaces within a router are found under the attribute **IP Routing Protocols...RIP Parameters...Process Parameters**. Under this attribute, the subattribute **Number of Rows** controls the total number of different RIP configuration definitions in this router. By default, one such configuration definition exists in every router, and the compound attribute **IPv4 – Any** contains such a definition for IPv4 unicast and multicast traffic.

In each configuration definition, the attribute **Address Family** controls the type of traffic the RIP definition will be applied to. In addition, the value of this attribute also determines the actual name of the RIP configuration definition attribute. For example, in the default configuration definition **IPv4 – Any**, if you change the value of the attribute **Address Family** to IPv4 – Unicast, then the name of the parent attribute will also change to **IPv4 – Unicast**.

Within each configuration definition, such as **IPv4 – Any**, the compound attribute **Process Parameters** contains parameters for specifying the current node's configuration of the RIP process. The attribute **Process Parameters**, as shown in Figure 11.6, consists of the following subattributes:

- **Start Time**—specifies the simulation time when the RIP process will start execution on this node. This attribute value is defined through a random probability distribution function. By default, this attribute value is set to uniform (5, 10), which means that the start time for RIP will be selected randomly from the interval 5 to 10 seconds using the uniform probability distribution function. You should not start RIP too early in a simulation because certain critical network functions must be completed before successful transport of RIP packets can occur. You can also specify the RIP start time for all the nodes in the network using a different technique, which is discussed in Section 11.2.4.
- **Timers**—a compound attribute that configures various RIP-specific timers that control the protocol's behavior. Specifically, you can configure the following RIP timers:

- **Update Interval (seconds)**—determines the time interval between two successive routing updates exchanged between neighboring routers. The default value of this attribute is 30 seconds.
- **Route Invalid (seconds)**—specifies the length of the time period after which the route is considered to be no longer valid. A route invalid timer is initialized when a new route is inserted into the routing table or whenever the route is updated. Upon expiry of this timer, the route is not immediately deleted, but it is marked as invalid by setting the route cost to infinity. The default value of this attribute is 180 seconds.
- **Flush (seconds)**—specifies the length of the time period after which the route is removed from the routing table. This timer works in a similar way to the route invalid timer, except the route is actually deleted when this timer expires. The value of the route flush timer should be greater than that of the route invalid timer. The default value of this attribute is 240 seconds.
- **Holddown (seconds)**—specifies the value of the holddown timer, which is started upon expiration of the route invalid timer. During the holddown period, the router discards all updates regarding the affected routes to minimize the effects of route flapping on the routing table. This attribute is currently not supported in OPNET and has no effect on the simulation performance.

- **Failure Impact**—specifies what information the router will remember when there is a failure and subsequent node recovery. This attribute accepts the following two values:
  - Clear Route Table—configures the router to clear the RIP routing table maintained at this node in the event of a node failure. The routing table is rebuilt after the node recovers from the failure.
  - Retain Route Table—will configure the router to remember its prefailure RIP routing table when it recovers from a failure. This is the default setting for the attribute **Failure Impact**.
- **Version**—specifies the RIP version used in the router as a whole. The version may be either Version 1 or Version 2 (default is Version 1). Individual interfaces may ignore this setting and they may send or receive routing updates using the RIP version specified via the RIP interface-specific attributes **Send Version** and **Receive Version**, respectively.
- **Send Style**—specifies how the interfaces will handle backward compatibility in the case when an interface is configured to use both RIPv1 and RIPv2 for sending routing updates (via the RIP interface-specific attribute **Send Version**). This attribute accepts two values:
  - Broadcast—the router will broadcast RIPv2 messages so that RIPv1 routers can receive routing updates (RFC 2453).
  - Broadcast & Multicast—the router will broadcast RIPv1 messages and will multicast RIPv2 messages.
- **Auto Summary**—controls whether RIP performs subnet filtering defined in RFC 1058. By default, this feature is enabled.

The rest of the router-wide attributes configure various advanced RIP features, which are not discussed in this book.

11.2.3   RIP INTERFACE-SPECIFIC CONFIGURATION ATTRIBUTES

To specify RIP interface-specific configuration attributes for an interface, you need to set values of the compound attribute **IP Routing Protocols…RIP Parameters… Interface Information…<interface name>**, where **<interface name>** identifies an interface of interest. As shown in Figure 11.7, RIP interface-specific configuration includes the following attributes:

- **Name**—specifies the interface name.
- **Status**—specifies whether RIP is running on this interface. However, this attribute has no effect on simulation execution. The value of this attribute cannot be modified at this level as it is set based on how the routing protocols are configured to run on various interfaces of each node.
- **Silent Mode**—specifies how RIP operates on this interface and accepts two possible values:
  - Disabled—this interface both sends and receives RIP updates.
  - Enabled—this interface does not send out RIP updates, although it still receives and processes them.
- **Cost**—specifies the cost associated with RIP updates received on this interface. The default cost for all interfaces is 1. This cost is added to the metrics value received in an update for each destination.

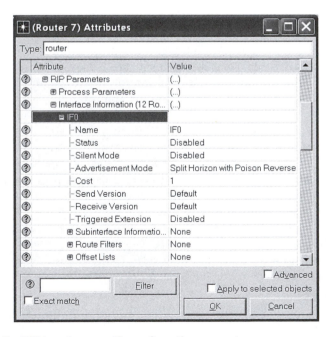

FIGURE 11.7   RIP interface-specific configuration parameters.

- **Send Version**—specifies the RIP version used to send update messages. The value of this attribute overrides the value of the router-wide attribute **Version**. This attribute accepts the following values:
  - Version1—RIPv1 updates are sent via broadcast.
  - Version2—RIPv2 updates are sent via multicast.
  - Version1&2—the router can send update messages for both RIPv1 and RIPv2 according to the setting of the router-specific attribute **Send Style**.
  - Default—uses either RIPv1 or RIPv2 update methods depending on the version of RIP specified via the router-wide attribute **Version**.
- **Receive Version**—specifies the type of RIP routing updates accepted by this interface. This attribute can be set to one of the following values:
  - Version1—only RIPv1 updates are accepted.
  - Version2—only RIPv2 updates are accepted.
  - Version1&2—both RIPv1 and RIPv2 updates are accepted.
  - Default—if the router-wide RIP configuration attribute **Version** is set to Version 1, then both RIPv1 and RIPv2 updates are accepted, otherwise (i.e., Version 2) only RIPv2 updates are accepted.
- **Advertisement Mode**—specifies how the RIP advertisement is performed on the current interface. This attribute accepts the following three values:
  - No Filtering—the router advertises all route entries on this IP interface, including those learnt through advertisements received on this interface.
  - Split Horizon—the router advertises all route entries on this IP interface, except those learnt through advertisements received on this interface. The **split horizon advertisement mode** reduces the impact of the **count to infinity problem** that plagues all distance-vector protocols such as RIP.
  - Split Horizon with Poison Reverse—the router advertises all route entries on this IP interface, while the routes learnt through advertisements received on this interface are advertised with a metric of infinity (i.e., 16). **Poison reverse** is an enhancement of the split horizon advertisement mode, which helps to speed up route convergence.
- **Triggered Extension**—This attribute specifies if the **triggered extension mode** is enabled or disabled on this interface. This attribute accepts only two values: Enabled and Disabled. If triggered extension mode is enabled, then no regular periodic updates are sent on that interface. The routing updates are sent only upon interface initialization and when the routing table in this router experiences a change. Specifically, after a routing table is changed, a routing update is sent on the interface after a time interval computed using the uniform probability distribution function with the outcome in the range of 1 to 5 seconds. If triggered extension mode is disabled for an interface, then routing updates are sent periodically based on the value of the update interval timer (i.e., the local router-wide RIP attribute **Update Interval (seconds)**). Notice that the **Triggered Extension** attribute must be enabled on all interfaces within the same subnet in order for the triggered extension mode to operate properly.
- **Subinterface Information**, **Route Filters**, and **Offset Lists**—Describe various advanced RIP features, which are not discussed in this book.

11.2.4 CONFIGURING RIP START TIME

There are two ways to specify the start time for RIP: (1) by setting the value of the **Start Time** attribute on individual routers (Section 11.2.2) or (2) by using the **Protocols** menu. The second method is a more convenient way of configuring the RIP start time as compared to directly changing the value of node attributes because it can be applied to a set of routers simultaneously. You need to perform the following steps to specify the RIP start time via the **Protocols** menu:

1. Select RIP-enabled routers in the network on which you would like to specify the RIP start time. This step can be omitted if you wish to configure the RIP start time on all the routers in the network.
2. Choose the **Protocols > RIP > Configure Start Time...** option, which opens the **RIP Start Time Configuration** window shown in Figure 11.8.
3. From the pull-down list of distributions, select the name of the probability distribution function that will be used to compute the RIP start time.
4. In the *Mean outcome* textbox, specify an input parameter for the probability distribution function. As shown in Figure 11.8, the constant probability distribution function takes a single parameter. Other distribution functions, such as uniform, may take more than one parameter in which case additional textboxes will appear.
5. Click on the corresponding radio button to specify whether you wish to apply the RIP start time configuration to all routers or only to the routers that you have selected in the first step.
6. Click the **OK** button to apply your configuration and to close the window, which will set the local router-wide attribute **Start Time** to the specified distribution function and mean outcome on the chosen routers (i.e., all routers in the network or those selected in step 1).

11.2.5 RIP SIMULATION EFFICIENCY MODE

To improve simulation efficiency, OPNET supports a set of attributes configuring which will speed up simulation execution. For example, in a network that

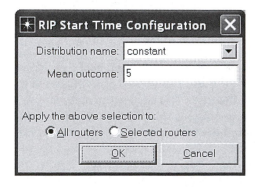

FIGURE 11.8   RIP Start Time Configuration window.

experiences no topological changes, it may be unnecessary to continue exchanging route update messages after a certain period of time, that is, when the routing tables have converged to a stable state. For this purpose, OPNET contains two global RIP-specific attributes that prevent routers from advertising routing updates after a certain time:

- **RIP Sim Efficiency**—An attribute that specifies whether RIP simulation efficiency mode is enabled or disabled. When set to `Enabled`, the routers in the network will stop sending RIP updates after the simulation passes the time value configured via the **RIP Stop Time (seconds)** attribute. In most simulations where RIP is not the object of study, the suppression of these updates improves simulation efficiency by reducing the number of events that must be processed by the simulation kernel. When set to `Disabled`, the routers will continue sending RIP routing updates throughout the simulation. This setting is useful when the network in your simulation study experiences topological changes or when RIP is the focus of your study.
- **RIP Stop Time (seconds)**—Specifies the time after which no more RIP routing updates will be generated in the network. This attribute is only applicable when **RIP Sim Efficiency** is set to `Enabled`. By default, OPNET sets **RIP Stop Time (seconds)** to 65 seconds.

As shown in Figure 11.9, you can configure these parameters via the **Inputs... Global Attributes...Simulation Efficiency** compound attribute accessible in the

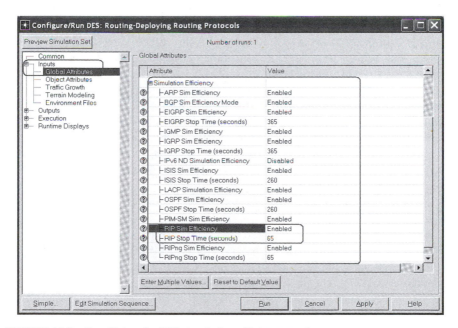

FIGURE 11.9   Specifying the RIP simulation efficiency mode.

Configure/Run DES window. You can open this window by selecting the **DES >** **Configure/Run Discrete Event Simulation…** option, pressing *Ctrl + R* keys, or by clicking on the *Running Man* icon (Section 4.3).

11.3   ROUTING WITH OSPF

11.3.1 INTRODUCTION TO OSPF

Open Shortest Path First (OSPF) is a link-state routing protocol. In OSPF, each router sends link-state advertisements (LSAs) to all other routers in its domain. The received link-state information is maintained on each router as a link-state database (LSDB), which is essentially a representation of the entire network topology. Upon receiving link-state information, each router uses Dijkstra's shortest path first (SPF) algorithm to independently calculate the shortest paths to all destinations. The transmission of LSAs is done via a modified flooding algorithm, which propagates them quickly to all routers but prunes duplicate updates. As a result, OSPF is quick to detect changes in topology such as link failures and converges very quickly to relatively loop-free routes as compared to RIP.

OSPF associates an externally specified cost metric with each link. The cost may represent link bandwidth or any other desired parameter such as distance, delay, reliability, and so on. When there are multiple paths of equal cost to a destination, OSPF includes all these paths in its routing table and in the *IP Forwarding Table*. This allows IP to do load balancing across equal-cost paths to a given destination.

An OSPF routing domain can be divided into several areas. Each area is typically a group of contiguous networks and attached hosts. Areas are identified by 32-bit integer numbers expressed either as simple decimal numbers or in octet-based dot-decimal notation similar to IPv4 addresses. By convention, area 0 (or 0.0.0.0) represents the core or backbone area that provides connectivity between all other areas. A router that belongs to multiple areas is called an Area Border Router (ABR), but each interface of the router can only belong to one area. All areas must be connected to the backbone area by an ABR or by a virtual link. LSAs are only propagated within the same area, but an ABR participates in the information exchange of each area it belongs to. An ABR also maintains separate topological databases for each of its areas. Sometimes routing information at area boundaries may be condensed by combining a group of routes into a single advertisement, reducing both the load on the router and the perceived complexity of the network. This process of grouping routes together is called route aggregation.

Since OSPF is a link-state routing protocol, OSPF routers establish neighbor relationships to exchange routing updates with other routers. The routers use Hello packets to discover and maintain these neighbor relationships. Hello messages also act as keep-alives to inform neighbors that a router is still functional. Routers that are part of a LAN such as an Ethernet select a designated router (DR) and a backup designated router (BDR). The main purpose of DR/BDR nodes is to reduce the amount of LSA traffic. Instead of exchanging link-state updates with one another, all the routers within the LAN send their LSAs to the DR that acts as a hub to reduce traffic between the routers. The DR acts on behalf of all the LAN routers to exchange LSAs with other routers within the same area. As the name implies, the BDR is a backup to the DR and it sends link-state updates only if the DR goes down.

Multiple OSPF processes can be configured on a router. Processes let you run different OSPF configurations on different interfaces of the same node. The interfaces of a router can be configured to run one of the processes configured on that router. Each OSPF process builds its own LSDB, although routes can be learned from other OSPF processes. In addition, both OSPF and RIP can be configured to build their routing tables by accepting routes learned by other routing protocols configured on this router. This technique is called route redistribution, and it is not discussed in this book.

OPNET supports OSPF version 2 (RFC 2328) for IPv4 and OSPF version 3 (RFC 5340) for IPv6. In this chapter, we describe the default version of OSPF, which is version 2, whereas OSPF version 3 is not discussed in this book.

11.3.2 OSPF Attributes

All parameters related to the configuration of OSPF are located under the compound attribute **IP Routing Protocols…OSPF Parameters**. As shown in Figure 11.10, the attribute **OSPF Parameters** consists of the following subattributes:

- **Processes**—Specifies the configuration of OSPF processes running in this router.
- **Interface Information**—Specifies the OSPF configuration on each IP interface within this router. Again, as with RIP, we will call this attribute

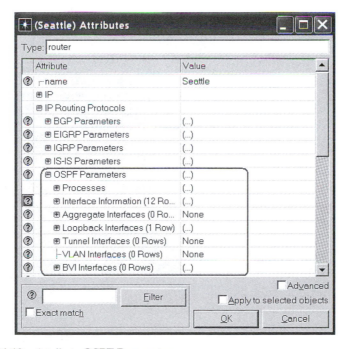

FIGURE 11.10   Attribute **OSPF Parameters**.

OSPF interface-specific to distinguish it from the physical interface-specific attribute of the same name under **IP…IP Routing Parameters**.
- **Aggregate Interfaces**, **Loopback Interfaces**, **Tunnel Interfaces**, **VLAN Interfaces**, and **BVI Interfaces**—Specify the OSPF configuration on various other types of interfaces. We do not discuss these attributes in this book.

11.3.3 Configuring OSPF Processes

To configure individual OSPF processes, you need to expand the attribute **IP Routing Protocols…OSPF Parameters…Processes** and set the attribute **Number of Rows** to the total number of OSPF processes you would like to configure. By default, OPNET sets the attribute **Number of Rows** to 1. Each OSPF process is specified through the following parameters:

- **Process Tag**—specifies a string that uniquely identifies this OSPF process.
- **VRF Name**—the name of the VPN Routing and Forwarding (VRF) table associated with this process. We do not discuss this attribute in the book.
- **Process Parameters**—configuration parameters for this OSPF process.
- **Address Family**—the type of traffic for which OSPF provides its routing services. By default, OSPF is configured to provide routing services to IPv4 unicast traffic.

We are primarily interested in the **Process Parameters** compound attribute, which specifies the OSPF process configuration. As shown in Figure 11.11, **Process Parameters** consists of the following subattributes:

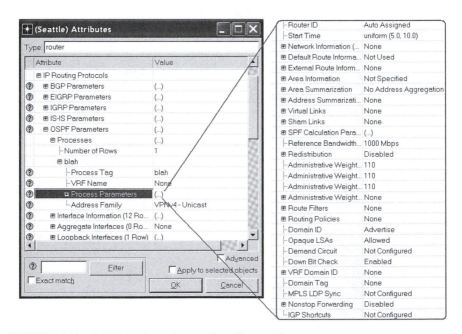

FIGURE 11.11   OSPF configuration attribute **Process Parameters**.

- **Router ID**—specifies a unique OSPF router identifier, which can be written as a positive integer value or in a dotted quad notation. If **Router ID** is set to `Auto Assigned`, then the value specified under **IP...IP Routing Parameters...Router ID** is used as the OSPF router identifier.
- **Start Time**—specifies the time when the OSPF process is started on this node. The value of this attribute is specified as a probability distribution function. OSPF should not be started too early in a simulation because certain critical network functions must be completed before successful transport of OSPF packets can occur. OPNET also allows you to specify the OSPF start time via the **Protocols** menu, as described in Section 11.3.10.
- **Network Information**—this attribute has no effect on the simulation.
- **Default Route Information**—specifies whether this router will advertise itself as a default router. Currently, OPNET supports only two possible settings:
  - `Not Used` and `Originate`—both values will configure the router not to advertise itself as a default router. The meanings of the two settings are different, but OPNET does not support the `Originate` setting, effectively resulting in the same behavior for both.
  - `Originate Always`—the router will always advertise itself as a default router.
- **External Route Information**—specifies a list of static external routes that this router may use within the OSPF routing domain. External routes provide this router with the path information to destinations that are reachable but which are not configured to run any dynamic routing protocols. Notice that routers in the network should not be able to compute specified external routes via dynamic routing protocols. Each static external route entry is specified as a separate row in the **External Route Information** table and consists of standard route information configurable via the following attributes:
  - **Destination Address** and **Destination Mask**—specify a range of destination addresses for this route entry.
  - **Cost**—represents the cost of reaching these destination addresses.
  - **Metric Type**—specifies how the value of the attribute **Cost** is interpreted. This attribute accepts two values: `Type 1`, which means the path cost is computed as a sum of the **Cost** attribute value and the cost of reaching the intermediate router and `Type 2`, which means that the **Cost** attribute value is used by itself to compute the path cost.
  - **Forwarding Address**—specifies the next hop address on the path to the destination.
- **Area Information**—specifies the areas configured for this router. While the area information can be edited in this attribute, it is generally not a good idea to do so as some of the information can be difficult to figure out. Instead, the areas should be configured using the **Protocols** menu option described in Section 11.3.8. The **Area Information** is a compound attribute, which contains a separate row for each definition of area information. Each row contains such attributes as **Area ID**, **Area Type**, **Default Route Information**, and some other advanced attributes.

- **Area Summarization**—specifies the address ranges this router advertises or hides for each connected area. We describe how to configure area summarization for border routers in Section 11.3.9.
- **Address Summarization**—creates aggregate addresses for OSPF. This compound attribute has each row representing a single address aggregate, which is defined via the following attributes:
  - **Address** and **Subnet Mask**—define the range of addresses, which will represent a single address aggregate.
  - **Advertise**—specifies whether or not the router advertises the routes that match the defined address aggregate (i.e., routes will not be advertised when this attribute is set to Disabled).
  - **Tag**—specifies a "match" value for controlling redistribution via route maps.

  This attribute models the summary-address command on Cisco routers.
- **SPF Calculation Parameters**—configures computation of the Shortest Path First (SPF) algorithm thorough three attributes. Attribute **Style** specifies how SPF computations will be performed. If set to Periodic, the router performs calculations periodically, provided there is a need for a calculation due to a change in the LSDB. If **Style** is set to LSA Driven, then the router performs the calculation when an LSA is received, but after waiting for a period of time specified by the attribute **Delay (seconds)**. In both cases, two consecutive SPF computations must be separated by at least the amount of time specified via the attribute **Hold Time (seconds)**.
- **Reference Bandwidth**—specifies the bandwidth value to be used to calculate the OSPF interface cost. Specifically, if the OSPF interface-specific attribute **Interface Information...<interface name>...Cost**, which specifies the cost of sending a packet on that interface (i.e., interface cost), is set to Auto Calculate, then the interface cost is computed as a ratio between the value of the **Reference Bandwidth** attribute and the available bandwidth on the link attached to this interface. This calculation method assigns a lower interface cost to links with higher bandwidth. Thus, if the attribute **Reference Bandwidth** is set to 100 Mbps, then an interface with a bandwidth of 10 Mbps would have an interface cost of 10, whereas one with a bandwidth of 1 Mbps would have a cost of 100.

The remaining attributes for configuring an OSPF process specify various advanced features, which are not described in this book. For more information about the available OSPF process configuration parameters, refer to OPNET documentation (i.e., available via the **Protocols > OSPF > Model User Guide** option) and a description of the individual **Process Parameters** attributes that is available via the blue circle with the question mark icon located to the left of each attribute. You may also find it useful to read *Cisco's OSPF Design Guide* (i.e., document ID: 7039) and related Cisco documentation, which often appear to have been used as a reference guide for OPNET's implementation of OSPF.

11.3.4 SPECIFYING OSPF CONFIGURATION ON ROUTER INTERFACES

The OSPF interface-specific attribute **Interface Information** (i.e., **IP...OSPF Parameters...Interface Information**) specifies the OSPF configuration on a router's interfaces. In fact, each row contains the OSPF configuration of a single physical interface on this router and the name of each row is that of the corresponding physical interface. You can specify the OSPF configuration on an interface by setting the values of the following attributes, as shown in Figure 11.12:

- **Name**—this is a read-only attribute, which specifies the name of the physical interface that is being configured.
- **Status**—specifies whether OSPF is enabled or disabled on this interface. The attribute cannot be directly modified: its value is already set based on the routing protocol deployed on this interface. If this interface is configured to run OSPF, then **Status** is set to Enabled, otherwise it is set to Disabled. Recall that the routing protocol deployed on an interface is specified via the physical interface-specific attribute **IP...IP Routing Parameters... Interface Information...<interface name>...Routing Protocol(s)**.
- **Silent Mode**—specifies whether this interface runs OSPF in passive mode. This interface will not send or receive any LSAs when this attribute is set to Enabled.
- **Type**—specifies the type of an underlying interface. This attribute accepts the following values:
  - Point To Point—for SLIP links.
  - Broadcast—for Ethernet and other types of interfaces that have broadcast capabilities.
  - Non-Broadcast—for multiaccess interfaces without broadcast capability, such as ATM or Frame Relay, which have their permanent virtual circuit (PVCs) set up in a full mesh fashion.

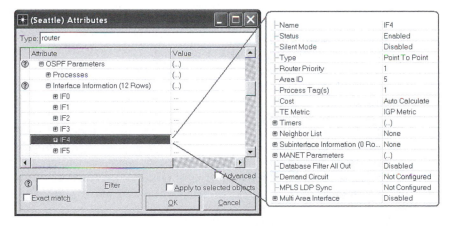

FIGURE 11.12 Attributes for configuring OSPF on the interfaces.

- **Point to Multipoint**—for multiaccess interfaces without broadcast capability that have their PVCs set up in a partial mesh fashion.
  - MANET—for wireless interfaces.
- **Area ID**—specifies the ID of the area to which this interface belongs. Area 0 is the core or backbone area. The area ID can be a simple integer such as 4 or it may be specified in four-number dotted notation, for example, Area 4 has ID 0.0.0.4.
- **Process Tag(s)**—specifies which OSPF process is configured to run on this interface. If this attribute is set to multiple values, then only the first value in the list is considered. This is a local OSPF process tag and therefore it must match those of its neighboring nodes.
- **Cost**—specifies the cost of sending a packet on this interface. You can specify the cost value as an integer number or set it to Auto Calculate. When set to Auto Calculate, the cost value is computed based on the bandwidth of the attached link and the value of the attribute **Reference Bandwidth** as described in Section 11.3.3. In this method, low bandwidth links are assigned a higher cost, and high bandwidth links are assigned a lower cost. In Section 11.3.6, we describe an alternative method for configuring the link cost for all or selected interfaces.
- **Timers**—specifies various OSPF-specific timers for this interface. These are described in Section 11.3.7.

11.3.5 CONFIGURING OSPF

The simplest way to use OSPF is to leave it in its default mode. Of course, you have to deploy OSPF on the desired router interfaces, as described in Section 11.1.1, because the default routing protocol for IPv4 is RIP. When OSPF is deployed in its default mode, there is a single area and all nodes belong to this default backbone area numbered 0. Also, there is only one OSPF process running on each node. You can use the following steps to configure OSPF for a more complex operational environment:

1. Assign IP addresses to all router interfaces (Section 9.3).
2. Deploy OSPF on desired router interfaces (Section 11.1.1).
3. Configure link costs (Section 11.3.6).
4. Configure OSPF timers (Section 11.3.7).
5. Configure OSPF areas (Section 11.3.8).
6. Configure ABRs (Section 11.3.9).
7. Configure start time (Section 11.3.10).
8. Configure simulation efficiency mode (Section 11.3.11).

Some of these configuration steps can be performed by using the **Protocols >
OSPF** options, which provide a menu of possible OSPF configuration operations. The advantage of using this technique is that most of these operations can be performed on all router interfaces or on selected interfaces in a single step.

11.3.6 CONFIGURING LINK COSTS FOR OSPF ROUTING

Link Cost (or Interface Cost) is specified for each interface and is used as the basis for the shortest path calculation. The Interface Cost represents the cost of routing an outbound packet via this interface. The inbound cost is not specified and is taken to be the outbound cost on the other side of the link. The Interface Cost can be specified either as an integer value or as computed based on the Reference Bandwidth and Link Bandwidth values. As described in Section 11.3.3, each OSPF process contains the configuration attribute **Reference Bandwidth**, which by default is set to 100 Mbps. The Link Bandwidth depends on the type of the link attached to the interface. During simulation, the Interface Cost is computed as the ratio between the value of the **Reference Bandwidth** attribute and Link Bandwidth. Thus, if the Link Bandwidth is 1 Mbps, the Interface Cost will be 100. You can specify Link Cost using one of the four methods described below:

1. Use the default setting for configuring link costs. In such a case, the simulation will automatically compute the Link Cost based on the values of the **Reference Bandwidth** attribute and the bandwidth of the link attached to this interface (the Link Bandwidth). The disadvantage of this method is that you do not explicitly control the Link Cost. Instead, the Link Cost depends directly on the allocated bandwidth on the corresponding link.
2. Use the **Protocols** menu to explicitly specify the cost of all or selected interfaces in the network. This is the easiest and preferred method of assigning link costs because it allows you to explicitly specify the cost of multiple links in a single operation. To specify Link Cost on multiple links, you need to perform the following steps:
   a. Select the links that you would like to configure with the same cost value. This step may be omitted if you would like to set Link Cost on all connected interfaces in the network to the same value.
   b. Select the **Protocols > OSPF > Configure Interface Cost** menu option, which opens the **OSPF Interface Cost Configuration** window shown in Figure 11.13.
   c. Select the checkbox *Set the interface cost explicitly to:* and enter the desired cost value.
   d. Specify whether you want the cost configuration to be specified on all or on selected links only by clicking either on the *All links* or on the *Selected links* radio button, respectively.

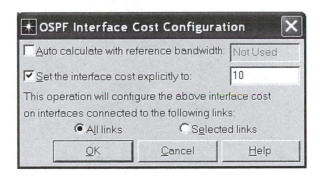

FIGURE 11.13   OSPF Interface Cost Configuration window.

e. Click **OK** to complete this operation.

f. You can repeat this process for other links, which may require a different cost value.

3. Specify the link cost by configuring the Reference Bandwidth and Link Bandwidth values. The first step is to set the Link Bandwidth for all links in the network, which can be achieved as follows:

a. Select the links that you would like to configure with the same Link Bandwidth value. This step may be omitted if you would like to set Link Bandwidth on all connected interfaces in the network to the same value.

b. Select the **Protocols > IP > Routing > Configure Interface Metric Information** option, which opens the **Configure Interface Metric Information** window shown in Figure 11.14.

c. Set the value of the textbox *Bandwidth (kbps)* to the desired Link Bandwidth. Note that the units of Bandwidth here are kbps so if you want a Link Bandwidth of 10 Mbps, you must enter 10,000.

d. Specify whether you want the Link Bandwidth to be set on all connected interfaces or on interfaces of selected links by clicking either on the *All connected interfaces* or on the *Interfaces across selected links* radio button, respectively.

e. Click **OK** to complete this operation.

f. You can repeat this process for other links, which may require a different Link Bandwidth value.

Next, you may want to set the Reference Bandwidth to a value different from its default value of 100 Mbps by performing the following steps:

a. Select the links that you would like to configure with the same Reference Bandwidth value. This step may be omitted if you would like to set Reference Bandwidth on all connected interfaces to the same value.

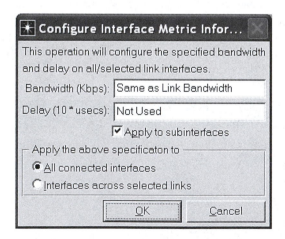

FIGURE 11.14   **Configure Interface Metric Information** window.

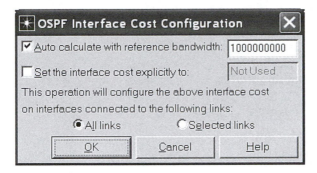

FIGURE 11.15   OSPF Interface Cost Configuration window.

b. Select the **Protocols > OSPF > Configure Interface Cost option**, which opens the **OSPF Interface Cost Configuration** window, as shown in Figure 11.15.

c. Select the checkbox *Auto calculate with reference bandwidth:* and in the adjacent textbox, input the desired value of Reference Bandwidth. Note that the box is prepopulated with the number 1,000,000,000, and the units are bits per second. This value corresponds to 1.0 Gbps, which is different from the default value.

d. Specify whether you want the Reference Bandwidth to be set on all connected interfaces or on interfaces of selected links by clicking either on the *All links* or on the *Selected links* radio button, respectively. Generally, it is a good idea to have the Reference Bandwidth set to the same value on all network links.

e. Click **OK** to complete this operation.

f. You can repeat this process for other links, which may require a different Reference Bandwidth value.

Now when the simulation runs, the link costs will be assigned based on the Reference Bandwidth and the Link Bandwidth values.

4. Manually assign Link Cost for each router interface. The Link Cost is specified via the OSPF interface-specific attribute **IP Routing Protocols... OSPF Parameters...Interface Information...<interface name>...Cost** available in all router models. You can manually edit this attribute for each router interface. Typically, you may want to perform this step only when you need to update the Link Cost value on only a few interfaces. You can also examine this attribute to verify that the cost has been properly set by one of the methods described above.

11.3.7 CONFIGURING OSPF TIMERS

OPNET supports configuration of the following four OSPF timers. You can view or configure the values of these OSPF timers via the OSPF interface-specific compound

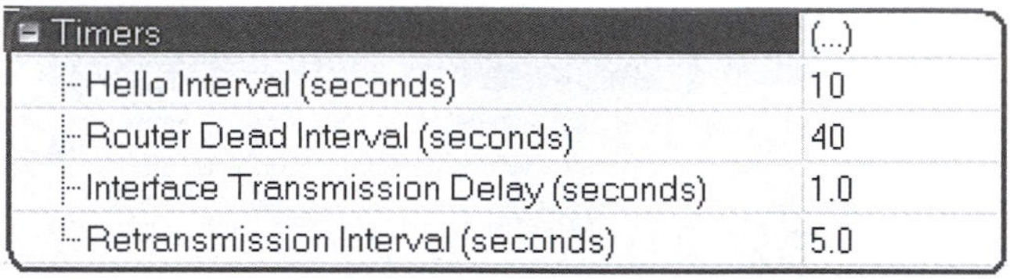

☐ Timers	(...)
├─Hello Interval (seconds)	10
├─Router Dead Interval (seconds)	40
├─Interface Transmission Delay (seconds)	1.0
└─Retransmission Interval (seconds)	5.0

FIGURE 11.16   Configuration attributes for the OSPF interface timers.

attribute **IP Routing Protocols…OSPF Parameters…Interface Information…
<interface name>…Timers**, as shown in Figure 11.16. You can configure the OSPF
timers for each individual interface in a router by specifying the following subattribute values:

- **Hello Interval (seconds)**—specifies the interval between two consecutive
  Hello messages generated from this interface. Note that this value should be
  the same for all interfaces connected to the same network. The default value
  for this attribute is 10 seconds for all interface types except for MANET
  interfaces, which by default set this attribute to 2 seconds.
- **Router Dead Interval (seconds)**—specifies the duration of time after
  which a neighboring router from which no Hello messages were received is
  considered down. For example, if a router has not seen any Hello messages
  from a neighboring router for a length of time greater than or equal to the
  value of the **Router Dead Interval (seconds)** attribute, then that neighboring router is considered inactive. Generally, this attribute should be set to
  some multiple of the Hello Interval and should be the same for all interfaces
  connected to the same network. The default value for this attribute is 40
  seconds (6 seconds for MANET interfaces).
- **Interface Transmission Delay (seconds)**—specifies an estimated amount
  of time it takes to transmit an LSA packet over this interface. To prevent
  indefinite circulation of the LSA in the flooding process, the value of this
  attribute is used to age an LSA each time it is forwarded by a router. By
  default, this attribute is set to 1 second.
- **Retransmission Interval (seconds)**—specifies the length of time between
  LSA retransmissions. By default, this attribute is set to 5 seconds.

You can also configure OSPF timers using the **Protocols** menu, as described below:

1. Select the links that you would like to configure with the same set of OSPF
   timer values. This step may be omitted if you would like to configure the
   OSPF timers the same way on all connected interfaces in the network.
2. Select the **Protocols > OSPF > Configure Interface Timers** option, which
   opens the **Configure OSPF Timers** window, as shown in Figure 11.17.
3. Specify the desired timer values in the corresponding textboxes.
4. Specify whether you want to apply the provided configuration to all connected interfaces or to interfaces of selected links by clicking either on the

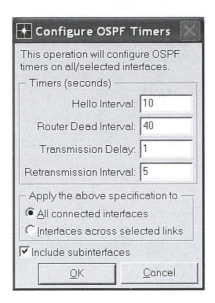

FIGURE 11.17   **Configure OSPF Timers** window invoked from the **Protocols** menu.

All connected interfaces or on the *Interfaces across selected links* radio button, respectively.
5. Click **OK** to complete this operation.
6. You can repeat this process for other links, which may require a different set of values for OSPF timers.

11.3.8 CONFIGURING OSPF AREAS

You can configure OSPF areas via the **Protocols** menu by performing the following steps:

1. Select all the links that belong to the same OSPF area.
2. Select the **Protocols > OSPF > Configure Areas** option, which opens the **OSPF Area Configuration** window, as shown in Figure 11.18.
3. In the *Area identifier* textbox, specify the desired OSPF area number. Notice that only the backbone OSPF area can be assigned area number 0, whereas non-backbone areas can be numbered 1, 2, 3, etc. (but not 0).
4. Click **OK** to complete this operation.
5. Repeat these steps for each OSPF area that you would like to configure. However, there is no need to configure the backbone area (Area 0) as all links not assigned to any other area will be assigned to Area 0 by default.

Once you have configured the OSPF areas, you can see a visual depiction of the areas by selecting the **View > Visualize Protocol Configuration > OSPF Area Configuration** menu option in the **Project Editor**. The pop-up window that appears

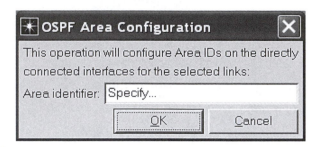

FIGURE 11.18   OSPF Area Configuration window.

lists the colors used to denote each OSPF area. You can change the color map if needed. Notice that only those interfaces that are configured to run the OSPF protocol can be assigned to an OSPF area and will be marked with the corresponding colors, when visualized.

11.3.9   Configuring OSPF Area Border Routers

For each Area Border Router (ABR) in OSPF, you can specify policies that determine how routing advertisements are sent through this ABR across the area boundaries in the network. Routing policies are specified via the **Area Summarization** attribute in the router models. For a given area and a range of addresses within it, you can specify whether the related routing advertisements will be distributed (i.e., `Advertise` setting) or hidden (i.e., `Hide` setting). If you choose the `Advertise` setting, then all routes to the given range of addresses will be sent to the rest of the network using a single advertisement. If you choose the `Hide` setting, then the address range of interest will remain hidden from the rest of the network and no advertisement will be forwarded. Finally, if you do not configure **Area Summarization**, then by default all applicable advertisements will be forwarded individually.

When an OSPF area contains multiple border routers connected to other areas, then it is common practice to configure some of such ABRs to *not* carry cross-area traffic. This is done by specifying policies for one (or more) of the ABRs to hide all subnet addresses that are internal to the given area. As a result, such a router will not advertise these addresses to the other areas to which it connects and will not receive traffic from other areas destined for these subnet addresses.

You can configure route advertisement policies for a router by setting the values of the compound attribute **IP Routing Protocols...OSPF Parameters... Processes...<Process Tag>...Process Parameters...Area Summarization**. This attribute consists of multiple rows, each of which specifies a single router advertisement policy. Each row consists of the following attributes:

- **Address** and **Subnet Mask/Prefix Length**—specify the address range for this area summarization.
- **Area ID**—specifies the ID of the area assigned to this address range. The specified area must be connected to this router.

- **Status**—specifies whether the router advertises or hides this address area from other areas. This attribute accepts two values only: `Advertise` and `Hide`.
- **Metric**—specifies the metric associated with the summary address. The summary metric is computed by default as the highest metric for all addresses being summarized. Specifying a metric value along with the summary overrides the default calculation.

11.3.10 CONFIGURE THE OSPF START TIME

In addition to configuring the OSPF start time via the attribute **Start Time**, described in Section 11.3.3, you can also use the **Protocols** menu as described below:

1. Select the OSPF-enabled routers in the network on which you would like to specify the OSPF start time. This step may be omitted if you wish to configure the OSPF start time on all the routers in the network.
2. Choose the **Protocols > OSPF > Configure Start Time...** option, which opens the **OSPF Start Time Configuration** window. This window is almost identical to that for configuring the RIP start time, as shown in Figure 11.8.
3. From the pull-down list of distributions, select the name of the probability distribution function, which will be used to compute the OSPF start time.
4. In the *Mean outcome* textbox, specify an input parameter for the probability distribution function.
5. Specify whether you wish to apply the specified OSPF start time configuration to all routers or on the routers that you have selected in the first step by clicking on the corresponding radio button.
6. Click the **OK** button to apply your configuration and to close the window.

This procedure is very similar to the one for configuring the RIP start time, described in Section 11.2.4. This is a more convenient way of setting the OSPF start time as compared to directly changing the corresponding attribute because it can be applied to all or a subset of routers.

11.3.11 OSPF SIMULATION EFFICIENCY MODE

Similarly to RIP, by default, OPNET has the attribute **OSPF Sim Efficiency** set to `Enabled`, which means that OSPF updates are only exchanged between nodes for the first 260 seconds of simulation run-time (i.e., the duration of the update interval which is configurable via the attribute **OSPF Stop Time (seconds)**). After that time, no more OSPF updates are generated. In most simulations where OSPF is not the object of study, the suppression of these updates improves simulation efficiency by reducing the number of events that must be processed by the simulation kernel. However, if you are interested in observing the behavior of OSPF over time, then you should consider disabling the OSPF simulation efficiency mode by executing the following steps:

1. Open the **Configure/Run DES** dialog box by selecting the **DES> Configure/Run DES** menu option, pressing *Ctrl + R*, or alternatively clicking on the *Running Man* icon.

2. In the **Configure/Run DES** dialog box (i.e., detailed view), expand the attribute **Inputs…Global Attributes…Simulation Efficiency**.
3. Specify the values of attributes **OSPF Sim Efficiency** and **OSPF Stop Time (seconds)**. You can now specify a value for **OSPF Stop Time (seconds)** to change the time at which OSPF updates will be stopped, or you can completely disable the Simulation Efficiency Mode by setting the value of the attribute **OSPF Sim Efficiency** to `Disabled`.
4. Click the **Apply** button to save the configuration changes without executing the simulation or click the **Run** button to execute the simulation with the new configuration.

11.4   COMMON ROUTING STATISTICS

OPNET provide both Global and Node Statistics for evaluating routing protocols. Specifically, each routing protocol has a separate statistic category that combines all Global and Node Statistics related to that protocol. Table 11.1 provides a summary of the available RIP and OSPF statistics.

TABLE 11.1
Summary of RIP- and OSPF-Related Global and Node Statistics

Category	Name	Description
Global Statistics OSPF	*Network Convergence Activity*	This statistic records a square wave that is 0 when there is no convergence activity anywhere in the network and 1 when there is convergence activity somewhere in the network. Convergence activity is defined as changes in the routing table.
	Network Convergence Duration (sec)	This statistic records the duration in seconds of convergence cycles for routing tables across the whole network.
	Total OSPF Protocol Traffic Sent (bits/sec), Total OSPF Protocol Traffic Sent (pkts/sec)	These statistics record the total amount of OSPF traffic sent across all connected interfaces on all nodes in the network. These statistics are available in units of bits/second and packets/second.
Global Statistics RIP	*Network Convergence Activity*	This statistic records a square wave that is 0 when there is no convergence activity anywhere in the network and 1 when there is convergence activity somewhere in the network. Convergence activity is defined as changes in the routing table.
	Network Convergence Duration (sec)	This statistic records the duration in seconds of convergence cycles for routing tables across the whole network.

	Traffic Received (bits/sec), Traffic Sent (bits/sec)	These statistics record the total amount of RIP traffic received and sent across all connected interfaces on all nodes in the network. These statistics are available in units of bits/second.
Global Statistics OSPF Advanced	*Database Description Traffic Sent (bits/sec), Database Description Traffic Sent (pkts/sec), Hello Traffic Sent (bits/sec), Hello Traffic Sent (pkts/sec), Link-State Acknowledgement (Multicast) Traffic Sent (bits/sec), Link-State Acknowledgement (Multicast) Traffic Sent (pkts/sec), Link-State Acknowledgement (Unicast) Traffic Sent (bits/sec), Link-State Acknowledgement (Unicast) Traffic Sent (pkts/sec), Link-State Request Traffic Sent (bits/sec), Link-State Request Traffic Sent (pkts/sec), Link-State Update (Multicast) Traffic Sent (bits/sec), Link-State Update (Multicast) Traffic Sent (pkts/sec), Link-State Update (Unicast) Traffic Sent (bits/sec), Link-State Update (Unicast) Traffic Sent (pkts/sec)*	These statistics record the total amount of traffic sent for various types of OSPF messages. These statistics are recorded across all connected interfaces on all the nodes in the network. These statistics are collected for the following types of OSPF messages: Database Description, Hello, Link-State ACK (multicast), Link-State ACK (unicast), Link-State Request, Link-State Update (multicast), and Link-State Update (unicast). For each OSPF message type, these statistics are recorded in units of bits/second and packets/second.
	LSA Acknowledgement Rate, LSA Installation Rate, LSA Origination Rate, LSA Retransmission Rate	These statistics record the rate at which various events concerning LSAs occur at all interfaces within an OSPF area. Specifically, these statistics are available for the following LSA event rates: • *Origination rate*—rate at which LSAs are originated. • *ACK rate*—rate at which ACKs are sent for LSAs. • *Retransmission rate*—rate at which LSAs are retransmitted when no ACK is received for them. • *Installation rate*—rate at which LSAs are installed in the database which happens when a new update is received.
	Retransmission Traffic Sent (bits/sec), Retransmission Traffic Sent (pkts/sec)	These statistics record the amount of retransmissions of database synchronization packets that were sent by all OSPF interfaces. These statistics are available in units of bits/second and packets/second.

(Continued)

TABLE 11.1 (*Continued*)
Summary of RIP- and OSPF-Related Global and Node Statistics

Category	Name	Description
Node Statistics OSPF	*Router Convergence Activity*	This statistic records a square wave which is 0 when there is no convergence activity at a specific router and is 1 when there is convergence activity at the router.
	Router Convergence Duration (sec)	This statistic records the duration in seconds it takes for the routing table at a given router to converge.
	Traffic Received (bits/sec), Traffic Sent (bits/sec)	These statistics record the total amount of traffic received and sent by all OSPF processes running on a given router. These statistics are available in units of bits/second.
Node Statistics OSPF Process	*Database Description Traffic Sent (bits/sec), Database Description Traffic Sent (pkts/sec), Hello Traffic Sent (bits/sec), Hello Traffic Sent (pkts/sec), Link-State Acknowledgement (Multicast) Traffic Sent (bits/sec), Link-State Acknowledgement (Multicast) Traffic Sent (pkts/sec), Link-State Acknowledgement (Unicast) Traffic Sent (bits/sec), Link-State Acknowledgement (Unicast) Traffic Sent (pkts/sec), Link-State Request Traffic Sent (bits/sec), Link-State Request Traffic Sent (pkts/sec), Link-State Update (Multicast) Traffic Sent (bits/sec), Link-State Update (Multicast) Traffic Sent (pkts/sec), Link-State Update (Unicast) Traffic Sent (bits/sec), Link-State Update (Unicast) Traffic Sent (pkts/sec)*	These statistics record the total amount of traffic sent for various types of OSPF messages. These statistics are recorded separately for each individual OSPF process on a given router. These statistics are collected for the following types of OSPF messages: Database Description, Hello, Link-State ACK (multicast), Link-State ACK (unicast), Link-State Request, Link-State Update (multicast), and Link-State Update (unicast). For each OSPF message type, these statistics are recorded in units of bits/second and packets/second.
	MANET Designated Router Status	This statistic is only available for MANET interfaces, and it records in the form of a square wave the status of MANET Designated Router (MDR), which may be either **MDR**, recorded as 2; **Backup MDR**, recorded as 1; or **other MDR**, recorded as 0.

TABLE 11.1 (*Continued*)
Summary of RIP- and OSPF-Related Global and Node Statistics

Category	Name	Description
	Traffic Sent (bits/sec), *Traffic Sent (pkts/sec)*	These statistics record the total amount of traffic generated by a single OSPF process on a given router. These statistics are available in units of bits/second and packets/second.
Node Statistics RIP	*Router Convergence Activity*	This statistic records a square wave that is 0 when there is no convergence activity at a specific router and 1 when there is convergence activity at the router.
	Router Convergence Duration (sec)	This statistic records the duration in seconds it takes for the routing table at a given router to converge.
	Traffic Received (bits/sec), *Traffic Received (pkts/sec),* *Traffic Sent (bits/sec),* *Traffic Sent (pkts/sec)*	These statistics record the total amount of traffic received and sent by all RIP processes running on a given router. These statistics are available in units of bits/second and packets/second.

11.5 VIEWING ROUTING TABLES

You can export various reports related to routing so they can be viewed after the simulation ends. You can export the routing tables created by RIP, OSPF, or other routing protocols deployed in the network. You can also export the OSPF LSDB, *IP Forwarding Table*, and other network related reports. You can configure the simulation to export the contents of these tables at designated times during the simulation and/or at the end of the simulation. In addition, you can configure the simulation to export routing tables of specific individual routers or multiple routers in the network. In this section, we will only discuss steps for generating reports related to the RIP routing protocol. The same general steps are applicable to the export of any of the other report tables.

You need to perform the following steps to configure individual routers to export RIP-related tables:

1. Select the routers from which you would like to export their RIP routing information.
2. Edit the attributes of any one of the selected routers.
3. Expand the attribute **Reports**, which contains several subattributes for configuring the router to export various report tables. By default, the router is configured to not export any information at all.
4. To export the *RIP Routing Table* at the end of the simulation, set the value of the attribute **RIP Routing Table** to Export at End of Simulation.

5. Alternatively, you may want to export the *RIP Routing Table* at other times during the simulation, in which case you need to expand the attribute **RIP Routing Table** and set its subattribute **Status** to Enabled. Figure 11.19 illustrates this process.
6. Expand the attribute **Export Time(s) Specification** and set the value of the attribute **Number of Rows** to the total number of times during the simulation you would like to export the *RIP Routing Table*.
7. Expand each created row and set its attribute **Time** to the simulation time when the *RIP Routing Table* should be exported. The attribute **Time** also accepts the value End of Simulation, which will configure the simulation to export this router's *RIP Routing Table* at the end of the simulation.
8. Select the checkbox *Apply to selected objects* to specify your configuration on all the routers that you have selected in the first step of this process.
9. Click the **OK** button to apply the configuration changes. Notice that the attributes **OSPF Routing Table** and **OSPF Link State Database** are responsible for exporting the *OSPF Routing Table* and OSPF LSDB, respectively.

You can view the exported router information, such as the *RIP Routing Table*, only after the simulation has completed execution, by performing the following steps:

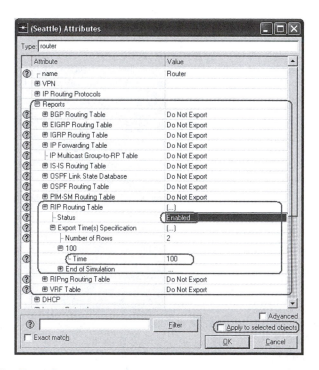

FIGURE 11.19 Exporting router reports.

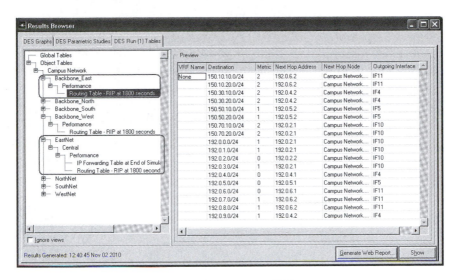

FIGURE 11.20   Preview of an exported *RIP Routing Table.*

1. Open the **Results Browser** window by using the **DES > Results > View Results** menu option.
2. Click on the *DES Run Tables* tab and expand the **Object Tables** attribute tree, as shown in Figure 11.20.
3. Expand the attribute **Performance** and then click on the available reports to display a preview of the exported table on the selected node. As illustrated in Figure 11.20, the *Preview* panel shows the *RIP Routing Table* for the node *Backbone_East* in *Campus Network,* which is the only router report that has been exported, whereas the node *Central* in the *EastNet* subnet of *Campus Network* contains both the *IP Forwarding Table* and *RIP Routing Table,* both exported at the end of the simulation.
4. Click the **Show** button to display the currently previewed report in a separate window.
5. You can also click the **Generate Web Report** button, which allows you to select which router reports from this scenario you would like to export and then saves the selected information as a set of HTML files in a desired location on your computer.

Figure 11.20 shows an example of an exported *RIP Routing Table* at *Backbone_ East* at time 1800 seconds from the RIPv2 scenario of the standard OPNET project RIP. We configured this scenario to export the *RIP Routing Table* on all the routers within the network at the end of the simulation, which runs for 30 minutes. As Figure 11.20 shows, each row in the exported *RIP Routing Table* contains the following information:

- **Destination**—the destination network reachable from the current router (i.e., *Backbone_East*).

- **Metric**—the number of hops to reach the destination network.
- **Next Hop Address**—the IP address of the next hop on the route to the destination network.
- **Next Hop Node**—a logical name of the next hop on the route to the destination network.
- **Outgoing Interface**—the name of the interface on which a packet would be transmitted to reach the next hop on its way to the final destination.

12 Data Link and Physical Layers

12.1 INTRODUCTION

OPNET supports a wide range of data link layer protocols and technologies including Ethernet, Asynchronous Transfer Mode (ATM), Fiber Distributed Data Interface (FDDI), Frame Relay (FR), Token Ring (TR), X.25, and so on. In addition, OPNET software also contains various models for simulating wireless communication, such as Wireless LAN (WLAN), WiMAX (802.16e), Mobile Ad Hoc Networks (MANET), Time Division Multiple Access (TDMA), Universal Mobile Telecommunications System (UMTS), and others. Each of these protocols and technologies typically has a set of simulation attributes for specifying a desired configuration. However, a complete description of all related parameters and configuration settings is beyond the scope of this book. Furthermore, the principles and patterns for configuring these technologies are often very similar. A reader with a working knowledge of any of the above protocols and with an understanding of OPNET basics can easily determine the meaning of the configuration attributes as well as design and configure a simulation model of a network that employs the corresponding data link layer or wireless communication technology. For this reason, instead of describing all available models and their attributes in detail, we only introduce the main principles behind deploying data link layer and wireless communication technologies in OPNET and provide an overview of a few most popular protocols and technologies. It is important to note that certain wireless protocol suites, such as WLAN, WiMAX, and UMTS, require separate module licenses. Thus, none of the features described in Sections 12.6 and 12.7 will be available unless the wireless module license is present. For a complete description of all supported data link layer and wireless communication protocols and technologies, refer to OPNET's product documentation.

The rest of this chapter is organized as follows: In Section 12.2, we describe general principles for deploying and configuring data link layer node and link models. In Section 12.3, we provide an overview of link models through link configuration attributes and available statistics for analyzing performance on the links. We follow with a description of Ethernet and TR data link layer technologies in Sections 12.4 and 12.5, respectively. The next two sections deal with wireless networks: Section 12.6 describes WLANs while Section 12.7 describes MANET technology. In Section 12.8 we illustrate how to define node mobility, and in Section 12.9 we show how to use the Wireless Deployment Wizard to create new wireless networks.

12.2 DEPLOYING AND CONFIGURING SIMULATION MODELS WITH DATA LINK LAYER TECHNOLOGIES

To deploy a particular data link technology requires placing node and link models that support that technology within the project workspace. You can identify the data link protocols supported by a node or a link model based on its name. Link, end node (e.g., workstations and servers), and hub models typically support a single data link layer technology, while switches and layer 3 devices can have interfaces that run different data link technologies. For example, models *ATM_E1, 1000Base_X_adv, FDDI, FR_link, TR16*, and *sl_x25_int* represent ATM, Ethernet, FDDI, FR, TR, and X.25 links, respectively. Similarly, *atm_server_int, ethernet_wkstn, fddi_cache_server, fr_wkstn_adv*, and *tr_unitx_wkstn* are examples of the end node models with different data link layer interfaces. Hub models use the following naming conventions: *<data link layer><number of ports>_hub_<suffix>*. For example, an advanced model of an Ethernet hub with 32 ports is named *ethernet32_hub_adv*. Routers and switches use similar naming conventions. However, since such devices support multiple interfaces, each potentially running a different data link layer technology, the model name separates each different interface type with an underscore symbol. For example, *eth2_fddi2_tr2_switch_adv* is an advanced model of a multipurpose switch that contains two Ethernet interfaces, two FDDI interfaces, and two Token Ring interfaces.

Link models contain no attributes for configuring data link layer properties. Instead, such configuration attributes are located in the node models. Typically, only advanced node models have attributes available for configuring data link layer characteristics, while the regular node models have these attributes hidden and set to the default values. For example, the *tr_server_adv* node model contains the compound attribute **Token Ring** for specifying TR configuration, while the *tr_server* model does not have this attribute available for configuration. On the other hand, both node models *fr_server* and *fr_server_adv* contain the compound attribute **Frame Relay** for specifying FR settings. Generally, when studying properties of the data link layer, it is a good idea to use advanced node models.

As you may recall, when building network topology, you can connect two nodes with a link only if each of these nodes has an interface or a port that supports the same data link layer technology. For example, you cannot directly connect *ethernet_wkstn* and *tr_server*. To set up a communication channel between these two nodes, you need to add an intermediate node, such as a switch or a router. As an example, for the intermediate node, you can use such node models as *eth2_fddi2_tr2_switch* or *ethernet_tr_slip8_gtwy*, each of which contains at least one Ethernet and one Token Ring interface. After the intermediate node is added to the network topology, you can connect *ethernet_wkstn* to the intermediate node using an Ethernet link model such as *100BaseT* and *1000BaseX* and finally connect the intermediate node to *tr_server* using such TR link models as *TR_adv* and *TR4*.

12.3 LINK MODEL ATTRIBUTES AND STATISTICS

OPNET contains several link categories, which include duplex, simplex, bus, and bus tap models. Typically, the model's name identifies the type of the link and the data link layer technology used to connect to the physical medium. For example, *eth_tap_adv* represents an advanced Ethernet bus tap link model. Similarly, *FR_T1_int* represents an intermediate FR duplex link model carrying data at T1 rate. You should note that the model's name does not indicate whether the link is simplex or duplex. However, the model description specifies whether the link is simplex or duplex. In addition, simplex links are represented as a line with an arrow at the end, indicating the direction of traffic flow on the link, while duplex link models are represented as a line without any arrows.

As shown in Figure 12.1, link models typically contain the following configuration attributes:

- **transmitter**, **transmitter a**, **transmitter b**—identify the transmitting module(s) in one or both nodes attached to this link. The **transmitter** attribute is available in simplex and bus tap models only because the data flow is unidirectional. The **transmitter a** and **transmitter b** attributes are only available in duplex link models and bus link models.
- **receiver**, **receiver a**, **receiver b**—identify the receiving module(s) in one or both nodes attached to this link. The **receiver** attribute is available in simplex and bus tap models only, while the **receiver a** and **receiver b** attributes are available in duplex link models only.
- **Propagation Speed**—specifies the propagation speed of this physical medium in units of meters per second. This attribute is only visible in advanced link models. The value of this attribute together with the distance between the nodes is used to determine the propagation delay on the link, when the attribute **delay** is set to `Distance Based`, which is the default value.

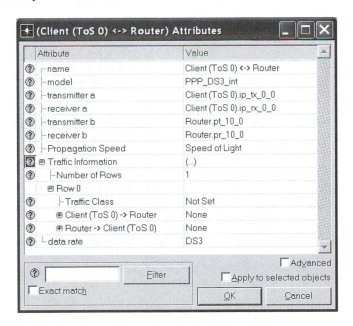

FIGURE 12.1 Configuration attributes for a typical link model.

TABLE 12.1
Summary of low-level point-to-point Link Statistics

Name	Description
bit error rate	This statistic records the average bit error rate on the link.
bit error rate per packet	This statistic records the average number of bit errors per packet sent over the link.
busy, *busy ->,* *busy <-*	These statistics are collected as Boolean values 0 or 1, which represent whether the link is free or is occupied due to packet transmission or reception. The arrows <- and -> indicate the direction of the data flow over the link for which the statistic is collected.
packet loss ratio	This statistic is also collected as Boolean values 0 or 1. A value of 0 signifies a successful packet transmission over this link, whereas a value of 1 means a packet transmission was not successful.

- **delay**—specifies the propagation delay on this link. Typically, this attribute is hidden and is set to the default value of `Distance Based`. You need to select the *Advanced* checkbox in order to view and/or modify this attribute. If needed, you can explicitly specify the propagation delay value to be experienced by the traffic traversing this link.
- **Traffic Information**—specifies the baseline traffic load on the link. Please refer to Section 6.6.3 for details regarding the configuration of this attribute.
- **data rate**—specifies the transmission rate on the link in units of bits per second. This attribute is available only in advanced link models. Certain link models, such as Ethernet links (e.g., 1000BaseX, 100BaseT, etc.), do not contain this attribute.

OPNET provides two types of Link Statistics:

- *low-level point-to-point*—a set of statistics that describe low-level characteristics of the physical channel. Table 12.1 contains a summary of the *low-level point-to-point* Link Statistics.
- *point-to-point*—a set of statistics that describe such link characteristics as queuing delay, throughput, and utilization. Table 12.2 contains a summary of the *point-to-point* Link Statistics.

12.4   ETHERNET

OPNET implements most features of the Ethernet protocol with the exception of bit serialization for data transfer to and from the physical channel. The Ethernet model implemented in OPNET is based on IEEE 802.3, IEEE 802.3u, and IEEE 802.3z standards. Specifically, OPNET supports the following Ethernet features: FIFO processing of transmission requests, computation of propagation delay based on the

TABLE 12.2
Summary of point-to-point Link Statistics

Name	Description
queuing delay (sec), *queuing delay (sec) ->,* *queuing delay (sec) <-*	These statistics record the instantaneous measurement of the packet's waiting time in the link's queue. The delay is measured from the time the packet enters the link's queue until the time when the packet finishes transmission.
throughput (bits/sec), *throughput (bits/sec) ->,* *throughput (bits/sec) <-,* *throughput (packets/sec),* *throughput (packets/sec) ->,* *throughput (packets/sec) <-*	These statistics record the average throughput on the link. Throughput is defined as the number of bits or packets successfully received or transmitted over the link per second.
utilization, *utilization ->,* *utilization <-*	These statistics record the link utilization as a value between 0% and 100%, where 100% means that the link's capacity is fully consumed. Link utilization is defined as the ratio between the amount of data carried on the link and the link's capacity.

length of the link, carrier sensing, collision detection, binary exponential back-off with the maximum number of retransmission attempts set to 16, interframe gap timing for deference, jam sequence transmission after collisions, 802.3 minimum and maximum frame sizes, and frame bursting for 1000BaseX Ethernet links operating in half-duplex mode, as well as full and half-duplex data transmission. OPNET also allows configuring port-based Virtual LAN on all generic bridge and switch models and provides support for Cisco's Fast EtherChannel technology. OPNET allows deployment of Ethernet models in bus and hub topologies only. Please refer to the product documentation for a full list of all Ethernet features supported in OPNET.

OPNET provides only a few configuration attributes for Ethernet. These attributes are only visible in the advanced node models such as *eth2_fddi2_tr2_switch_adv,* *ethernet16_switch_adv,* and *ethernet_wkstn_adv.* All Ethernet configuration attributes are collected under the compound attribute **Ethernet...Ethernet Parameters**. In the node models that support multiple Ethernet ports, the word **Parameters** is followed by the port number in parenthesis. As shown in Figure 12.2, you can specify the Ethernet configuration through the following attributes:

- **Address**—specifies the unique MAC address for this Ethernet interface. By default, this attribute is set to the value `Auto Assigned`, but if needed, you can explicitly set this attribute to the desired MAC address value.
- **Frame Bursting**—specifies whether or not Frame Bursting is enabled on this interface. Frame Bursting is only applicable to Gigabit Ethernet (IEEE 802.3z). This attribute value is ignored in all other scenarios. This attribute accepts only two values: `Enabled` and `Disabled`. When enabled, the node is allowed to transmit additional frames on this interface without

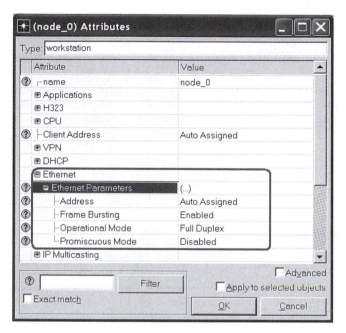

FIGURE 12.2   Ethernet configuration attributes.

restarting procedures for channel contention. The node may continue transmitting until there is no more data or until the burst timer expires, whichever occurs first.

- **Operational Mode**—specifies whether this interface operates in Full Duplex (i.e., can send and receive frames simultaneously) or Half Duplex (i.e., can only perform either transmission or receiving of the data frames at a time) mode. This attribute accepts only two values: `Full Duplex` and `Half Duplex`.

- **Promiscuous Mode**—specifies whether the promiscuous mode is enabled on this interface. This attribute accepts only two values: `Enabled` and `Disabled`. When enabled, the node accepts and forwards to the upper layer all the packets regardless of their destination MAC address. Promiscuous mode is typically enabled on all Ethernet interfaces of bridge and switch nodes.

OPNET also contains a set of Node Statistics for evaluating Ethernet performance. These statistics are aggregated under the **Ethernet** category. Notice that the **Ethernet** Node Statistics category is hidden in simulation models where there are no nodes with Ethernet interfaces. Table 12.3 contains a summary of **Ethernet** Node Statistics.

12.5   TOKEN RING

OPNET includes support for the TR MAC networking protocol. Provided TR models operate at transmission rates of 4 or 16 Mbits/second in networks organized in ring or hub topology. OPNET's TR implementation is based on the IEEE 802.5 standard.

TABLE 12.3
Summary of Ethernet Node Statistics

Name	Description
Burst Duration (sec)	This statistic records the amount of time the Ethernet interface was in the burst mode. This statistic is computed for each burst as an interval starting from the time the first frame of the burst started transmission until the time when the last frame of the burst ended its transmission. This is applicable only to the Gigabit Ethernet.
Burst ON/OFF	This statistic records the burst status of the Ethernet interface as a sequence of Boolean values: 0, when there is no burst transmission and 1, when there is burst transmission. This is applicable only to the Gigabit Ethernet.
Burst Size (packets)	This statistic records the number of packets sent during each burst transmission. This is applicable only to the Gigabit Ethernet.
Collision Count	This statistic records the total number of collisions experienced by the data transmitted from this Ethernet interface.
Delay (sec)	This statistic records the delay experienced by the frames arriving at this Ethernet interface.
Load (bits), *Load (bits/sec),* *Load (packets),* *Load (packets/sec)*	These statistics record the amount of data sent from the upper layers to the Ethernet layer on this interface. When the units are bits or packets, the total amount of data sent is recorded, whereas when the units are bits/second or packets/second, the average rate at which data is transmitted per second is recorded.
Traffic Received (bits), *Traffic Received (bits/sec),* *Traffic Received (packets),* *Traffic Received (packets/sec)*	These statistics record the amount of data received over the link and sent to the upper layers from the Ethernet layer on this interface. When the units are bits or packets, the total amount of data received is recorded, whereas when the units are bits/second or packets/second, the average rate at which data is received per second is recorded.
Transmission Attempts	This statistic records the number of transmission attempts by this station before the packet is successfully received at its destination.

However, not all of the TR features have been implemented in OPNET because the model was primarily designed for simulation and performance estimation only. Specifically, OPNET does not explicitly model ring initialization and recovery processes. The interface between the network management (NMT) and MAC is not implemented as well. There is no support for error detection and error reporting to NMT. To improve simulation performance, the TR model implements the token acceleration feature, which blocks token passing during the time periods when there is no data transmission. When data transmission restarts, the token is injected back into the TR LAN.

The TR model implements an interface between the MAC and Logical Link Control (LLC) sublayers as well as an interface between the MAC and Physical (PHY) layers. In addition, it also explicitly models such features as definable priority levels for LLC PDUs, a 24-bit assured minimum latency, the effects of station latency and propagation delay, restack operation, and the token holding timers (THT).

OPNET provides only a few configuration attributes for the TR. These attributes are only available in the advanced node models such as *eth2_fddi2_tr2_switch_adv*, *tr16_bridge_adv*, and *tr_wkstn_adv*. All TR configuration attributes are collected under the compound attribute **Token Ring...Token Ring Parameters**. In node models with multiple TR ports, the word **Parameters** is followed by the port number in parentheses. As shown in Figure 12.3, you can specify TR configuration through the following attributes:

- **Address**—specifies the unique TR MAC address on the current TR interface. By default, this attribute is set to `Auto Assigned`, but if needed, you can explicitly specify the desired MAC address value.
- **Hop Propagation Delay (seconds)**—specifies the time taken by a frame to travel from one TR interface to the next.
- **Operation Mode**—specifies whether this interface operates in `Switched` (i.e., can send and receive frames simultaneously) or `Shared` (i.e., can only perform either transmission or receiving of the data frame at a time) mode. This attribute accepts only two values: `Switched` and `Shared`.
- **Promiscuous Mode**—specifies whether the promiscuous mode is enabled on this interface.

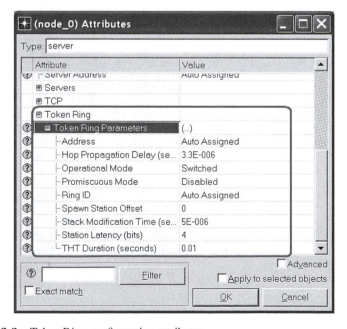

FIGURE 12.3   Token Ring configuration attributes.

TABLE 12.4
Summary of Token Ring Node Statistics

Name	Description
Delay (sec)	This statistic records the delay experienced by the frames arriving on this Token Ring interface.
Load (bits), *Load (bits/sec),* *Load (packets),* *Load (packets/sec)*	These statistics record the amount of data sent from the upper layers to the Token Ring layer on this interface.
Traffic Received (bits), *Traffic Received (bits/sec),* *Traffic Received (packets),* *Traffic Received (packets/sec)*	These statistics record the amount of data sent to the upper layers from the Token Ring layer on this interface.

- **Ring ID**—specifies the identity of the ring to which this TR interface belongs.
- **Spawn Station Offset**—specifies the value used to select the TR interface, which creates and injects the token into the ring.
- **Stack Modification Time (seconds)**—specifies the time required for a TR MAC to modify its stack.
- **Station Latency (bits)**—specifies the delay, in units of bits, introduced to a packet by each TR MAC it visits on its path to the destination. The bit value is converted into time based on the connected interface speed (e.g., 4 Mbps or 16 Mbps).
- **THT Duration (seconds)**—specifies the maximum amount of time a TR interface may use the token before releasing it to the next TR station. This value is often referred to as Token Holding Time.

OPNET also contains a set of Node Statistics for evaluating TR performance. These statistics are aggregated under the **Token Ring** category. Notice that the **Token Ring** Node Statistics category is hidden in simulation scenarios where there are no nodes with TR interfaces. Table 12.4 contains a summary of the **Token Ring** Node Statistics.

12.6 WIRELESS LANS

OPNET software includes the WLAN model suite, but you need a wireless module license to run simulations that employ this suite. All the features described in this section on WLAN and in the next section on MANET are only available if you have this license, and they may not be visible if the license is not present.

The implementation of the WLAN model suite is based on the IEEE 802.11, 802.11a, 802.11b, 802.11g, and 802.11e standards, and it supports such features as

- Carrier sense multiple access and collision avoidance (CSMA/CA) distributed coordinating function (DCF) access scheme with exponential back-off mechanism.
- The point coordination function (PCF) access scheme.
- Hybrid Coordination Function (HCF) that supports Enhanced Distributed Channel Access (EDCA) with four Access Categories (ACs) for prioritized contention-based access, Transmission Opportunity (TXOP), Frame Bursting, and EDCA Parameter Set distribution by Access Point (AP).
- Threshold-based Request-to-Send/Clear-to-Send (RTS/CTS) exchange for reliable data transmission.
- Optional data frame fragmentation.
- AP functionality. Even though all WLAN nodes have AP capabilities, only WLAN bridges, switches, or routers can connect the base station subsystem (BSS) to external networks.
- Roaming capability, where a WLAN node can scan for other APs and, if the signal strength with the current AP drops below an acceptable level, the node can switch to another AP with a stronger signal.
- MAC-level acknowledgments: normal ACK, Block-ACK, and no-ACK.
- Interoperability between 802.11b, 802.11g, and/or 802.11e nodes.
- Physical layer technologies: frequency-hopping spread spectrum (FHSS), infrared (IR), direct-sequence spread spectrum (DSSS), orthogonal frequency-division multiplexing (OFDM), and Extended Rate PHY-OFDM.
- Differential phase shift keying (DPSK), binary phase shift keying (BPSK), quadrature phase-shift keying (QPSK), quadrature amplitude modulation (QAM), and complementary code keying (CCK) modulation techniques.
- Data rates of 1, 2, 5.5, 6, 9, 11, 12, 18, 24, 36, 48, and 54 Mbps.
- Terrain Modeling. By default, OPNET uses the free space propagation model. However, if you have licenses for the Terrain Modeling Module (TMM), then you can use other models such as Longley-Rice and Terrain Integrated Rough Earth Model (TIREM). You can manage terrain modeling through the **Topology > Terrain** menu option, which appears only if the TMM license is present.

However, other features such as background traffic, frequency hopping, power save mode, authentication and security, node failure and recovery, variable transmission rate in WLAN nodes, certain 802.11e features, and a few others are not supported in OPNET's WLAN model suite. Please refer to the OPNET product documentation for a full list of features and limitations of the WLAN model.

12.6.1 WLAN Configuration Attributes

The **Object Palette Tree** keeps most commonly used WLAN node models in the *wireless_lan, wireless_lan_adv*, and *wlan* object palettes. There are several distinct types of WLAN nodes:

- Wireless workstation and server models such as *wlan_server* and *wlan_wkstn_adv*. These node models contain all layers of the TCP/IP reference model stack.

- Wireless terminal station models such as *wlan_station_adv*. These node models contain only the link layer and physical layer WLAN functionality along with the capability to directly generate and receive traffic, using explicit packet generation (see Section 6.5). These models were designed for simulation studies that focus primarily on evaluation of WLAN MAC and physical layers.
- Wireless routers and bridges (e.g., *wlan_fddi2_tr2_router, wlan_eth_bridge*), which contain both wireless and wired (i.e., Ethernet, TR, SLIP, etc.) interfaces.

WLAN node configuration settings are combined under the compound attribute **Wireless LAN**, which consists of two subattributes, as shown in Figure 12.4:

- **Wireless LAN MAC Address**, which specifies a unique WLAN MAC address. By default, this attribute is set to `Auto Assigned`, but if needed, you can specify a desired WLAN MAC address value.
- **Wireless LAN Parameters**, which specifies the physical characteristics of this WLAN interface.

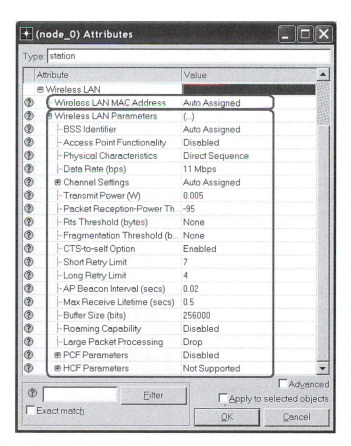

FIGURE 12.4   WLAN configuration attributes.

Wireless LAN Parameters is also a compound attribute and it consists of the following subattributes:

- **BSS Identifier**—specifies the identity of the BSS to which this WLAN node belongs. By default, this attribute is set to Auto Assigned, in which case all the nodes within the current subnet belong to the same BSS. If you set this attribute to a non-default value in any one node in the network, then you need to explicitly specify the BSS id in all the other nodes in the network as well. In the case of roaming nodes, this attribute value specifies the initial BSS id. As the node moves, it can become assigned to a different BSS.
- **Access Point Functionality**—specifies whether AP functionality is enabled or disabled on this node. OPNET documentation identifies the following rules for specifying AP functionality:
  - Each BSS can have at most one AP. A BSS may have no APs.
  - If a BSS is connected to other networks or it deploys the PCF, then it is required to have a node with AP functionality enabled.
  - A WLAN MAC interface on a bridge, switch, or router must have AP functionality enabled. An exception to this rule is a WLAN interface in a router, which is part of a WLAN backbone.
  - Enabling AP functionality in a node creates additional simulation processing overhead. It is recommended to disable this option when not explicitly required in the simulation.
- **Physical Characteristics**—identifies the physical layer technology deployed on this interface. This attribute accepts the following values: Frequency Hopping (currently not supported), Direct Sequence, Infra Red, OFDM (802.11a), and Extended Rate PHY (802.11g).
- **Data Rate (bps)**—specifies the transmission rate on this WLAN interface. The set of acceptable data rate values depends on the deployed physical layer technology as configured via the attribute **Physical Characteristics**.
- **Channel Settings**—is a compound attribute, which specifies the frequency band used by the transmitter and receiver on this WLAN interface. Please refer to the attribute description for additional details.
- **Transmit Power (W)**—this attribute determines the transmission range of this WLAN node. It specifies the signal transmit power of this node in units of watts.
- **Packet Reception-Power Threshold (dBm)**—specifies the minimum signal power value, which makes the arriving packet accepted by the receiver. Any packet with a power value lower than the value of this attribute is considered noise and can cause interference. Only packets with a power greater than or equal to the threshold value are accepted and processed by this WLAN interface.
- **Rts Threshold (bytes)**—specifies the minimum frame size, including WLAN MAC header (28 bytes), which triggers RTS/CTS exchange. Any packet whose size exceeds this attribute value must wait for a successful RTS/CTS exchange completion before being transmitted. This attribute accepts the value None, which indicates that RTS/CTS frame exchange is not used on this node.

- **Fragmentation Threshold (bytes)**—specifies the maximum size of a frame that can be transmitted without fragmentation. Each frame of size larger than the value of this attribute is fragmented into chunks of size **Fragmentation Threshold (bytes)**, except for possibly the last fragment. However, all data packets larger than the MAC Service Data Unit (MSDU) size, which is 2304 bytes, are discarded unless the attribute **Large Packet Processing** is set to `Fragment`.
- **CTS-to-self Option**—determines if the CTS-to-self protection mechanism is enabled on this WLAN interface. This attribute is applicable only to a WLAN MAC operating in 802.11g mode. If this attribute is set to `Disabled`, then a regular RTS/CTS frame exchange is executed, except for 802.11g MAC broadcast transmissions, which use the CTS-to-self mechanism regardless of this attribute's value.
- **Short Reply Limit**—specifies the maximum number of transmission attempts for frames of size smaller than or equal to the **RTS Threshold** value. The interface gives up its attempt to transmit the frame and discards it, if the frame fails to transmit after **Short Reply Limit** number of tries.
- **Long Reply Limit**—specifies the maximum number of transmission attempts for frames of size greater than the **RTS Threshold** value.
- **AP Beacon Interval (seconds)**—specifies the Target Beacon Transmission Time (TBTT) value, which determines the time period between successive beacon transmissions by the AP.
- **Max Receive Lifetime (seconds)**—specifies how long, starting from the first fragment arrival, the node has to wait for all fragments to arrive before giving up and discarding a partially received frame.
- **Buffer Size (bits)**—specifies the maximum buffer size for storing the data arriving from upper layers. When the buffer is full, all arriving packets from the upper layer are discarded until more room becomes available in the buffer.
- **Roaming Capability**—specifies if this node has roaming capability enabled. Roaming is not supported in an ad hoc BSS and in a BSS with PCF enabled.
- **Large Packet Processing**—specifies how packets larger than the maximum MSDU size are processed. This attribute accepts two values:
  - `Drop`—the default value indicating that all packets of size larger than the maximum MSDU are discarded.
  - `Fragment`—indicates that the node will attempt to fragment large packets. The node will fragment a large packet only if the attribute **Large Packet Processing** is set to `Fragment` and fragmentation is enabled on this WLAN interface (e.g., attribute **Fragmentation Threshold (bytes)** is set to a value other than `None`).
- **PCF Parameters**—this is a compound attribute, which specifies configuration of PFC mode. By default, this attribute is set to `Disabled`.
- **HCF Parameters**—this is a compound attribute, which specifies configuration of the WLAN QoS mechanism of HFC mode defined in the IEEE 802.11e standard. By default, this attribute is set to `Not Supported`.

OPNET also contains four global WLAN attributes, as shown in Figure 12.5. These attributes are available only if the simulation contains WLAN object models; otherwise, they are hidden and not available for configuration. These attributes, which are accessible under the *Global Attributes* tab of the **Configure/Run DES** window, are the following:

- **Closure Method (non-TMM)**—specifies which method will be used to determine if communication is possible between two WLAN nodes when the TMM, which requires a separate software license, is not available. This attribute accepts the following two values:
  - No Occlusion—specifies that every pair of wireless nodes should be able to communicate, that is, the transmission path between any two nodes in the simulation is never blocked. This is the default setting in WLAN.
  - Earth Line-of-Sight—the closure between any two nodes is computed using Line-of-Sight with a spherical Earth model. When using this setting, make sure to properly configure node altitude to avoid a situation where communication between two nodes fails because of blocking due to the Earth's curvature. OPNET recommends configuring all WLAN nodes with a nonzero altitude value when using the Earth Line-of-Sight closure method.
- **WLAN AP Connectivity Check Interval (seconds)**—specifies the length of the period between two successive attempts by the WLAN node to verify its connectivity with the AP and to scan for alternative APs, if

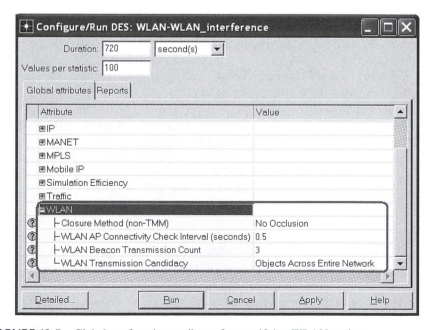

FIGURE 12.5   Global confiuration attributes for specifying WLAN settings.

necessary. This attribute is applicable only when beacon efficiency mode is disabled (i.e., the attribute **WLAN Beacon Transmission Count** is set to `Periodic`) and roaming is enabled in WLAN nodes.

- **WLAN Beacon Transmission Count**—specifies if beacon efficiency mode is enabled or not. This attribute accepts the following values:
  - A positive integer value—indicates that beacon efficiency mode is enabled (i.e., no beacons will be generated). This value specifies the number of beacons the AP will generate before discontinuing beacon transmission. Roaming is still possible with this setting. The nodes examine the signal strength from nearby APs, without explicit broadcast of beacons, to determine when switch to another AP is needed.
  - `Periodic` value—indicates that beacon efficiency mode is disabled and the AP will generate a beacon periodically based on the configuration value of the node attribute **AP Beacon Interval (seconds)**.

  Please refer to the attribute documentation for additional information regarding this attribute.

- **WLAN Transmission Candidacy**—this attribute allows enabling simulation efficiency, which will eliminate any communication and/or interference between nodes that belong to different subnets. In effect, only nodes within the same subnet will be able to communicate with one another and influence transmissions (i.e., introduce interference) of nodes within the same subnet. This attribute accepts the following two values:
  - `Objects in Same Subnet`—indicates that simulation efficiency is enabled and that the nodes can communicate with and influence communication of the other nodes within the same subnet only. A node outside a subnet has no effect on communication within that subnet.
  - `Objects Across Entire Network`—indicates that this simulation efficiency is disabled and the simulation executes as normal. In this case, a transmission from a node may cause interference at nodes in other subnets if those nodes are within the interference range of this node. This is the default value for this attribute.

In addition, OPNET contains a **Protocols** menu option for configuring physical characteristics and the data rate of either all or only selected WLAN nodes in the simulation. You can perform this configuration step by selecting the **Protocols > Wireless LAN > Configure PHY and Data Rate...** option from the **Project Editor**'s pull-down menu, which opens the **Configure WLAN PHY and Data Rate** window as shown in Figure 12.6. In this window, you can specify the desired physical layer technology and data rate to be configured in the network. Once you configure the desired values, you can click on either **All WLAN nodes** or **Selected WLAN nodes** radio button to indicate whether your configuration setting will be applied to all or to only selected WLAN nodes in the network. Clicking the **OK** button updates the value of the corresponding attributes in the specified WLAN nodes.

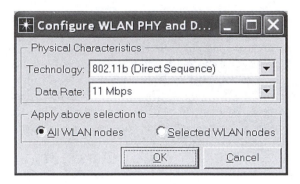

FIGURE 12.6   **Configure WLAN PHY and Data Rate** window.

12.6.2   WLAN Statistics

OPNET provides Global and Node Statistics for the evaluation of WLAN performance. Both Global and Node Statistics contain two statistic categories each for analyzing WLAN performance:

*   **Wireless LAN**—contains statistics about overall WLAN performance. A summary of available Global and Node Statistics is provided in Tables 12.5 and 12.6, respectively.
*   **WLAN (Per HCF Access Category)**—contains WLAN statistics collected for each HCF AC separately. These statistics are identical to regular WLAN statistics and hence we omit their description from the book.

The OPNET product documentation provides an excellent overview of the basic steps for creating and deploying WLAN networks. In addition, an example project called WLAN contains several scenarios that illustrate configuration and evaluation of WLAN networks under various settings.

12.7   MANET

MANET is a group of WLAN nodes arranged in an ad hoc fashion, which communicate with one another forming a network. In some cases, MANET nodes may connect to a wired network through a gateway. OPNET supports the following MANET routing protocols: Open Shortest Path First version 3 (OSPFv3), Ad Hoc On Demand Distance Vector (AODV), Dynamic Source Routing (DSR), Geographic Routing Protocol (GRP), and Temporally Ordered Routing Algorithm (TORA). However, OPNET does not support multiple MANET routing protocols on the same node and provides very limited support for route redistribution between routing protocols. For a complete list of supported MANET routing protocol features, please refer to OPNET's product documentation.

TABLE 12.5
Summary of Global WLAN Statistics

Name	Description
Data Dropped (Buffer Overflow) (bits/sec)	This statistic records the total amount of data that was received from the upper layer and then dropped by all WLAN nodes in the network due to upper layer buffer overflow (buffer size controlled via the node attribute **Buffer Size (bits)**) or due to data packets exceeding the maximum MSDU size.
Data Dropped (Retry Threshold Exceeded) (bits/sec)	This statistic records the total amount of data that was received from the upper layer and then dropped by all WLAN nodes in the network due to repeatedly failed retransmissions (i.e., exceeded the corresponding short retry or long retry threshold value).
Delay (sec)	This statistic records the end-to-end delay experienced by all the packets successfully received by all WLAN interfaces in the network.
Load (bits/sec)	This statistic records the total amount of data submitted by the upper layer for transmission by the WLAN layer on all the nodes in the network.
Media Access Delay (sec)	This statistic records the medium access delay experienced by the packets submitted for transmission on all WLAN interfaces in the network. This value is computed as the interval from the time the packet was inserted into the transmission queue until the time when the packet was sent to the physical layer for the first time.
Network Load (bits/sec)	This statistic is computed on a per-BSS basis. It represents the amount of data from the higher layer that was received, accepted, and queued for transmission by the entire WLAN BSS.
Retransmission Attempts (packets)	This statistic records the total number of retransmission attempts by all WLAN nodes in the network.
Throughput (bits/sec)	This statistic records the amount of data forwarded from WLAN layers to higher layers in all WLAN nodes of the network.

The **Object Palette Tree** keeps most commonly used MANET-capable node models under the *MANET* object palette. There are three distinct types of MANET-capable nodes:

1. WLAN servers and workstations, for example, *wlan_server* and *wlan_wkstn*. These node models support standard applications such as HTTP, FTP, and Remote Login and can run any MANET routing protocol.
2. MANET stations such as *manet_station_adv*. These models explicitly generate network traffic (via the compound attribute **MANET Traffic Generation Parameters**), can be a source, an intermediate node, or a destination for the traffic stream, and can run any MANET routing protocols.
3. WLAN routers and MANET gateway node models such as *manet_gtwy_wlan_ethernet_slip4* primarily operate as an AP from the MANET to an IP network.

TABLE 12.6
Summary of Node WLAN Statistics

Name	Description
AP Connectivity	This statistic records the id of the AP to which the current node is connected. If the node is not connected to any AP, then the value − 1 is recorded. When searching for a satisfactory AP in scanning mode, the node is considered to be unconnected.
Backoff Slots (slots)	This statistic records the number of slots the station needed to back-off while contending for the medium before it was able to transmit the frame.
Control Traffic Rcvd (bits/sec), *Control Traffic Rcvd (packets/sec),* *Control Traffic Sent (bits/sec),* *Control Traffic Sent (packets/sec)*	These statistics record the amount of control traffic received or sent on this WLAN interface. Control traffic includes such frames as RTS, CTS, ACK, CF-End, CF-End + CF-ACK, Block-ACK Request, and Block-ACK.
Data Dropped (Buffer Overflow) (bits/sec), *Data Dropped (Buffer Overflow) (packets/sec)*	These statistics record the amount of data traffic discarded by this WLAN interface due to the upper layer buffer being fully occupied or due to the upper layer packet size exceeding the maximum MSDU size.
Data Dropped (Retry Threshold Exceeded) (bits/sec), *Data Dropped (Retry Threshold Exceeded) (packets/sec)*	These statistics record the total amount of data that was received from the upper layer and then dropped by this WLAN interface due to repeatedly failed retransmissions (i.e., exceeded the corresponding short retry or long retry threshold value).
Data Traffic Rcvd (bits/sec), *Data Traffic Rcvd (packets/sec)*	These statistics record successfully received WLAN data traffic on this WLAN interface from the physical layer. When these statistics are reported in units of bits/second, the physical and the MAC header sizes are included in the computation of the total amount of traffic received. These statistics record all the data received on the WLAN interface regardless of the destination address.
Data Traffic Sent (bits/sec), *Data Traffic Sent (packets/sec)*	These statistics record the amount of data transmitted by this WLAN interface onto the physical layer. When these statistics are reported in units of bits/second, the physical and the MAC header sizes are included in the computation of the total amount of traffic sent.

Delay (sec)

This statistic records the end-to-end delay experienced by all the packets successfully received on this WLAN interface from the physical layer and forwarded to the upper layer.

Load (bits/sec),
Load (packets/sec)

These statistics record the total amount of data submitted by the upper layer to this WLAN interface for transmission.

Management Traffic Dropped (bits/sec),
Management Traffic Dropped (packets/sec)

These statistics record the amount of non-data traffic that was discarded by the WLAN MAC. These statistics record loss of management frames transmitted as a result of 802.11e Block-ACK scheme execution. Thus, these statistics will be recorded only if the current WLAN interface supports the 802.11e protocol and has Block-ACK scheme enabled.

Management Traffic Rcvd (bits/sec),
Management Traffic Rcvd (packets/sec)

These statistics record the total amount of management traffic (i.e., beacon and Block-ACK frames) received from the physical layer on this WLAN interface. These statistics record successful arrival of management frames even if they are not destined to this interface. When these statistics are reported in units of bits/second, the physical and the MAC header sizes are included in the computation of the total amount of traffic received.

Management Traffic Sent (bits/sec),
Management Traffic Sent (packets/sec)

These statistics record the total amount of management traffic (i.e., beacon and Block-ACK frames) transmitted by this WLAN interface onto the physical medium. When these statistics are reported in units of bits/second, the physical and the MAC header sizes are included in the computation of the total amount of traffic sent.

Media Access Delay (sec)

This statistic records the medium access delay experienced by the packets submitted for transmission to this WLAN interface. This value is computed as the interval from the time the packet was inserted into the transmission queue until the time the packet was sent to the physical layer for the first time.

Queue Size (packets)

This statistic records the total number of packets buffered in the transmission queue of this WLAN interface.

Retransmission Attempts (packets)

This statistic records the total number of retransmission attempts (i.e., until the packet is successfully transmitted or is discarded due to reaching the limit of the short or long retry threshold) on this WLAN interface.

Throughput (bits/sec)

This statistic records the total amount of data that was successfully received from the physical medium and forwarded from this WLAN interface to the upper layer.

You can specify MANET configuration through node and global attributes available via the **Configure/Run DES** dialog box. The MANET node configuration is specified through the following compound attributes as shown in Figure 12.7:

- **AD-HOC Routing Parameters**—specify the MANET routing protocol deployed on this node as well as configuration settings for that protocol. Each MANET routing protocol contains another compound subattribute such as **AODV Parameters**, **DSR Parameters**, and **GRP Parameters** depending on which protocol is deployed. We omit description of the routing protocol configuration attributes due to space limitation. Please refer to the OPNET product documentation for a complete description of the supported MANET routing protocol features and available configuration parameters.
- **Wireless LAN**—specifies the physical characteristics of the WLAN interface. Please refer to Section 12.6 for a description of these attributes.
- **MANET Traffic Generation Parameters**—specifies characteristics of the traffic generated by this node. Please refer to Section 6.6.3 for a description of these attributes.

Global attributes for specifying the MANET configuration, as shown in Figure 12.8, primarily address various aspects of routing protocol specification applied to the simulation scenario as a whole. The meaning of these configuration attributes is either self-explanatory or protocol specific and we omit their description from this book.

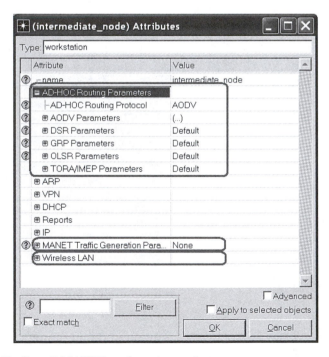

FIGURE 12.7   Overall MANET configuration attributes.

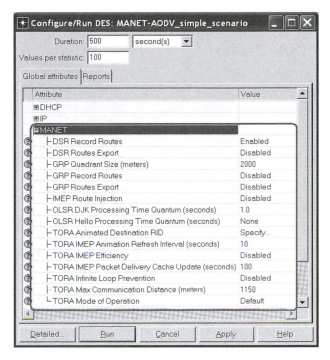

FIGURE 12.8   Global configuration attributes for specifying MANET settings.

In addition, OPNET also provides an option for displaying or hiding DSR and GRP routes. You can access this feature via the **Protocols > MANET > DSR > Display DSR Routes.../ Hide DSR Routes** and **Protocols > MANET > GRP > Display GRP Routes.../ Hide GRP Routes** menu options.

OPNET provides Global and Node Statistics for analyzing MANET and MANET routing protocols. We omit description of the available statistics for the evaluation of MANET routing protocols. However, a summary of **MANET** statistics is available in Table 12.7. Both Global and Node Statistics include a category called **MANET**, which contains statistics for the evaluation of MANET performance. MANET statistics are only available for collection if a simulation scenario contains at least one MANET node model.

The OPNET product documentation provides an excellent overview of the basic steps for deploying the MANET simulation model. In addition, an example project called MANET contains several scenarios, which illustrate configuration and evaluation of MANET networks under various settings.

12.8   SPECIFYING NODE MOBILITY

Any WLAN node can be either fixed or mobile. Fixed nodes remain stationary throughout the simulation, while mobile nodes can move around as the simulation progresses. You can specify the movement of the node during simulation via attribute **trajectory**, which contains a list of preset movement trajectories. If needed, you can

TABLE 12.7
Summary of MANET Statistics

Type	Name	Description
Global Statistics **MANET**	*Delay (secs)*	This statistic records the end-to-end delay experienced by any MANET packet traveling through the network. The end-to-end delay is computed as the time elapsed between the creation of the MANET packet at the source and its destruction at the destination.
	Traffic Received (bits/sec), *Traffic Received* *(packets/sec)*	These statistics record the amount of traffic received by all MANET traffic destinations in the entire network.
	Traffic Sent (bits/sec), *Traffic Sent (packets/sec)*	These statistics record the amount of traffic sent by all MANET traffic sources in the entire network.
Node Statistics **MANET**	*Delay (secs)*	This statistic records the end-to-end delay experienced by MANET packets arriving at this node.
	Traffic Received (bits/sec), *Traffic Received* *(packets/sec)*	These statistics record the amount of traffic received by this node from all other MANET nodes in the network.
	Traffic Sent (bits/sec), *Traffic Sent (packets/sec)*	These statistics record the amount of traffic sent by this node to all other MANET nodes in the network.

specify your own trajectory of the node movement through the **Topology > Define Trajectory...** menu option or clear the deployed trajectory through the **Topology > Clear Trajectory Assignment...** option.

12.8.1 DEFINING A NODE TRAJECTORY

To define a new trajectory, perform the following steps:

1. Select the **Topology > Define Trajectory...** option from the **Project Editor**'s pull-down menu.
2. Specify the trajectory name in the **Define Trajectory** window (Figure 12.9) that appears. Notice that you cannot define a new trajectory without giving a name to the trajectory. You can also specify the amount of time the node will wait before starting its movement, the initial node altitude, as well as the initial roll, pitch, and yaw of the node. However, typically the latter attributes remain set to their default values. It is also a good idea to select the checkbox called *Coordinates are relative to node's position.* Selecting this option will ensure that the starting point in the node's trajectory will be relative to the

node's initial position. As a result, each node, which is assigned the same trajectory will follow a path with the same geometric and temporal characteristics but starting from a different, that is, its own, location in the workspace.

3. Click the **Define Path** button to specify the first segment of the trajectory path. This opens a **Trajectory Status** window (Figure 12.10) and sets a blue line with the word `start` next to it, as a mouse point. You can modify the speed or duration of node movement for this trajectory segment as well as define the units used to measure the distance via the **Trajectory Status** window.

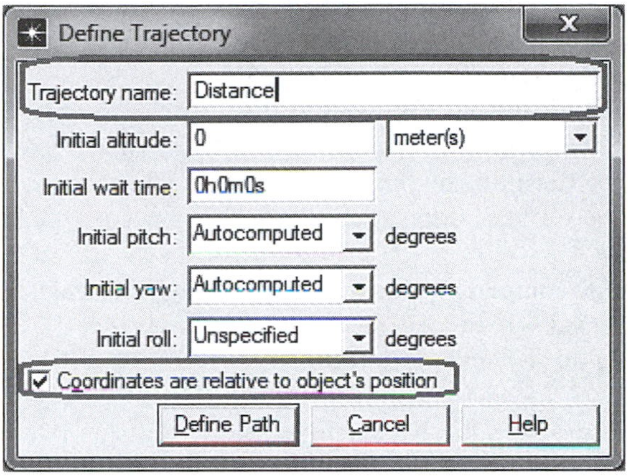

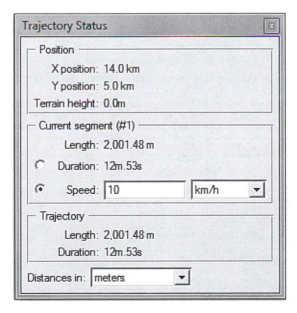

FIGURE 12.9   Define Trajectory window.

FIGURE 12.10   Trajectory Status window.

4. Click on a location in the project workspace where you would like your trajectory to start.
5. Drag the trajectory through the workspace until you reach the place where this trajectory segment should end and click on that location. A red arrow that defines the path for this trajectory segment will appear in the workspace. In addition, the **Segment Information** window (Figure 12.11) will open as well.
6. In the **Segment Information** window, you can specify the node speed as it traverses this trajectory segment as well as the node's altitude, roll, pitch, and yaw as it reaches the end of this segment. You can also specify how long the node should remain stationary at this location.
7. After you have defined the current trajectory segment, you can perform one of the following actions:
    a. Click the **Undo** button to redefine this trajectory segment. This operation can be performed on any trajectory segment except the initial one.
    b. Click the **Continue** button to define the next segment in this trajectory.
    c. Click the **Cancel** button to terminate this trajectory definition without saving it.
    d. Click the **Complete** button to finish and save this trajectory definition. The created trajectory definition is saved as a text file with extension .trj in the default model directory.

Once the new trajectory has been defined, you can assign it to a desired mobile node by setting the **trajectory** attribute of the node to the name of the newly defined trajectory. You will need to find the name of the desired trajectory in the list of all defined trajectories, which appears when you attempt to set the value of this attribute. Also recall that the attribute **trajectory** is available only in mobile nodes. The same trajectory can be assigned to multiple nodes, if desired. In such a case, if the trajectory was defined with coordinates relative to the node's position, then each node to which this trajectory is assigned will start from its own initial position and follow a path that has the same geometric and temporal characteristics as the defined

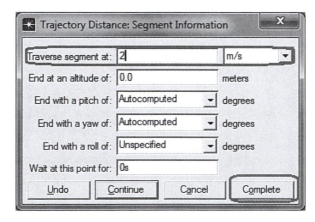

FIGURE 12.11   **Segment Information** window.

trajectory. On the other hand, if the trajectory was defined with absolute coordinates, then all such nodes will follow an identical path irrespective of their initial starting positions.

12.8.2 CONFIGURING A MOBILITY PROFILE

Instead of defining a trajectory with specified coordinates (either absolute or relative), you can choose to configure a node to move randomly during the simulation according to a mobility profile. The **Topology > Random Mobility...** set of menu options allows you to deploy or clear mobility profiles as well as reuse a trajectory that was created using the mobility profile.

Specifically, there are the following three menu options:

- **Set Mobility Profile...** opens the **Configure Mobility Profile on Selected Nodes...** window, which allows you to specify which mobility profile you would like to deploy on selected nodes. Notice that you must select one or more mobile nodes prior to executing this operation. By default, you can configure the selected nodes with one of the following three mobility profiles: `Default Random Waypoint`, `Random Waypoint (Record Trajectory)`, or `Static`.
- **Clear Mobility Profile...** opens a **Clear mobility profiles...** window, which allows you to remove a mobility profile either from all the nodes in the network or from only those selected prior to executing this operation. OPNET removes a mobility profile by setting the attribute **trajectory** on the corresponding nodes to the value `NONE`.
- **Set Trajectory Created from Random Mobility...** allows you to reuse a trajectory that was recorded during the previous simulation run. This option is useful when you would like to rerun a simulation using the same trajectory that was created using a mobility profile during the previous run of the simulation. Notice that in order for this operation to work, you first need to deploy a mobility profile, which records the created trajectory (i.e., the profile attribute **Record Trajectory** is set to `Enabled`) and execute the simulation at least once so that the trajectory is recorded.

In addition to the three default mobility profiles, namely `Default Random Waypoint`, `Random Waypoint (Record Trajectory)`, and `Static`, you can also define your own mobility profile. When you select one of the default profiles with the **Set Mobility Profile...** menu option, a *Mobility Config* node is automatically added to the workspace if it is not part of the simulation scenario yet. Alternatively, you can explicitly add a *Mobility Config* node to the simulation scenario. In either case, you can then define a new mobility profile by adding a new row in the compound attribute **Random Mobility Profile** of the *Mobility Config* node and setting its attributes (Figure 12.12) to desired values. The meaning of these attributes is self-explanatory and thus is omitted here. The number of random mobility profiles available for assignment to wireless nodes corresponds to the number of

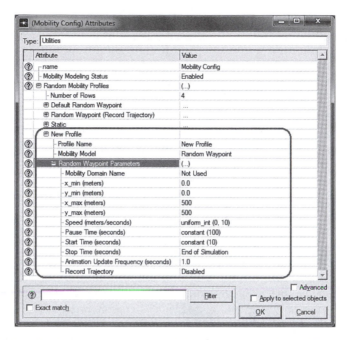

FIGURE 12.12   Configuration attributes for defining a new mobility profile.

profile entries in the *Mobility Config* node. The profiles added by you become available in the **Configure Mobility Profile on Selected Nodes...** window in addition to the three default profiles.

12.9   USING THE WIRELESS DEPLOYMENT WIZARD

OPNET contains a Wireless Deployment Wizard, which significantly simplifies the task of creating and deploying wireless network scenarios. Both WLANs and MANETs can be created with this wizard. To start the Wireless Deployment Wizard, select the **Topology > Deploy Wireless Network...** menu option, which opens the window as shown in Figure 12.13. The deployment process consists of specifying wireless network details through a series of configuration screens. Each screen contains a brief description of its purpose. As you are using the wizard to deploy your wireless network, you will have several buttons become available to you when applicable:

- **Quit** button—terminates the wireless network deployment without saving any changes.
- **Back/Next** button—moves to the previous or next configuration screen, respectively.
- **Help** button—opens a help window, which provides a brief description of the options available in the current configuration screen.
- **Finish** button—completes the configuration process and deploys the specified wireless network in the project workspace.

FIGURE 12.13   **Welcome** window of the Wireless Deployment Wizard.

Now, let us review the procedure for using the Wireless Deployment Wizard:

1. Start the Wireless Deployment Wizard by selecting the **Topology > Deploy Wireless Network...** option.
2. Click the **Continue** button to proceed to the next screen, which allows you to specify whether to load the wireless network specification from a file or to use the wizard to create a brand new network definition.
3. If you open a previously saved configuration file, then all the configuration screens will be preset with the values loaded from that file. Otherwise, all the configuration screens will contain the default values. Regardless of whether you load the network specification from a file or create a new one using the wizard, you still will need to go through all the configuration screens in order to deploy your wireless network.
4. The first configuration screen allows you to specify the location of your wireless network within the project workspace.
5. The next configuration screen, as shown in Figure 12.14, allows you to specify the wireless technology employed in your network. You can select the wireless technology through a pull-down menu, which contains such values as WLAN (Ad-hoc), WLAN (Infrastructure), WiMAX, and TDMA (Distributed). You should use WLAN (Ad-hoc) to create a MANET network and WLAN (Infrastructure) to create a WLAN with an infrastructure of base stations. As you select the wireless technology from the pull-down list, the set of configuration attributes will change.

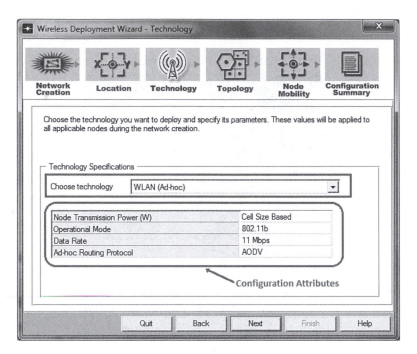

FIGURE 12.14 **Technology** configuration screen of the Wireless Deployment Wizard.

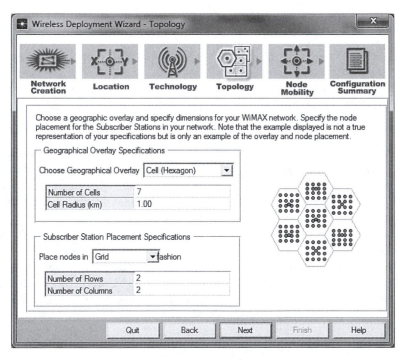

FIGURE 12.15 **Topology** configuration screen of the Wireless Deployment Wizard for specifying node placement.

If desired, you can set these attributes to different values. All the nodes in the deployed wireless network will have the corresponding attributes set to the values specified in this configuration screen.

6. The next configuration screen allows you to specify the placement of the nodes within your network (Figure 12.15). You can specify such information as the size of the area where your wireless network will be deployed as well as the node placement. This is a crucial aspect of the wireless network configuration because the network performance depends on the communication range. For example, placing nodes too far away from one another may result in poor performance because the received signal will be too weak. On the other hand, placing nodes too close to one another may result in every node being able to talk to all the other nodes causing a lot of interference or may prevent you from observing the routing protocol behavior since all routes in the network will be one hop long. Please note that the appearance and the available configuration values depend on your selections in the previous screens.

7. The next configuration screen, as shown in Figure 12.16, allows you to specify the type and the number of the node models to be used in your wireless network.

8. After that you can configure how the nodes in your wireless network will move during the simulation. You can configure your nodes to move randomly according to one of the mobility profiles or specify the actual movement trajectory for the nodes. As shown in Figure 12.17, by changing the

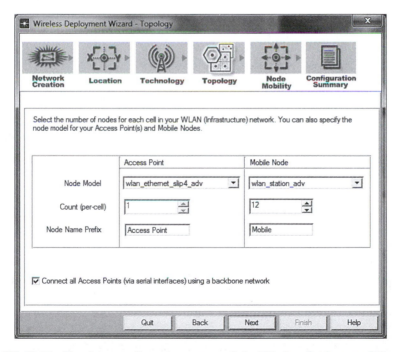

FIGURE 12.16 Topology configuration screen of the Wireless Deployment Wizard for specifying the type and the number of nodes.

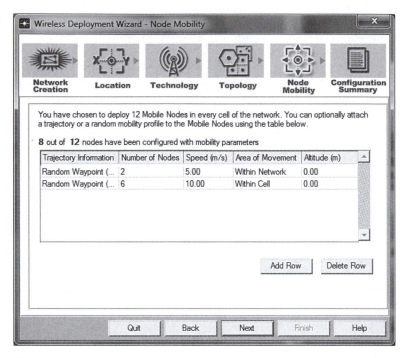

FIGURE 12.17   **Node Mobility** configuration screen of the Wireless Deployment Wizard.

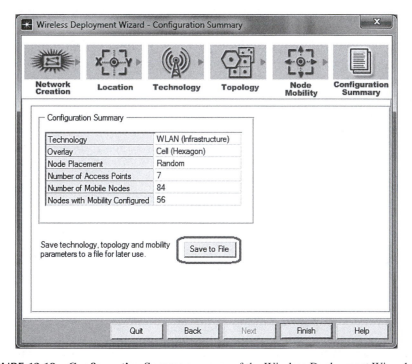

FIGURE 12.18   **Configuration Summary** screen of the Wireless Deployment Wizard.

values of the corresponding cell in the mobility configuration table, you can also specify how many nodes in your wireless network will be moving, their speed and altitude, as well as their area of movement.

9. The final configuration screen (Figure 12.18) provides a brief summary of the created network topology and allows you to save the created network into a file for future reuse. The wireless network topology file is saved into your default model directory as an XML file. However, if needed, you can save the file into an alternative location.

Laboratory Assignment #1: Introduction to OPNET

Recommended reading for this laboratory assignment: Chapters 1 through 4.

L1.1   INTRODUCTION

In this laboratory assignment, you will perform your first complete simulation study. The goal of this laboratory assignment is to provide an exercise that will allow you to practice working with various basic features of OPNET software. Specifically, in this first laboratory assignment you will perform such basic OPNET operations as setting up an OPNET simulation project, creating network topology, configuring individual objects, collecting simulation statistics, running the simulation, managing scenarios, and comparing collected statistics. If you are familiar with these features, then you may skip this assignment.

In this introductory laboratory assignment, we include additional configuration details to help novice OPNET users navigate through the challenges of creating their first OPNET simulation. In the remaining laboratory assignments, we assume that you are already familiar with the basic functionality of OPNET software, and therefore we will omit many such details. Instead, in subsequent assignments we will refer you to the relevant sections in the main chapters of this book, which describe the corresponding configuration steps.

L1.2   CREATING SIMULATION PROJECT AND SCENARIO

Before performing this configuration step, consider reviewing Section 1.6.2, which describes how to create projects and scenarios with the **Startup Wizard**.

Using the **Startup Wizard**, create a new project and an empty scenario (i.e., option `Create Empty Scenario` within the **Initial Topology** window) named `Assignment 01` and `Initial_Network`, respectively. When configuring the initial scenario setup, choose the option `Logical` in the **Network Scale** window and select the *internet_toolbox* model family in the **Select Technologies** window.

L1.3   CREATING NETWORK TOPOLOGY

We recommend that you review Sections 2.3 and 2.4, which describe steps for working with the **Object Palette Tree** and for creating network topology, respectively. These steps are necessary for completing the next portion of this laboratory assignment.

Your first step is to create the network topology shown in Figure L1.1 by dragging into the project workspace the following object models: *ethernet_ip_station*, *ip32_cloud*, *ethernet_slip8_gtwy*, and *ppp_server*. Notice that the default view of the **Object Palette Tree** does not contain the object *ethernet_ip_station*. You need to search the **Object Palette Tree** to locate the model family to which this object belongs (Sections 2.3.2 and 2.4.1).

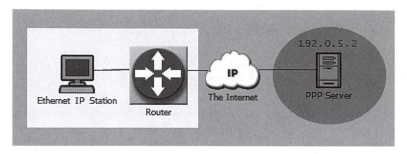

FIGURE L1.1 Network topology for this assignment.

Once the desired node models have been placed within the project, you need to connect them using link models. Specifically, connect *ethernet_ip_station* and *ethernet_slip8_gtwy* using the *10BaseT_int* link model, *ethernet_slip8_gtwy* and *ip32_cloud* using the *PPP_56K* link model, and *ip32_cloud* and *ppp_server* using the *PPP_DS1* link model (Section 2.4.2).

After the basic network topology has been created, verify that you have properly connected the nodes in the simulated network. Refer to Section 2.6.2 for a detailed description of the configuration steps for verifying link connectivity.

L1.4 CONFIGURING NETWORK TOPOLOGY

Once the basic network topology has been created and the link connectivity has been verified, you can start configuring the individual nodes. Specifically, you will need to perform the following steps:

1. Name all the node objects in the simulated network. You can change the name of an object by either selecting the **Set Name** option or by changing the value of the node's attribute **name** as described in Sections 3.2.2 and 3.3, respectively. Name all the node objects in the network as shown in Figure L1.1.
2. Configure the loss and latency experienced by the packets that travel through the Internet by setting the values of the attributes **Performance Metrics...Packet Discard Ratio** and **Performance Metrics...Packet Latency (secs)** in The Internet node to 1.0% and 50 milliseconds, respectively.
3. Next, specify the IP address and the subnet mask values in the node PPP_server by setting attributes **Address** and **Subnet Mask** to 192.0.5.2 and 255.255.255.0, respectively. The attributes **Address** and **Subnet Mask** are part of the compound attribute **IP...IP Host Parameters...Interface Information**.
4. Now, configure the traffic generation source in the Ethernet IP Station node by setting the attribute **IP...Traffic Generation Parameters** to the following values:
   a. **Packet Inter-Arrival Time (seconds)** = constant (0.5)
   b. **Packet Size (bytes)** = constant (2000)
   c. **Destination IP Address** = 192.0.5.2
   d. **Start Time (seconds)** = 100.0

Notice that the attribute **IP...Traffic Generation Parameters** may consist of multiple instances (Section 3.3.3), that is, each row represents a separate traffic source. In this simulation scenario, we are only interested in a single traffic source originating from the `Ethernet IP Station` node, and therefore will require only one attribute instance.

5. Finally, create annotation for the simulated network. Open the **Annotation Palette** (Section 2.9) and place a filled rectangular shape around the `Ethernet IP Station` and `Router` nodes. Also place a filled circular shape around the `PPP Server` node and place a text object that contains the string `192.0.5.2` inside the circular shape as shown in Figure L1.1. You can set a filling color for each added object such as yellow for the rectangular shape and blue for the circular shape.

This completes the process of creating the simulation model of the network. We provide a summary of the network configuration in Table L1.1.

TABLE L1.1
Summary of the Network Topology

Object Name	Object Model	Configuration Details
`Ethernet IP Station` Node	*ethernet_ip_station* node object	Attribute **IP...Traffic Generation Parameters** • **Packet Inter-Arrival Time (seconds)** = `constant (0.5)` • **Packet Size (bytes)** = `constant (2000)` • **Destination IP Address** = `192.0.5.2` • **Start Time (seconds)** = `100.0`
`Router` Node	*ethernet_slip8_gtwy* node object	NONE
`The Internet` Node	*ip32_cloud* node object	Attribute **Performance Metrics** • **Packet Discard Ratio** = `1.0%` • **Packet Latency (secs)** = `0.05`
`PPP Server` Node	*ppp_server* node object	Attribute **IP...IP Host Parameters...Interface Information** • **Address** = `192.0.5.2` • **Subnet Mask** = `255.255.255.0`
`Ethernet IP Station <-> Router` Link	*10BaseT_int* link object	NONE
`Router <-> The Internet` Link	*PPP_56K* link object	NONE
`The Internet <-> PPP Server` Link	*PPP_DS1* link object	NONE

L1.5   CONFIGURING AND RUNNING THE SIMULATION

In this section of the laboratory assignment, you will configure the simulation to collected desired statistics and run the developed simulation model. Your first step here is to configure the simulation to collect the following statistics:

1. Global Statistics in category **IP** (Section 4.2.4)
   a. **Traffic Dropped (packets/sec)**
2. Node Statistics in category **IP** to be collected individually for all nodes (Section 4.2.3)
   a. **Traffic Dropped (packets/sec)**
   b. **Traffic Received (packets/sec)**
   c. **Traffic Sent (packets/sec)**
3. Link Statistics in category *point-to-point* (Section 4.2.3)
   a. **Throughput (bits/sec)** →
   b. **Throughput (bits/sec)** ←
   c. **Throughput (packets/sec)** →
   d. **Throughput (packets/sec)** ←

Next, you need to configure your simulation as follows (Section 4.3):

- **Duration** = 500 seconds
- **Seed** = 128
- **Values per statistic** = 100

Now execute the simulation (Section 4.3.7). When the execution is complete, examine the number of DES Log entries generated (Section 4.8). If the total number of log entries is less than 5, then it is likely that the simulation was configured correctly and you can proceed to the next step of examining the collected results. Otherwise, open the DES Log and check the log entries generated by the simulation. Rerun the simulation after you have identified and fixed any configuration errors.

L1.6   EXAMINING COLLECTED RESULTS

Finally, examine the statistics collected during the simulation (Section 4.5.1). In this portion of the assignment, you will display four sets of simulation statistic results. The first set of results shows the amount of IP traffic dropped at each node in the network. The second set examines the relationship between the amount of traffic sent by Ethernet IP Station, received by PPP Server, and dropped by The Internet. The third set compares the amount of traffic sent by the IP layer in the Ethernet IP Station node and the traffic rate in packets per second forwarded on the link between the Ethernet IP Station and Router nodes. Finally, you will also examine the throughput in bits per second on the link between the Ethernet IP Station and Router nodes.

TABLE L1.2
Configuration Summary of Statistic Graphs

Set	Statistics	Graph Panel Configuration
1	**IP...Traffic Dropped (packets/sec)** for the statistic types: • Global Statistics • Ethernet IP Station Node Statistic • Router Node Statistic • The Internet Node Statistic • PPP Server Node Statistic	Options: • Stacked Statistics • As Is
2	Ethernet IP Station Node Statistic **IP...Traffic Sent (packets/sec)** PPP Server Node Statistic **IP...Traffic Received (packets/sec)** The Internet Node Statistic **IP...Traffic Dropped (packets/sec)**	Options: • Overlaid Statistics • As Is
3	Ethernet IP Station Node Statistic **IP...Traffic Sent (packets/sec)** Link Ethernet IP Station <-> Router statistic • **point-to-point...throughput (packets/sec) ->**	Options: • Overlaid Statistics • As Is
4	Link Ethernet IP Station <-> Router statistic • **point-to-point...throughput (bits/sec) ->**	Options: • As Is

Table L1.2 provides a summary of the statistic results that you need to display. Display each set of statistics in a separate analysis panel. Use the collected results to answer the following questions:

Questions

Q1. Examine the analysis panel that displays the statistic set 1.
  a. Which nodes in the simulated network discard packets? Which nodes in the simulated network do not discard any packets? Why? (Hint: review the network configuration of the simulation).
  b. What is the difference between Global and Node Statistics? Do you observe any differences between the **IP...Traffic Dropped (packets/sec)** graphs for the Global and Node Statistics which discards packets? Why or why not?

Q2. Examine the graph panel that displays the statistic set 2.
  a. How does the amount of traffic sent by Ethernet IP Station compare to the traffic received by PPP Server and the traffic dropped by The Internet? Explain.

Q3. Examine the graph panel that displays the statistic set 3.
  a. What is the Ethernet IP Station sending rate in packets per second reported by the simulation? What is the throughput rate in packets per

second on the link between Ethernet IP Station and Router as reported by the simulation? Are these values the same or different? Why?

b. According to your configuration of the traffic generation source, Ethernet IP Station generates one packet each 0.5 seconds, which corresponds to a sending rate of 2 packets per second. What is the Ethernet IP Station sending rate in packets per second as reported by the simulation? Are these numbers the same or different? Why? What is your hypothesis as to what happens at Ethernet IP Station? (Hint: What is the packet size generated by Ethernet IP Station and what is the MTU size in the Ethernet? What happens when the upper layer packet is too large to be transmitted over the physical medium?)

Q4. Examine the graph panel that displays the statistic set 4.

a. What is the transmission rate in bits per second on the link between Ethernet IP Station and Router? Do the results reported by the simulation correspond to the traffic source configuration? (Hint: assume that IP and Ethernet headers are 20 and 26 bytes, respectively.)

L1.7 DUPLICATING SCENARIOS, RERUNNING THE SIMULATION, AND COMPARING COLLECTED RESULTS

In this part of the assignment, you will create three copies of the current scenario (Section 1.7) and modify the scenario configuration as follows:

1. Duplicate the Initial_Network scenario, naming the new scenario Packet_Size_1000bytes. Modify the configuration of the traffic generation source in the Ethernet IP Station node so that each newly generated packet is 1000 bytes instead of 2000 bytes. Do not execute the simulation yet.

2. Duplicate the Initial_Network scenario, naming the new scenario 500_values_statistic. Modify the simulation configuration by setting the attribute **Values per statistic** to 500. Save the simulation configuration without running the simulation (Section 4.3).

3. Duplicate the Initial_Network scenario, naming the new scenario 5000_values_statistic. Modify the simulation configuration by setting the attribute **Values per statistic** to 5000. Save the simulation configuration without running the simulation (Section 4.3).

Using the **Manage Scenarios** menu (Section 1.7.2), (re)collect the statistics for all the scenarios within the Assignment 01 project. Once all scenarios have been executed, display the collected statistics as follows:

- Display the Ethernet IP Station Node Statistic **IP...Traffic Sent (packets/sec)** for scenarios Initial_Network and Packet_Size_1000bytes in a single graph panel using the *Overlaid Statistics* and *As Is* options (Section 4.4).

- Display The Internet Node Statistic **IP...Traffic Dropped (packets/ sec)** for scenarios Initial_Network, 500_values_statistic, and 5000_values_statistic in a single graph panel using the *Overlaid Statistics* and *As Is* options (Section 4.4).

Now, answer the following questions:

Questions

Q5. What do you observe in the first graph, which displays the traffic sent by the Ethernet IP Station node? Are there any differences in the packet sending rate as reported by the results collected in the Initial_Network and Packet_Size_1000bytes scenarios? Why? Notice that you did not change the packet interarrival rate at the node Ethernet IP Station. Do these results confirm or refute the hypothesis developed for Question 3(b) in Section L1.6?

Q6. What do you observe in the second graph, which displays the traffic discarded by The Internet node? Is the number of packets discarded the same or different in each scenario? Why or why not? (Hint: recall how statistics are collected in OPNET and how the attribute **Values per statistic** influences the reported results [Sections 4.1.1 and 4.2.7]).

Laboratory Assignment #2: Simple Capacity Planning

Recommended reading for this laboratory assignment: Chapters 1 through 4.

L2.1 INTRODUCTION

In this laboratory assignment, you will examine the influence of link capacity on network performance as well as practice configuring traffic demands and specifying multiple values for a single attribute. Specifically, you will evaluate the performance of the given network under various conditions and recommend an optimal configuration.

Consider the following situation: ABC Inc., a small private company, is in the process of expanding and would like to add another office on the other end of town. The company plans to double the size of the new office in the future. In this laboratory assignment, you will help ABC Inc. determine the best option for provisioning the links connecting their offices to the Internet.

L2.2 MODELING ABC INC.'S NETWORK

Create a new project and empty scenario named `Assignment 02` and `ABC_Network`, respectively (Section 1.6.2). Create the network topology as shown in Figure L2.1. Make sure that you use node and link models as specified in Table L2.1 (following the instructions in Section 2.4). Once you have created the network topology, verify link connectivity (Section 2.6.2).

Next, you need to add and configure traffic demands (Section 6.6). Add and configure four *ip_traffic_flow* demand models as follows:

- All demands should be configured to have 1% of traffic modeled as explicit traffic (attribute **Traffic Mix**).
- All demands should start transmitting data at time 100 seconds.
- All demands should continue transmitting data until the end of simulation.
- Both `Main Office → DB Server` and `New Office → DB Server` IP demands transmit data at constant rates of 1200 kbps and 100 packets per second.
- Both `Main Office → E-mail Server` and `New Office → E-mail Server` IP demands transmit data at constant rates of 800 kbps and 10 packets per second.

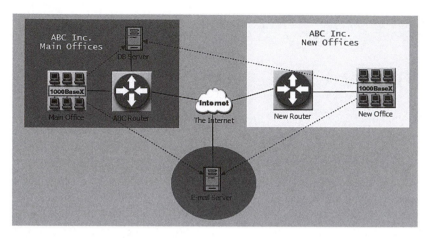

FIGURE L2.1   Network topology for this assignment.

TABLE L2.1
Summary of ABC Inc.'s Network Topology

Object Name	Object Model
DB Server	*ethernet_server* node object
E-mail Server	*ppp_server* node object
Main Office New Office	*1000BaseX_LAN* node object
ABC Router New Router	*ethernet4_slip8_gtwy* node object
The Internet	*ip32_cloud* node object
Main Office <-> ABC Router DB Server <-> ABC Router New Office <-> New Router	*1000BaseX* link object
ABC Router <-> The Internet New Router <-> The Internet	*PPP_DS1_int* link object
The Internet <-> E-mail Server	*PPP_DS3* link object

L2.3   EVALUATING ABC INC.'S NETWORK

Configure your simulation to collect the following statistics:

- All Demand Statistics (Section 6.8).
- All Link Statistics in the category *point-to-point* (Section 4.2.3).

Execute the simulation for 1 hour (Section 4.3.7) and then examine the following collected statistics (Section 4.5):

- Traffic sent by each of the traffic demands.
- Utilization on all the links connected to The Internet node.
- Traffic received by each of the demand destinations.
- End-to-end delay experienced by the demand packets. Note that if you did not configure your demands to generate traffic explicitly (i.e., did not set the **Traffic Mix** attribute to a value greater than 0%), then this Demand Statistic will report no data.

Questions

Q1. What was the sending rate by each demand? Did the simulation results correspond to the demand configuration?

Q2. What was the utilization on the links connected to the Internet? Why? (Hint: What are the capacities of 1000Base-T, DS-1, and DS-3 links? What are the transmission rates of each demand?)

Q3. What was the rate at which traffic arrived at each demand's destination? Why?

Q4. What was the packet end-to-end delay for each of the demands? Why?

Q5. In your opinion, what is the problem with the current network configuration? Which link in the network is the bottleneck?

L2.4   COMPARING APPLICATION PERFORMANCE

Duplicate the original scenario and create a new one called Eliminating Bottleneck (Section 1.7). In the new scenario, set the capacity of the bottleneck link to the **smallest** value that will result in the bottleneck link utilization of around 60%. Note that if the link object does not contain the attribute **data rate**, which represents the link's capacity, then you need to change the link model to one that contains the suffix _int in its name. To change the model associated with the object, you first need to select the *Advanced* checkbox in the **Edit Attributes** window of the object and then change the value of the attribute **model** (Section 3.4.5). After you have updated the capacity of the bottleneck link, rerun the simulation and examine the collected statistics.

Questions

Q6. What should be the new capacity of the bottleneck link?

Q7. What was the utilization on the links connected to the Internet? Why?

Q8. What was the rate at which traffic arrived at each demand's destination? Why?

Q9. What was the end-to-end delay experienced by the traffic of each demand? Why?

L2.5   IDENTIFYING THE OPTIMAL BANDWIDTH/COST RATIO

Duplicate the scenario Eliminating Bottleneck and create a new one called Best Option (Section 1.7). Assume that ABC Inc. doubled the staff at their new offices, which resulted in twice the amount of traffic generated from these offices. To model such behavior, double the transmission rate of the traffic demands originating from the New Office node.

Currently, the Internet service provider (ISP) offers the following options for connecting to the Internet:

- T1 line for $50 per month
- 4 Mbps line for $100 per month
- 10 Mbps line for $150 per month
- 50 Mbps line for $500 per month

Promote the attribute **data rate** on the bottleneck link and then set the value of the promoted attribute to the Internet access rates available through the ISP (i.e., T1 or DS1, 4 Mbps, 10 Mbps, and 50 Mbps [Section 3.5]). Rerun your simulation and examine the bottleneck link utilization and the end-to-end delay experienced by the packets of the IP demands originating from the New Office node.

Questions

When answering the following questions, consider the four different capacities specified on the bottleneck link.

Q10. What was the link utilization on the bottleneck link? What was the packet end-to-end delay for the IP demand traffic between New Office and E-mail Server? Based on these results, which of the link capacities should ABC Inc. choose to connect their new offices to the Internet?

Q11. What was the packet end-to-end delay for the IP demand traffic between New Office and DB Server? Why?

Laboratory Assignment #3: Introduction to Standard Applications

Recommended reading for this laboratory assignment: Chapters 1 through 5 and 7.

L3.1   INTRODUCTION

The primary goal of this laboratory assignment is to introduce you to the modeling of standard applications in OPNET. Specifically, in this assignment you will configure several standard applications, specify and debug the created user profile definitions, deploy the specified applications in the network, and finally examine application performance.

Consider the following situation, where a private R&D company called Researchers Inc. is spread over three sites:

- `Main Site` consists of 50 employees who primarily use e-mail, web browsing, and voice conferencing to perform their daily duties.
- `RnD Site` consists of 100 employees who primarily use remote login, e-mail, and web searching to perform their daily tasks.
- `Server Farm` contains servers for e-mail, web, and remote login applications.

Your assignment is to create a simulation model of the Researchers Inc. network and examine the performance of the applications deployed in the network.

L3.2   MODELING RESEARCHERS INC.'S NETWORK TOPOLOGY

In this portion of the laboratory assignment, you will create a model of the Researchers Inc. network by performing the following steps:

1. Create a new project and an empty scenario named `Assignment 03` and `Researchers Network`, respectively (Section 1.6.2).
2. Create the network topology as shown in Figure L3.1. Make sure that you use node and link models as specified in Table L3.1 (following instructions of Section 2.4).
3. Configure your `Main Site` and `RnD Site` LAN objects to contain 50 and 100 workstations, respectively. You can specify the number of workstations in a LAN object by setting the attribute **LAN...Number of Workstations** to the desired value.
4. Once you have created the network topology and configured the LAN objects, verify link connectivity (Section 2.6.2).

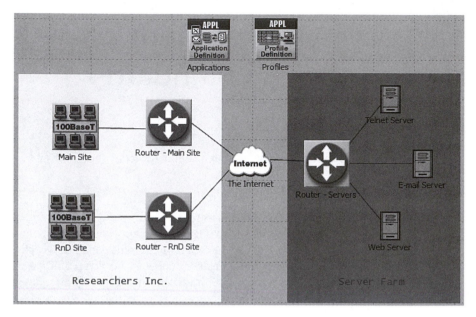

FIGURE L3.1   Network topology for this assignment.

TABLE L3.1
Summary of Researchers Inc.'s Network Topology

Object Name	Object Model
Telnet Server	*ethernet_server* node object
E-mail Server	
Web Server	
Router-Servers	*ethernet4_slip8_gtwy* node object
Router-Main Site	
Router-RnD Site	
Main Site	*100BaseT_LAN* node object
RnD Site	
The Internet	*ip32_cloud* node object
Main Site <-> Router-Main Site	*100BaseT* link object
RnD Site <-> Router-RnD Site	
Telnet Server <-> Router-Servers	
E-mail Server <-> Router-Servers	
Web Server <-> Router-Servers	
The Internet <-> Router-Main Site	*PPP_DS1_int* link object
The Internet <-> Router-RnD Site	
The Internet <-> Router-Servers	

L3.3   CONFIGURING AND DEPLOYING APPLICATIONS
IN RESEARCHERS INC.'S NETWORK

In this portion of the laboratory assignment, you will configure and deploy applications in the Researchers Inc. network by performing the following steps:

1. Configure individual applications (Section 5.3.2).
   - Add the **Application Config** node into your project workspace, if it has not been added yet.
   - Define five standard applications according to the configuration details provided in Table L3.2.
2. Define user profiles (Section 7.2). For now, let us assume that each Researchers Inc. employee executes applications sequentially, that is, one after another. Based on this assumption, you will create two profiles:
   - Main Site Employees, which runs the following three applications: Web Browsing, E-mail, and VoIP.
   - RnD Employees, which runs the following three applications: Web Research, Telnet, and E-mail.
3. Configure each profile to run applications in Serial (Ordered) **Operation Mode** while leaving the remaining profile configuration attributes set to their default values.
4. Deploy the created profiles according to the following rules (Section 7.4):
   - Main Site Employee profile is configured to have Main Site serve as the source of the profile, whereas objects Web Server, E-mail Server, and RnD Site operate as servers or destinations for Web Browsing, E-mail, and VoIP applications, respectively.
   - RnD Employees profile is configured to have RnD Site serve as the source of the profile, whereas objects Web Server, Telnet Server, and Email Server operate as servers or destinations for Web Research, Telnet, and E-mail applications, respectively.
5. Configure both LAN objects to have the deployed profiles executed on all nodes within the LAN. You can specify this requirement by setting the

---

TABLE L3.2
Application Configuration Settings

Application Name	Application Model Attribute	Application Model Attribute Value
Web Browsing	**Http**	Heavy Browsing
Telnet	**Remote Login**	High Load
Web Research	**Http**	Searching
E-mail	**Email**	Medium Load
VoIP	**Voice**	GSM Quality Speech

attribute **Applications...Application: Supported Profiles...<profile>...
Number of Clients** to `Entire LAN`, where **<profile>** is either **Main Site
Employee** or **RnD Employee**.

6. Configure your simulation to collect all Global Statistics for **Email**, **HTTP**,
   Remote Login, and **Voice** applications (Sections 4.2.4 and 5.6).
7. Execute the simulation for 5 minutes (Section 4.3).

Questions

Q1. How many DES Log entries were generated upon simulation completion
   (Section 4.8)?

Q2. Open the **Log Viewer** window (**DES > Open DES Log**) and examine the
   DES Log entry that deals with application setup (Section 4.8). What problem was reported by that log entry?

Q3. Examine the Global Statistic **Traffic Sent (bytes/sec)** for each application.
   Do the simulation results confirm or refute the DES Log warning? Why or
   why not?

Q4. Which of our original assumptions and subsequent configuration steps led
   to this issue and how could this issue be corrected?

L3.4   FIXING THE FIRST CONFIGURATION PROBLEM

One way to fix the problem above is to change how the applications within the profile
are executed. The current configuration allows only the first application within the
profile to start. The remaining application never executed because the first application
is configured to run until the end of simulation, while applications within the profile
are executed in `Serial (Ordered)` fashion, meaning that the next application can
start only after the previous application has completed its execution (Section 7.6).

Clearly, our original assumption that applications are executed one after another
is incorrect. In practice, it is common for a user to run the web browser, e-mail,
Telnet, and other applications simultaneously while switching between applications
as needed. Perform the following steps to model such behavior:

1. Create a duplicate scenario called `Fixing Problem 01` (Section 1.7).
2. Change the operation mode from `Serial (Ordered)` to `Simultaneous` for
   both `Main Site Employee` and `RnD Employee` profiles (Section 7.2).
3. Configure the simulation to also collect link *point-to-point* statistics
   (Section 4.2.3).
4. Rerun the simulation.

Questions

Q5. Examine the sending rate of each application. Did the above profile
   configuration change solve the problem discovered in the previous section?

Q6. Examine the download response time for the `E-mail` application,
   page response time for web applications, response time for the `Telnet`

application, and the end-to-end delay experienced by the VoIP packets. Display these statistics using the average function in a single analysis panel (Section 4.5). Did the displayed statistic graphs of the application response times and delays demonstrate a trend? Why or why not?

Q7. Do you think the delays experienced by the applications are acceptable to the company employees or not? Why or why not?

Q8. Examine utilizations on the links that connect the network routers with the Internet. Based on the observed link utilizations, which application causes congestion in the network?

L3.5   FIXING THE SECOND CONFIGURATION PROBLEM

The simulation statistics collected in the previous section report that links Router–Main Site <-> The Internet and Router–RnD Site <-> The Internet are utilized 100%, whereas the link between the Internet and the server farm is utilized less than 20%. This suggests that the VoIP application is the culprit that causes network congestion. After further discussion with the Researchers Inc. staff, you discover that the Main Site actually sets up voice conferences only once a day. Usually, this conference call lasts only for about an hour and is conducted from a single meeting room at about 2:00 PM in the afternoon.

To model this behavior, you will need to modify your application deployment, change the definition of the Main Site Employee profile, and create an additional profile to model the conference call. Specifically you need to perform the following configuration steps:

1. Duplicate the current scenario Fixing Problem 01, creating a new one called Fixing Problem 02 (Section 1.7).
2. Clear the application deployment in the network (Section 7.4.8).
3. Remove the VoIP application from the definition of the Main Site Employee profile (Section 7.2).
4. Create a new profile called Conference Call, which runs only the VoIP application. The VoIP application should start 6 hours after the start of the profile. It should run for only one hour and it should repeat once. You can leave the remaining attributes set to their default values. Notice that the configuration attributes accept the value of time in units of seconds and that the profile's operation mode has no influence on the simulation because the profile consists of a single application (Section 7.2).
5. Redeploy all the profiles in the network. Configure the Conference Call profile to have Main Site as the source and RnD Site as destination for the VoIP application (Section 7.4).
6. Modify the configuration of the Main Site LAN object to have the profile Conference Call run on a single client (i.e., attribute **Applications... Application: Supported Profiles...Conference Call...Number of Clients**).
7. Set the duration of simulation to 8 hours (Section 4.3).
8. Rerun the simulation.

Questions

Q9. Examine the download response time for the E-mail application, page response time for web applications, response time for the Telnet application, and the end-to-end delay experienced by the VoIP packets. How do these values compare to the results collected in the scenario Fixing Problem 01? Do you think that such delays are acceptable for the application users? Why or why not?

Q10. Examine the sending rate for each application. How do these values compare to the results collected in the scenario Fixing Problem 01? Which application generates the most traffic? Validate the observed results through the application configuration (i.e., examine each application setting and compute the amount of data generated by each application).

Q11. Examine the utilizations on the links that connect the network routers to the Internet. Did the new profile configuration eliminate the bottlenecks in the network?

Laboratory Assignment #4: HTTP Performance

Recommended reading for this laboratory assignment: Chapters 1 through 5 and 7.

L4.1 INTRODUCTION

In this laboratory assignment, you will examine the performance of various versions of the HTTP protocol. In particular, you will study the influence of persistent, non-persistent, with pipelining, and without pipelining, variations of HTTP on the performance of a web application.

Consider the following scenario: Jason is a graduate student who conducts research over the Internet. Jason's computer is connected to the Internet via a T1 line. Jason uses a web browser that runs HTTP version 1.1. A regular workday for Jason looks as follows: Jason comes to his office, starts his computer, and logs onto the network, which takes about 100 seconds. Then Jason runs his web browser, which takes about 5–10 seconds to start up. Usually, Jason visits a new page every 10 seconds. Generally, each web page consists of several large images. Most of the web pages that Jason visits reside on the same server. Jason stops browsing after 2 hours.

L4.2 CREATING THE SIMULATION MODEL

In this portion of the assignment, you will create a simulation model of Jason's network by performing the following steps:

1. Create a new project and an empty scenario named `Assignment 04` and `HTTP_1_1`, respectively (Section 1.6.2).
2. Create the network topology as shown in Figure L4.1 using node and link models specified in Table L4.1 (Sections 2.3 and 2.4).
3. Configure `The Internet` cloud to discard 1.0% of all arriving packets and introduce an additional 200 millisecond delay to all traffic traveling through it (see Laboratory Assignment #1 for configuration details).
4. Verify link connectivity (Section 2.6.2).
5. Configure the web applications as follows (Section 5.3):
   a. Name Jason's web browsing application `Web`.
   b. Configure Jason's application to run over HTTP 1.1 (by editing the attribute **HTTP Specification** [Section 5.4.4]).
   c. Configure Jason's application to request a new web page approximately every 10 seconds. Use the exponential distribution to simulate such behavior (via the attribute **Page Interarrival Time (seconds)**).

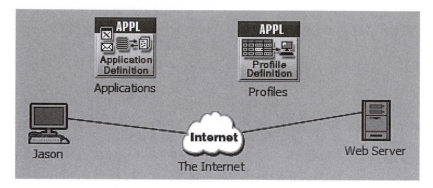

FIGURE L4.1   Network topology for this assignment.

TABLE L4.1
Summary of Network Topology

Object Name	Object Model
Jason	*ppp_wkstn* node object
Web Server	*ppp_server* node object
The Internet	*ip32_cloud* node object
Jason <-> The Internet Web Server <-> The Internet	*PPP_DS1_int* link object

---

    d. Configure the application to access web pages, each of which consists of a single text object of size 1000 bytes and seven large images embedded within it (use the compound attribute **Page Properties**).

    e. Leave the remaining attributes set to their default values.

6. Create a profile called Web Research to represent Jason's behavior as described in Section L4.1. (Section 7.2):

    a. Jason only runs the web browser on his computer (i.e., the profile contains only one application called Web).

    b. Jason starts the web browser 5–10 seconds after he logs onto the computer (i.e., the application starts 5–10 seconds after the start of the profile). Use the uniform distribution to model the application start time offset.

    c. Jason browses the web for 2 hours (i.e., the application runs for 2 hours). Use the constant distribution to model the duration of the application.

    d. It takes 100–110 seconds to boot the computer up (i.e., the profile starts 100–110 seconds after the start of simulation). Use the uniform distribution to model the profile start time.

    e. The simulation terminates as soon as Jason turns the computer off (i.e., the profile runs until the end of simulation).

7. Deploy the profile Web Research so that node Jason is its source and node Web Server is the server for the Web application (Section 7.4).
8. Configure the simulation to collect the following statistics:
   a. All Global Statistics for the HTTP application (Sections 4.2.4 and 5.6).
   b. Node Statistic **Active Connection Count** from category **TCP** on all nodes (Section 4.2.3).
   c. Any additional statistics that may be of interest to you.
9. Execute the simulation for 8000 seconds (Section 4.3).

Questions

Q1. What were the reported page and object response times? How do these values compare? Why?
Q2. Examine the maximum number of active TCP connections reported on the HTTP client and server nodes. Which property of HTTP version 1.1 explains the observed behavior and why?

L4.3   HTTP 1.0 VERSUS HTTP 1.1

In this portion of the laboratory assignment, you will compare the application performance when running HTTP version 1.0 and HTTP version 1.1. Before creating a new simulation scenario, write down your hypothesis as to what you expect to observe. Specifically, explain how the page and object response times and the number of active TCP connections will change while running HTTP version 1.0 instead of HTTP 1.1. Verify your hypothesis by creating another simulation scenario that runs the web browser over HTTP 1.0, executing it, and then examining the collected results:

1. Duplicate the current scenario naming the new one HTTP_1_0 (Section 1.7).
2. Change the Web application to run HTTP version 1.0 instead of version 1.1 (Section 5.4.4).
3. Execute the simulation scenario HTTP_1_0 (Section 4.3).

Questions

If you configured the simulation correctly, then you should observe the following:

- The page response time for HTTP version 1.1 will be smaller than the page response time for HTTP version 1.0.
- The object response time for HTTP version 1.1 will be larger than the object response time for HTTP version 1.0.

Review the specifications for HTTP 1.1 and HTTP 1.0 and explain the observed behavior (i.e., you can examine the assigned values for the compound attribute

HTTP Specification within the definition of the HTTP application). Specifically, answer the following questions:

Q3. Examine the page and object response time statistics collected in scenarios HTTP_1_1 and HTTP_1_0. What did you observe? Why is the page response time for HTTP 1.1 smaller than that for HTTP 1.0 while the object response time is larger for HTTP 1.1 than for HTTP 1.0?

Q4. Examine the number of active TCP connections established by the web application in each scenario. Which version of HTTP had fewer active TCP connections? Why?

Note: Use statistics collected at the Web Server node when examining the number of active TCP connections created by HTTP traffic and do not be too concerned with the absolute value for the number of active TCP connections. The reported statistic values may appear to be incorrect. This is primarily due to the nature of OPNET's mechanism for collecting statistics and the duration of the time period that a TCP connection remains open. Each time a node opens or closes a TCP connection, OPNET increments or decrements by 1 the value of the corresponding statistic variable. At the simulation level, the number of active TCP connections statistic is recorded as a sum of the statistic variable values over a period of time, typically larger than a second. On the client side, the TCP connections remain open for a very short period of time (i.e., a small fraction of a second, just long enough to retrieve the object). As a result, the number of open connections on the client side is often recorded as zero. On the other hand, the server side keeps the TCP connection open for a longer period of time, which is usually recorded properly by OPNET's statistic collection mechanism. In addition, TCP connections on the server side that handle different HTTP requests may remain open at the same time, resulting in the reported number of active TCP connection being larger than the actual number of requested page objects.

L4.4   HTTP WITH AND WITHOUT PIPELINING

In this portion of the laboratory assignment, you will examine how HTTP pipelining features influence the performance of web applications. Prior to working on this part of the assignment, write down what you expect to observe. Specifically, explain how the page and object response times and the number of active TCP connections will change when using HTTP version 1.1 without pipelining. Verify your hypothesis by creating another simulation scenario that runs the web browser over HTTP 1.1 without pipelining, executing it, and then examining the collected results:

1. Duplicate scenario HTTP_1_1 naming the new one HTTP_1_1_no_ pipelining (Section 1.7).

2. Change the Web application to run HTTP version 1.1 with the total number of pipelined requests set to 1 (Section 5.4.4).

3. Execute the simulation scenario `HTTP_1_1_no_pipelining` (Section 4.3).
4. Compare the collected statistics for all three scenarios. Consider displaying the collected response time statistics using the `average` option (Section 4.5.4).

Questions

If you configured the simulation correctly, then you should observe the following:

- The object response time for HTTP 1.1 without pipelining will be the smallest.
- The page response time for HTTP 1.1 without pipelining will be the largest.

Review the definitions of persistent HTTP connection and pipelining and explain the observed behavior. Specifically, answer the following questions:

Q5. Why does HTTP 1.1 without pipelining have the smallest object response time?
Q6. Why does HTTP 1.1 without pipelining have the largest page response time?
Q7. How does the number of active TCP connections differ for HTTP version 1.1 with and without pipelining? Why? Note that to answer this question, you may want to examine the statistics collected on the client side only.

L4.5   SIMPLE WEB PAGE

In the last portion of this laboratory assignment, you will examine how different HTTP configurations influence the performance of a web application that requests only simple one-object web pages. Before starting this part of the assignment, write down your hypothesis as to how the object response time, the page response time, and the number of active TCP connections will differ for HTTP version 1.0, HTTP version 1.1 with pipelining, and HTTP version 1.1 without pipelining. Verify your hypothesis by performing the following steps:

1. Duplicate scenario `HTTP_1_1` naming the new one `HTTP_1_1_simple` (Section 1.7).
2. Change the `Web` application definition by setting the web page to consist of a single 1000 bytes object (Section 5.4.4).
3. Repeat steps 1 and 2 for scenarios `HTTP_1_0` and `HTTP_1_1_no_pipelining`, naming the new scenarios `HTTP_1_0_simple` and `HTTP_1_1_no_pipelining_simple`, respectively.
4. Execute the newly created scenarios (Section 4.3).
5. Compare the collected statistics for all three scenarios. Consider displaying the collected statistics using the `average` option (Section 4.5).

Questions

If you have configured the simulation correctly, then you should observe the following:

- The page response time for HTTP 1.1 (both with and without pipelining) will be smaller than that for HTTP 1.0.
- The object response times in all cases will be almost identical.

Review the configuration setup and the definitions of persistent and non-persistent HTTP connections. Explain the observed behavior by answering the following questions:

Q8. How does the number of active TCP connections differ for HTTP version 1.1 with and without pipelining when requesting a simple web page? Why?

Q9. Why are the HTTP 1.1 page response times smaller than that for HTTP 1.0?

Q10. Why are the object response times for HTTP 1.1 and 1.0 nearly the same?

Laboratory Assignment #5: Modeling Custom Applications

Recommended reading for this laboratory assignment: Chapters 1 through 7.

L5.1 INTRODUCTION

In this laboratory assignment, you will develop and then compare the performance of two custom applications:

- An application that provides encrypted communication with authentication
- An application that provides only plain text communication

In particular, you will examine how the added level of security influences the overall application performance.

Consider the following scenario: Alice develops an application for communicating with Bob. Alice would like to communicate securely over the network, but she is not sure if the added layer of security will hinder her application's performance by introducing too much delay. Prior to starting program development, Alice would like to conduct a simulation study that compares the performance of two possible versions of her application. The first version of her application, which we call `Secure AB Talk`, consists of two distinct parts: (1) authentication and session key discovery and (2) encrypted message exchange. The second version of her application, which we call `AB Talk`, only does plain text message exchanges.

The authentication and session key discovery portion of the `Secure AB Talk` application works as described below. We will denote message M encrypted with key K as $K\{M\}$.

1. Alice sends a message to the Key Distribution Center (KDC), a trusted third-party authority, requesting a session key for her communication with Bob.
2. KDC generates a session key, K_{AB}, for communication between Alice and Bob.
3. KDC sends a message $K_{AP}\{K_{AB}, T_B\}$ to Alice, where:
   a. K_{AP}—Alice's public key.
   b. T_B—ticket generated by KDC to be sent to Bob. $T_B = K_{BP}\{K_{AB}, \text{``Alice''}\}$.

    c. K_{BP}—Bob's public key.

    d. *"Alice"*—Alice's identity.

4. Alice receives the above message from KDC and decrypts it using her private key, providing Alice with K_{AB} and T_B.

5. Alice sends a message $\{T_B, K_{AB}\{S_A\}\}$ to Bob where S_A is some secret number.

6. Bob receives the message from Alice and decrypts T_B using his private key, thereby obtaining K_{AB}.

7. Bob uses the retrieved key K_{AB} to decrypt Alice's secret number S_A. This proves Alice's identity, because only someone who knows Alice's private key could have obtained K_{AB} and the T_B ticket containing Alice's identity in step 4.

8. Bob encrypts and sends Alice a message $K_{AB}\{S_A+1\}$. Bob encrypts the value S_A+1 to protect against a replay attack.

9. Alice decrypts Bob's message using the session key K_{AB}, verifying Bob's identity, because only Bob could have known the value S_A and the session key K_{AB}.

    The algorithm described above is loosely based on the Kerberos authentication service, and Figure L5.1 presents its visual summary.

    The encrypted data exchange part of the `Secure AB Talk` application consists of an asynchronous exchange of encrypted messages between Alice and Bob. The `AB Talk` application consists of a single part, which models the exchange of plain text messages between Alice and Bob. The plain text data exchange is also done asynchronously, meaning that Alice can send a message to Bob prior to receiving a request and vice versa. In this model, we assume that encryption does not change the size of the message being transmitted. However, encryption is a computationally intensive task and therefore will increase the message processing time.

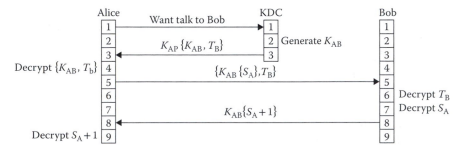

FIGURE L5.1 `Secure AB talk` authentication protocol.

L5.2 CREATING THE SIMULATION MODEL

In this portion of the assignment, you will create a simulation model of the network used by Alice and Bob for communication.

1. Create a new project and an empty scenario named `Assignment 05` and `AB_Talk`, respectively (Section 1.6.2).
2. Create the network topology as shown in Figure L5.2 using the node and link models specified in Table L5.1 (Sections 2.3 and 2.4). Make sure to use advanced models (i.e., suffix _*adv*) for the `Alice` and `Bob` nodes.
3. Verify link connectivity (Section 2.6.2).
4. Configure `The Internet` cloud to discard 0.0% of all arriving packets and to introduce an additional 100 millisecond delay to all traffic traveling through it (see Laboratory Assignment #1 for configuration details).

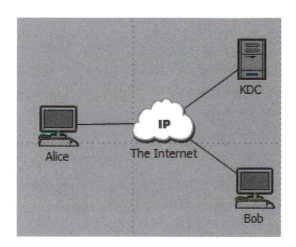

FIGURE L5.2   Network topology for this assignment.

TABLE L5.1
Summary of Network Topology

Object Name	Object Model
Alice, Bob	*ppp_wkstn_adv* node object
KDC	*ppp_server* node object
The Internet	*ip32_cloud* node object
Alice <-> The Internet KDC <-> The Internet Bob<-> The Internet	*PPP_DS1_int* link object

TABLE L5.2
Summary of Phase Configuration: Part 1

Task Name	Phase Name	Start After	Source	Destination
Authentication	Talk to KDC	Application Starts	Alice	KDC
	Talk to Bob	Previous Phase Ends	Alice	Bob
	Verify Bob	Previous Phase Ends	Alice	None
Secure Transmission	Alice to Bob	Application Starts	Alice	Bob
	Bob to Alice	Application Starts	Bob	Alice
Plain Text Transmission	Alice to Bob	Application Starts	Alice	Bob
	Bob to Alice	Application Starts	Bob	Alice

TABLE L5.3
Summary of Phase Configuration: Part 2

Task Name	Phase Name	Request	Response
Authentication	Talk to KDC	Init time = 0 seconds Number of requests = 1 Request size = 2 kbytes	Processing time $\cong$ 0.5 seconds Number of replies = 1 Response size = 5 kbytes
	Talk to Bob	Init time $\cong$ 0.4 seconds Number of requests = 1 Request size = 4 kbytes	Processing time $\cong$ 1 second Number of replies = 1 Response size = 1 kbytes
	Verify Bob	Init time $\cong$ 0.2 seconds Num requests = 0	N/A
Secure Transmission	Alice to Bob	Init time $\cong$ 0.2 seconds Number of requests = 1000 Inter-request time $\cong$ 1.2 second Request size = 1–10 kbytes	N/A
	Bob to Alice	Init time $\cong$ 0.2 seconds Number of requests = 1000 Inter-request time $\cong$ 1.2 second Request size = 1–10 kbytes	N/A
Plain Text Transmission	Alice to Bob	Init time = 0 seconds Number of requests = 1000 Inter-request time $\cong$ 1 second Request size = 1–10 kbytes	N/A
	Bob to Alice	Init time = 0 Number of requests = 1000 Inter-request time $\cong$ 1 second Request size = 1–10 kbytes	N/A

Next, you will configure the authentication, encrypted transmission, and plain text transmission tasks for modeling the `Secure AB Talk` and `AB Talk` applications.

- Add the *Task Config* object and name it `Tasks` (Section 6.2.1).
- Define application tasks (Section 6.2.2) according to the following rules, which are also summarized in Tables L5.2 and L5.3:
  - `Authentication` task:
    - Phase `Talk to KDC` models steps 1–3 of the authentication process. This phase starts at the same time as the application and it operates as follows:
      - A single message of size 2 kbytes is sent by Alice to KDC. Alice does not spend any initialization time for generating this message.
      - The KDC server spends about 2 seconds processing Alice's request before sending a single response message of size 5 kbytes back to Alice.
    - The phase `Talk to Bob` models steps 4–8 of the authentication process. This phase starts after the `Talk to KDC` phase completes, and it operates as follows:
      - Alice spends about 0.8 seconds to process KDC's response before sending a message of size 4 kbytes to Bob.
      - Bob spends about 1 second processing Alice's request before sending a response message of size 1 kbytes back to Alice.
    - The phase `Verify Bob` models the last step of the authentication process, during which Alice spends 0.2 seconds processing Bob's response message. This phase starts after the `Talk to Bob` phase completes.
  - `Secure Transmission` task:
    - The phase `Alice to Bob` models secure communication from Alice to Bob, which consists of encrypting messages and then sending them from Alice to Bob. Let us assume that the sizes of encrypted messages are uniformly distributed between 1 and 10 kbytes and that it takes about 0.2 seconds to encrypt a message (i.e., before sending a message, Alice spends 0.2 seconds encrypting it). Alice generates a new message to Bob about once every second. The total time between two consecutive requests sent by Alice consists of the inter-message generation time (1 second) and the time spent encrypting the message (0.2 seconds). Let us assume that Alice generates about 1000 messages throughout the course of her communication with Bob and that each of the messages fits into a single application packet. Since we model communication from Bob to Alice as a separate phase, there are no response messages generated during this phase. Configure this phase to start after the previous task completes (i.e., you should set the attribute **Start Phase After** to the value `Application Starts`).
    - The phase `Bob to Alice` models secure communication from Bob to Alice, which consists of encrypting messages and then

sending them from Bob to Alice. Configuration of this phase
is identical to that of the Alice to Bob phase, except the
source and destination nodes in this phase are switched. The
phase Bob to Alice starts at the same time as the Alice
to Bob phase.

- The Plain Text Transmission task also consists of two phases:
  Alice to Bob and Bob to Alice. Configuration of these phases
  is identical to those of the Secure Transmission task, except that
  the packets are not encrypted and are sent in plain text. As a result, both
  phases can send data without any additional encryption delay (i.e., the
  initialization time should be set to 0 and the inter-request time should
  be set to 1 second).

When applicable (e.g., setting an attribute to a non-zero value), use the exponential distribution to specify the initialization time, the inter-request time, and the
inter-packet time. Use the constant distribution to specify the size of the request and
response messages in the Authentication phase and use the uniform distribution with a minimum outcome of 1 kbytes and a maximum outcome of 10 kbytes to
specify the request size during the transmission phases. Use the constant distribution
to specify the number of request messages.

Once all the tasks have been specified, you will define two custom applications: Secure AB Talk and AB Talk (Section 6.3). The Secure AB Talk
custom application consists of two tasks, Authentication and Secure
Transmission executed in serial order, whereas the AB Talk custom application consists of a single task: Plain Text Transmission.

Next, you will define two profiles, Secure AB Talk User and AB Talk
User, supporting the Secure AB Talk and AB Talk custom applications,
respectively. Configure each profile to run an application only once during simulation. Deploy the defined user profiles in the network (Sections 7.2 and 7.4).

Finally, configure the simulation to collect all Global Statistics for **Custom
Application** (Sections 4.2.4 and 5.6) and then execute the simulation for 2000
seconds (Section 4.3).

Questions

Q1. What are the response times of each of the custom application phases?

Q2. How do the response times of the Secure Transmission and Plain
Text Transmission tasks compare to each other? Are these results
consistent with the reported response times of individual phases?

Q3. How do the response times of the Secure AB Talk and AB Talk applications compare? Which application requires more time to complete? Why?

Q4. What is the response time of the Authentication task? How
does the Authentication task influence the overall application
performance?

L5.3   APPLICATION PERFORMANCE VERSUS SHORTER APPLICATION TIME

In this portion of the assignment, you will examine how the duration of the application influences the overall performance.

- Duplicate the current scenario naming the new one AB_Talk_Short (Section 1.7).
- Change the definition of the data transmission phases in the Secure Transmission and Plain Text Transmission tasks to generate only five messages per phase, instead of 1000 (Section 6.2.3).
- Update the Secure AB Talk User and AB Talk User profiles to run their respective applications for 15 seconds and have unlimited repeatability (Section 7.2).
- Execute the simulation and examine the collected results.

Questions

Q5. What are the average response times of each of the custom application phases and tasks?

Q6. How do the average response times of the Secure AB Talk and AB Talk applications compare? Which application requires more time to complete? Why?

Q7. What is the average response time of the Authentication task? How does the Authentication task influence the overall application performance? What percentage of the Secure AB Talk application response time was spent on authentication?

L5.4   APPLICATION PERFORMANCE VERSUS SENDING APPLICATION MESSAGE VIA MULTIPLE PACKETS

In this portion of the assignment, you will examine what happens when a single application message is split into multiple packets, each of which has to be encrypted. Perform the following steps to simulate such a scenario:

- Duplicate the scenario AB_Talk naming the new one AB_Talk_Packets.
- Change the data transmission configuration (i.e., change the value of the attribute **Source->Dest Traffic** for both phases) in the Secure Transmission task as follows:
  - Set the initialization time to 0 seconds (constant distribution) because the application does not encrypt the message; instead, it encrypts individual packets within the message.
  - Set the inter-request time to 1 second.
  - Set the number of packets per request to 5 (constant distribution).

- Set the inter-packet request time to 0.21 seconds (exponential distribution): 0.2 seconds to encrypt a packet and 0.01 seconds to create individual packets by fragmenting the original message.
- Change the data transmission configuration in the Plain Text Transmission task as follows:
  - Set the number of packets per request to 5 (constant distribution).
  - Set the inter-packet request time to 0.01 seconds (exponential distribution): no encryption is needed; inter-packet delay is due to message fragmentation.
- Execute the simulation and examine the collected results.

Questions

Q8. What are the response times of each of the custom application phases and tasks?

Q9. Draw a table that compares the response times of the data transmission phases collected in the AB_Talk and AB_Talk_Packets scenarios. What differences do you observe? Why?

Q10. How do the application response times collected in the AB_Talk and AB_Talk_Packets scenarios compare? Why?

Laboratory Assignment #6: Influence of the Maximum Transmission Unit on Application Performance

Recommended reading for this laboratory assignment: Chapters 1 through 5 and 7.

L6.1 INTRODUCTION

In this laboratory assignment, you will study how the Maximum Transmission Unit (MTU) size influences application performance. Specifically, you will examine how varying the MTU value in the network layer changes the response time of an FTP application. Recall that MTU size determines how the application data is fragmented by IP. In particular, the fragment size influences the application response time through the following two factors:

1. If application data is broken into many small fragments, then the total amount of traffic carried through the network is more than when the data is broken into fewer larger fragments. The increase in traffic is due to additional header information carried by individual fragments (i.e., each IP fragment carries network and data link layer headers). The application response time increases when the overall amount of traffic carried by the network goes up (i.e., it takes longer time to transmit the same amount of application data). Thus, the application response time may increase when application data is fragmented into many small pieces.

2. On the other hand, the overall end-to-end delay for transmitting a large fragment is more than for transmitting a sequence of small fragments because of the "pipelining" effect over multiple hops. In the time it takes to transmit one large fragment, several small fragments can be transmitted by the source, received by the next hop, and forwarded further into the network. Fragmenting data into smaller chunks supports "pipelining," which may reduce the response time experienced by an application. Thus, the application response time may decrease when application data is fragmented into many small pieces.

Clearly these two factors have opposing effects and there should be some "optimal" MTU size, which will minimize the application response time. In this laboratory assignment, you will derive a formula for computing optimal MTU size and compare your computations with the results of a simulation study. Consider the network topology shown in Figure L6.1 in which the client node uploads 50,000-byte documents onto the server every 10 seconds. In this assignment, you will vary MTU size on the client and server node networks from 500 bytes to 3000 bytes in 500 byte increments, and then you will examine how MTU size influences FTP upload response time.

L6.2 CREATING THE SIMULATION MODEL

In this portion of the assignment, you will create a simulation model of the network for this laboratory assignment.

1. Create a new project and an empty scenario named `Assignment 06` and `Two_Hop_Network`, respectively (Section 1.6.2).
2. Create the network topology as shown in Figure L6.1 using node and link models specified in Table L6.1 (Sections 2.3 and 2.4).
3. Verify link connectivity (Section 2.6.2).
4. Add *Application Config* and *Profile Config* node objects (Sections 5.3.1 and 7.2.1).
5. Configure an FTP application to generate 50,000-byte file uploads every 10 seconds (Section 5.4.3).
6. Configure an FTP user profile that runs the defined FTP application once for the duration of the simulation (Section 7.2).

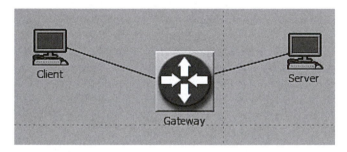

FIGURE L6.1   Network topology for the `Two_Hop_network` scenario.

TABLE L6.1
Node and Link Models for the Network Topology

Object Name	Object Model
Client Server	*ppp_wkstn* node object
Gateway	*ethernet4_slip8_gtwy* node object
Client <-> Gateway Server <-> Gateway	*PPP_DS1_int* duplex link object

---

7. Deploy the defined FTP user profile in the network so that `Client` and `Server` nodes operate as source and server for the FTP application, respectively (Section 7.4).

Next you will configure statistics to be collected in this simulation, specify MTU values (i.e., promote the MTU attribute) for `Client` and `Server` nodes, and execute the simulation.

8. Configure the simulation to collect all Global Statistics for the FTP application (Sections 4.2.4 and 5.6).
9. Open the `Client` node **Edit Attributes** menu, right click on attribute **IP... IP Host Parameters...Interface Information...MTU**, and set that attribute value to `promoted` (Section 3.5.1).
10. Repeat the same process for the `Server` node.
11. Specify the value of the promoted attribute to vary from 500 bytes to 3000 bytes in 500 byte increments. You will need to use the wildcard option to specify this range of values for the MTU attribute on both the `Client` and `Server` nodes (Sections 3.5, 4.3.5, and 4.3.6).
12. Execute the simulation for 400 seconds (Section 4.3).

Questions

Q1. What were the FTP upload response times for each of the MTU values?

Q2. Which MTU value resulted in the smallest FTP upload response time? Why? (Hint: think in terms of how fragmentation influences the application response time as discussed in Section L6.1).

Q3. Assume that the size of the data link layer header is 7 bytes and the size of the IP header is 20 bytes. Also assume that $T_{transmission}$, the time to transmit **D** bytes of application data, can be approximated according to Equation L6.1 shown below. To simplify your computation, we ignore the sizes of the FTP and TCP headers (which are small compared to the **D** bytes of application data) and also approximate the number of packets needed to send the data as a ratio between the data size (**D** bytes) and the MTU size.

$$T_{transmission} = \left[\frac{D}{MTU}\right] \times (H + MTU) \div R \approx \frac{D(H + MTU)}{MTU * R} \tag{L6.1}$$

where **D** is the amount of application data, **MTU** is the Maximum Transmission Unit size, **H** is the sum of the sizes of the data link layer and IP headers (27 bytes), and **R** is the transmission rate.

Assume that the propagation and processing delays are negligible. Using Equation L6.1, derive a formula for the end-to-end delay computed from the time the first bit of the data was transmitted by the source node until the time the last bit of the data arrived at the destination. Assume that the data is transmitted through a packet-switched network with **N** hops between the source and the destination, where one hop corresponds to the traversal of one link. (Hint: the total time it takes for data **D** to reach the destination is equal to the time needed to transmit data **D** by the source node on the first hop plus the time it takes for the last data packet to reach the destination by traversing the remaining **N-1** hops on the path. You can assume that all the packets have the same size.)

Q4. From the formula derived for Q3, compute the optimal MTU size. (Hint: to compute the optimal MTU size, take the derivative with respect to MTU of the equation computed for Q3 and equate the result to 0. Solve the obtained equation for MTU.)

Q5. Using the derived equation, compute the optimal MTU size for transmitting 50,000 byte application data over a network with two hops from the source to the destination (i.e., **N** = 2), which is the case with the network used in your simulation scenario. Assume that the total size of the network and data link layer headers is 27 bytes. Compare the obtained optimal MTU value with the results produced by the simulation. Do the simulation results agree with your computations? Why or why not?

L6.3   INCREASING NUMBER OF HOPS IN THE NETWORK

You will now repeat the experiment by increasing the number of hops in the network to four and examining how the interaction between the application response time and MTU size is influenced by the increased number of hops.

1. Duplicate the current scenario, naming the new one `Four_Hop_Network` (Section 1.7).
2. Add two more gateway nodes between `Client` and `Server`.
3. Connect the new nodes via DS1 links to form the topology shown in Figure L6.2.
4. Execute the simulation and examine the collected results.

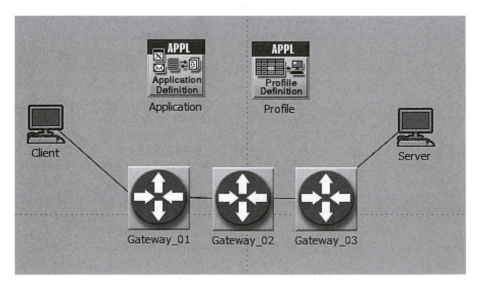

FIGURE L6.2   Network topology for the `Four_Hop_Network` scenario.

Questions

Q6. What were the FTP upload response times for each of the MTU values in the new scenario?

Q7. Which MTU value resulted in the smallest FTP upload response time? How do these results compare with the data collected in Section L6.2? Why?

Q8. Using the optimal MTU size formula derived in Section L6.2, compute the MTU value that should result in the smallest FTP upload response time for this simulation scenario. Note that the number of hops is now 4. How does the computed value compare with the simulation results? Do the simulation results support the computed optimal MTU value?

Laboratory Assignment #7: Transport Protocols: TCP versus UDP

Recommended reading for this laboratory assignment: Chapters 1 through 5, 7, and 8.

L7.1 INTRODUCTION

In this laboratory assignment, you will compare the performance of applications running over different transport protocols. In particular, you will examine how the application performance is influenced by the TCP and UDP protocols. Furthermore, you will hone your skills in creating complex OPNET simulation models, setting up standard applications, and developing user profiles.

Consider the following situation: Tim, "the Web Dude," developed a super secret (ala "James Bond") spy application for encoding and transmitting information securely over the network without anyone even being aware that some hidden information is being transmitted. The application encodes information as a series of images and is called Hermes, after the messenger of the gods in Greek mythology. For an outsider who monitors traffic generated by Hermes, it will appear that a bunch of small images are being transferred, whereas the images are actually carrying some hidden information.

Hermes can be deployed in environments with various loss characteristics. Tim's employer identified the following environment types where Hermes will operate:

- 0.00% loss (transfer over a loss-free LAN)
- 0.00% loss (transfer over the Internet with no data loss)
- 0.5% loss (transfer over the Internet with little data loss)
- 1.00% loss (transfer over the Internet with some data loss)
- 5.00% loss (transfer over the Internet with significant data loss)

Hermes encodes a single block of secret information in 10 consecutive images of size 10 kbytes generated every 100 milliseconds. In order for Hermes to operate normally, all 10 images should be received within 5 seconds.

Tim, "the Web Dude," hired you as a consultant to identify the most appropriate transport protocol for Hermes to be used within the above environments. Unfortunately, Tim cannot divulge the details of Hermes's implementation due to it being top secret. However, Tim provides you with a sufficient amount of detail to create a simulation model:

- Hermes is an FTP-like client–server application.
- The client generates requests for data downloads only.

- The client generates requests every 100 milliseconds.
- The requested file size is 10,000 bytes.
- The client runs the Hermes application as follows:
  - The client generates a set of requests for 10 files.
  - The client waits for 9 seconds before generating another set of requests.
  - The client repeats steps 1 and 2 for 100 seconds, then pauses for 100 seconds and starts over.
  - The client may repeat steps 1 through 3 as many times as necessary.

L7.2   HERMES APPLICATION IN AN ENVIRONMENT WITH NO DATA LOSS

In this portion of the assignment, you will create a simulation model of a network environment in which the Hermes application operates.

1. Create a new project and an empty scenario named Assignment 07 and Hermes with No Loss, respectively (Section 1.6.2).
2. Create the network topology as shown in Figure L7.1 using the node and link models specified in Table L7.1 (Sections 2.3 and 2.4).
3. Verify link connectivity (Section 2.6.2).
4. Add *Application Config* and *Profile Config* node objects.
5. Create two identical FTP applications and let us call them Hermes Local and Hermes over the Internet, which generate 10,000-byte file downloads every 100 milliseconds (Section 5.4.3).
6. Configure a profile and let us call it Local Use that employs the Hermes Local application as follows (Section 7.2):

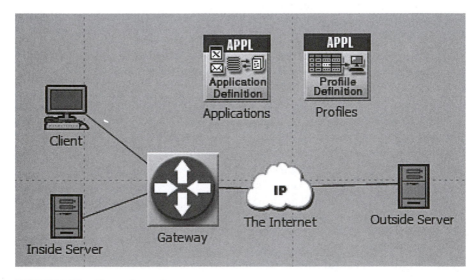

FIGURE L7.1   Network topology.

a. Specify application characteristics within the profile as follows:
   i. Set the application start time offset to 0 seconds to indicate that the application starts at the same time as the profile.
   ii. Since the Hermes application generates requests every 100 milliseconds, set the application duration to 1 second to simulate requesting 10 files.
   iii. To simulate the Hermes application waiting for 9 seconds before sending the next batch of requests (i.e., step 2), set the application inter-repetition time to 9 seconds, number of repetitions to Unlimited, and the repetition pattern to Serial.
b. To model the Hermes application pausing for 100 seconds before repeating steps 1 and 2, configure the profile characteristics as follows:
   i. Set the profile operational mode to Serial.
   ii. Set the profile start time to 100 seconds.
   iii. Set the duration of the profile to 100 seconds.
   iv. Set the profile repeatability to Unlimited.
c. Such a configuration will cause the profile to run for 100 seconds, then wait for 100 seconds, then start again, and repeat until the end of the simulation.
   i. To simplify data analysis, use only the constant distribution when setting time offsets and other profile or application characteristics.
7. Duplicate the profile Local Use and change its configuration as follows:
   a. Change the name of the new profile to Use over the Internet.
   b. Change the application employed by the new profile to Hermes over the Internet.
   c. Leave the remaining profile attributes unchanged.
8. Deploy the defined profiles in the network so that the Client operates as a source for the Hermes application, whereas the nodes Inside Server and Outside Server are the FTP servers for the Hermes application in the Local Use and Use over the Internet profiles, respectively (Section 7.4).

TABLE L7.1

Summary of Network Topology

Object Name	Object Model
Client	*ethernet_wkstn_adv* node object
Inside Server	*ethernet_server_adv* node object
Outside Server	*ppp_server_adv* node object
Gateway	*ethernet4_slip8_gtwy* node object
Client <-> Gateway Inside Server <-> Gateway	*100BaseT* link object
Gateway <-> The Internet The Internet <-> Outside Server	*PPP_DS1_int* link object

Next, you will configure the latency experienced by the packets that travel through the Internet, configure the statistics to be collected in this simulation, specify transport protocol values for the `Client`, `Inside Server`, and `Outside Server` nodes, and execute the simulation.

1. Set packet latency in `The Internet` node to 100 milliseconds.
    i.   You need to set the value of the node's attribute **Performance Metrics... Packet Latency (secs)** to `0.1`.
2. Configure the simulation to collect the following statistics (Sections 4.2.3 and 5.6):
    i.   All **Client FTP** Node Statistics.
    ii. All *point-to-point* Link Statistics.
3. Configure the end nodes `Client`, `Inside Server`, and `Outside Server` to allow the Hermes application to run over different transport protocols by setting the value of the node attribute **Applications: Transport Protocol Specification...FTP Transport** to `promoted` (Section 3.5.1).
4. Specify the values of the promoted attribute to `TCP` and `UDP`. You may need to use the wildcard option to specify the values for **FTP Transport** on all end nodes (Sections 3.5, 4.3.5, and 4.3.6).
5. Execute the simulation for 3000 seconds.

Questions

Once the simulation completes, examine how the TCP and UDP protocols influence the application response time and resource consumption. You should examine the download response time of the Hermes application, as well as utilization and throughput on the links connected to the servers in the following four cases:

- Hermes over TCP in the LAN.
- Hermes over UDP in the LAN.
- Hermes over TCP traveling through the Internet, with no data loss.
- Hermes over UDP traveling through the Internet, with no data loss.

Q1. Create a table that contains the maximum, minimum, and average download response times of the Hermes application for each of the above cases. Explain the observed results based on your understanding of TCP, UDP, and the physical configuration of the network.
   - When used in the LAN only, which protocol, TCP or UDP, results in a shorter response time for the Hermes application? Why?
   - When transmitting data over the Internet, which protocol, TCP or UDP, results in a shorter response time for the Hermes Application? Why?
   - When running over UDP, which network environment, LAN or the Internet, results in a shorter average download response time for the Hermes application? Why?
   - When running over TCP, which network environment, LAN or the Internet, results in a shorter average download response time for the Hermes application? Why?

Q2. For each of the four cases, examine the utilization and throughput on the links connected to the server nodes. Explain the observed results based on your understanding of TCP, UDP, and the physical configuration of the network.

- When used in the LAN only, which protocol, TCP or UDP, results in lower link utilization and throughput? Why?
- When transmitting data over the Internet, which protocol, TCP or UDP, results in lower link utilization and throughput? Why?
- When running over UDP, which network environment, LAN or the Internet, results in lower link utilization and throughput? Why?
- When running over TCP, which network environment, LAN or the Internet, results in lower link utilization and throughput? Why?

Q3. Based on the results of the simulation, when using Hermes in a network environment without packet loss, which transport protocol would you recommend to Tim and why?

L7.3   HERMES APPLICATION IN AN ENVIRONMENT WITH DATA LOSS

In real life, the Internet seldom provides lossless data delivery. In this part of the assignment, you will examine response time and resource consumption of the Hermes application in an Internet environment that loses data with a certain probability.

1. Duplicate the current scenario naming the new one `Hermes with Loss` (Section 1.7).
2. Promote the attribute **Performance Metrics...Packet Discard Ratio** in `The Internet` node.
3. Set the value of the promoted attribute **Packet Discard Ratio** to `0.5%`, `1.0%`, and `5.0%`.
4. Execute the simulation and examine the collected results.

Questions

When examining the simulation results, you may ignore the collected data for the cases when the Hermes application sends traffic through the local network. Instead, examine the application response time, as well as the utilization and throughput on the link between `The Internet` and `Outside Server` nodes for the cases when Hermes sends traffic through the Internet. For the purpose of all questions in this section, let us assume that Hermes recovers encoded information in the case of occasional packet loss. Usually to recover from a single packet loss, Hermes needs an additional 0.5 seconds of processing delay to complete its data recovery computations.

Q4. Display the graph that illustrates the application response time for the Hermes application that runs over UDP through a network with 0.5%, 1.0%, and 5.0% packet loss. Which of the packet loss values produces the shortest response time? Why? Considering the additional processing delay associated with data reconstruction in the case of a packet loss, which value of the **Packet Discard Ratio** attribute results in the shortest application response time?

Q5. How do the response times of the Hermes application that runs over UDP change in the network environment with loss as compared to that without loss? Why?

Q6. Display the graph that illustrates the application response time for the Hermes application that runs over TCP through the network with 0.5%, 1.0%, and 5.0% packet loss. Considering the additional processing delay associated with data reconstruction in the case of packet loss, which value of the **Packet Discard Ratio** attribute results in the shortest application response time? How does the response time of the Hermes application that runs over TCP change in the network environment with loss as compared to that without loss? Why?

Q7. How do the link utilization and throughput change in the network environment with loss as compared to that without loss when Hermes runs over UDP? Why?

Q8. How do the link utilization and throughput change in the network environment with loss as compared to that without loss when Hermes runs over TCP? Why?

Q9. Based on these observations, which transport protocol would you recommend to Tim for the Hermes application in the environment that experiences data loss? Why?

Laboratory Assignment #8: TCP Features

Recommended reading for this laboratory assignment: Chapters 1 through 5, 7, and 8.

L8.1 INTRODUCTION

In this laboratory assignment, you will investigate techniques for fine-tuning TCP configuration. In particular, you will investigate how such TCP features as Nagle's algorithm, Maximum Segment Size (MSS), receiver window size, and various TCP flavors influence application performance.

L8.2 NAGLE'S ALGORITHM

Nagle's algorithm addresses the issue of small packet size. Instead of immediately transmitting small data packets, each of which would carry 40 bytes of TCP and IP header information, Nagle's algorithm suggests buffering the data until the application provides more data to transmit or until all of its outstanding packets have been acknowledged. In this portion of the assignment, you will examine how Nagle's algorithm influences the application delay and the amount of traffic sent over the wire.

1. Create a new project and an empty scenario named Assignment 08 and Nagles Algorithm, respectively (Section 1.6.2).
2. Create the network topology as shown in Figure L8.1 using node and link models specified in Table L8.1 (Sections 2.3 and 2.4).
3. Verify link connectivity (Section 2.6.2).
4. Add *Application Config* and *Profile Config* node objects.
5. Create a Remote Login application, let us call it Telnet, that generates 1 byte commands to and from the terminal. To simplify the data analysis, use a constant distribution to specify the size of the command. Promote the attribute **Inter-Command Time (seconds)** by clicking the button **Promote**. Refer to Section 5.4.6 for additional information about the Remote Login application.
6. Configure a profile, let us call it Telnet User, that runs the Telnet application with the default setting (Section 7.2).
7. Deploy the defined profile in the network so that Client operates as source and Server is a Remote Login server for the Telnet application (Section 7.4).
8. Specify the node configuration as follows:
    a. Set packet latency in The Internet node to 100 milliseconds by specifying the value of the attribute **Performance Metrics...Packet Latency (secs)**.

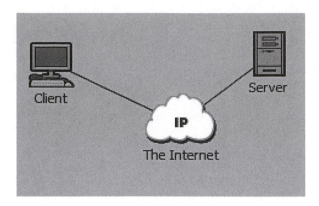

FIGURE L8.1   Network topology.

TABLE L8.1
Summary of Network Topology

Object Name	Object Model
Client	*ppp_wkstn* node object
Server	*ppp_server* node object
The Internet	*ip32_cloud* node object
All links in the network	*PPP_33K* link object

---

b.   Promote the attribute **TCP...TCP Parameters...Nagle Algorithm** at both Client and Server nodes.

9. Configure the simulation to collect the following statistics:

   a.   **Response Time (sec)** from the **Client Remote Login** Node Statistics category.

   b.   All *point-to-point* Link Statistics.

10. Specify the values of the promoted attributes as follows:

   a.   Set the value of the **TCP Parameters...Nagle Algorithm** attribute to Disabled and Enabled.

   b.   Set the value of the attribute **Inter-Command Time** to exponential (0.1) and exponential (10).

11. Run the simulation for 1 hour.

Questions

Examine the application response time and display a graph for the following four situations:

1. Nagle's algorithm is disabled and inter-command time is set to 0.1 seconds.

2. Nagle's algorithm is enabled and inter-command time is set to 0.1 seconds.
3. Nagle's algorithm is disabled and inter-command time is set to 10 seconds.
4. Nagle's algorithm is enabled and inter-command time is set to 10 seconds.

Q1. How does the application response time vary when Nagle's algorithm is enabled and disabled? Why?

Q2. Which of the above four scenarios resulted in the lowest and the highest application response times? Why?

Examine the throughput on the link between Client and The Internet, display a graph that compares the link throughput in the above four scenarios, and answer the following questions:

Q3. How does the link throughput vary when Nagle's algorithm is enabled and disabled? Why?

Q4. Which of the above four scenarios resulted in the lowest and the highest throughput on the link between Client and The Internet? Why?

L8.3   INFLUENCE OF THE END-TO-END DELAY ON NAGLE'S ALGORITHM

In this part of the assignment, you will examine how packet latency in the Internet influences the performance of Nagle's algorithm.

1. Duplicate the current scenario and name the new one Nagles Algorithm with Delay.
2. Promote the attribute **Performance Metrics...Packet Latency (secs)** in The Internet node.
3. Specify the values of the promoted attributes as follows:
   a.   Set the value of the **TCP...TCP Parameters...Nagle Algorithm** attribute to Disabled and Enabled.
   b.   Set the value of the attribute **Inter-Command Time** to exponential (0.1).
   c.   Set the value of the attribute **Performance Metrics...Packet Latency (secs)** to 0.1 and 0.001 seconds.
4. Run the simulation for 1 hour.

Questions

Examine the application response time and link throughput statistics collected during the simulation. Display the same set of graphs as in the previous part of the assignment and answer the following questions:

Q5. How does packet latency influence application response time when Nagle's algorithm is enabled and disabled? Why?

Q6. How does the application response time vary when the value of packet latency changes? Do you think that the reported results are consistent with the scenario's configuration? Why or why not?

Q7. How does packet latency influence link throughput when Nagle's algorithm is enabled and disabled? Why?

Q8. Which value of packet latency resulted in the lowest throughput on the link? Why?

L8.4   INFLUENCE OF TCP'S MSS SIZE ON APPLICATION PERFORMANCE

In this part of the assignment, you will examine the influence of MSS on application performance.

1. Create a new scenario named TCP MSS.
2. Create the network topology as shown in Figure L8.1 using the node models specified in Table L8.1. However, use the PPP_DS1 link model to connect the nodes in the network.
3. Verify link connectivity.
4. Add *Application Config* and *Profile Config* node objects.
5. Create an FTP application, let us call it FTP, which issues a get command to download a 1 Mb file every 3 minutes. Use a constant distribution to specify the file size and the inter-request time (Section 5.4.3).
6. Configure a profile, let us call it FTP User, that runs the FTP application with the default setting (Section 7.2).
7. Deploy the defined profile in the network so that Client operates as source and node Server is a server for the FTP application (Section 7.4).
8. Specify the node configuration as follows:
   a. Set packet latency in The Internet node to 100 milliseconds.
   b. Promote the attribute **TCP...TCP Parameters...Maximum Segment Size (bytes)** at both Client and Server nodes.
   c. Ensure that the size of the receive buffer on both end nodes is set to the default value of 8760 bytes.
9. Configure the simulation to collect the following statistics:
   a. **Download Response Time (sec)** from the **FTP** Node Statistics category.
   b. **Segment Delay (sec)** from the **TCP** Node Statistics category.
   c. All *point-to-point* Link Statistics.
10. Set the promoted attribute **TCP Parameters...Maximum Segment Size (bytes)** to the following values: Auto-Assigned, 500, 1000, 1400, 1500, 2000, and 5000.
11. Run the simulation for 10 minutes.

Questions

Examine the simulation results for the download response time, the TCP segment delay, and point-to-point throughput on the link between Client and The Internet.

Q9. Display the graph that illustrates the download response time versus MSS. Which MSS value resulted in the lowest download response times and why? Which MSS value resulted in the largest download response times and why? How does the download response time for MSS of 1500 bytes and 1400 bytes compare? Provide an explanation for the observed behavior.

Q10. Display the graph that illustrates TCP segment delay versus MSS. Which MSS value resulted in the lowest TCP segment delay and why? Which value resulted in the largest TCP segment delay and why?

Q11. Display the graph that illustrates TCP throughput on the link between `Client` and `The  Internet` versus MSS. Which MSS value resulted in the lowest link throughput and why? Which value resulted in the largest link throughput and why?

L8.5   INFLUENCE OF TCP'S RECEIVE BUFFER SIZE
ON APPLICATION PERFORMANCE

In this part of the assignment, you will examine the influence of receive buffer size on application performance. Recall that in OPNET, the attribute **Receive Buffer (bytes)** specifies the total amount of space available at the receiver to store arriving data before it is forwarded to the upper layers. This value is different from the advertised receive window (**rwnd**), which is the amount of free space currently available in the receive buffer to store arriving data.

1. Duplicate the `TCP MSS` scenario and name the new scenario `TCP Receive Buffer`.
2. Modify the node configuration as follows:
   a. Set the value of the attribute **TCP Parameters...Maximum Segment Size (bytes)** to `Auto-Assigned` at both `Client` and `Server` nodes.
   b. Promote the attribute **TCP Parameters...Receive Buffer (bytes)**.
3. Modify the configuration of the global attributes as follows:
   a. Delete previously promoted attribute **TCP Parameters...Maximum Segment Size (bytes)** from the **Object Attributes** table.
   b. Add and then set the promoted attribute **Receive Buffer (bytes)** to the following values: `8760`, `32768`, `65535`, and `Default`.
4. Run the simulation for 10 minutes.

Questions

Examine the simulation results for the download response time and point-to-point throughput on the link between `Client` and `The Internet`.

Q12. Display the graph that illustrates download response time versus receive buffer size. Which receive buffer size value resulted in the lowest download response times and why? Which receive buffer size value resulted in the largest download response times and why?

Q13. Based on the observed results and the fact that, by default, OPNET sets the receive buffer size to four times the value of MSS, guess the default value of the receive buffer size.

Q14. Display the graph that illustrates the TCP throughput on the link between Client and The Internet versus receive buffer size. Which receive buffer size value resulted in the lowest link throughput and why? Which value resulted in the largest link throughput and why? Do you think that reported link throughput results are consistent with your guess of the default receive buffer size? Why or why not?

L8.6 TCP CONGESTION CONTROL

In this portion of the assignment, you will examine how various TCP flavors behave in the presence of packet loss. Specifically, you will examine the performance of the Tahoe, Reno, New Reno, and SACK TCP flavors. To perform this study, you need to create a new OPNET scenario as follows:

1. Create a new scenario and name it TCP Congestion Control.
2. Create the network topology as shown in Figure L8.2 using the node models specified in Table L8.2. You can find the *Packet Discarder* node model in the *utilities* folder of the **Object Palette**.
3. Verify link connectivity.
4. Add *Application Config* and *Profile Config* node objects.

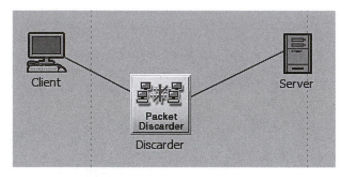

FIGURE L8.2   Network topology for the TCP congestion control scenario.

TABLE L8.2
Summary of Network Topology for TCP Congestion Control Scenario

Object Name	Object Model
Client	*ppp_wkstn* node object
Server	*ppp_server* node object
Discarder	*Packet Discarder* node object
All links in the network	*PPP_DS1* link object

5. Create an FTP application, let us call it FTP, which issues a get command to download a 2 Mb file every 5 minutes. Use a constant distribution to specify the file size and the inter-request time (Section 5.4.3).

6. Configure a profile, let us call it FTP User, that runs the FTP application as follows:
   a. Set the application start time offset to 1 second (use a constant distribution).
   b. Set the profile start time to 100 seconds (use a constant distribution).
   c. Set application and profile repeatability to Once at Start Time.

7. Deploy the defined profile in the network so that Client operates as source and node Server is a server for the FTP application.

8. Specify the node configuration as follows:
   a. Promote the attribute **TCP...TCP Parameters** at both Client and Server nodes.
   b. Configure the Discarder node to drop packets as follows:
      i.   Discard 1 packet between time 104 and 104.5 seconds.
      ii. Discard 1 packet between time 106 and 106.5 seconds.
      iii. Discard 3 packets between time 108 and 108.5 seconds.

9. Configure the simulation to collect the following statistics:
   a. **Download Response Time (sec)** from the **FTP** Node Statistics category.
   b. **Congestion Window Size (bytes)** from the **TCP Connection** Node Statistics category. Make sure that this statistic is collected using all values mode (Section 4.2.7).
   c. **Retransmission Count** from the **TCP Connection** Node Statistics category.
   d. All *point-to-point* Link Statistics.

10. Set the promoted **TCP...TCP Parameters** attribute to the following values: Tahoe, Reno, New Reno, and SACK.

11. Run the simulation for 130 seconds.

Questions

Q15. Provide a brief description of each TCP flavor used in this study.

Q16. Examine the scenario configuration and for each of the TCP flavors specify, in the form of a table, the values of the OPNET configuration attributes **Fast Retransmit**, **Fast Recovery**, and **Selective ACK (SACK)**. The table should contain three columns with the following headings: TCP flavor, the attribute name, and the attribute value. Briefly describe how fast retransmit, selective acknowledgment, and the Reno and New Reno variations of fast recovery operate.

Q17. Examine the collected statistics for and draw a graph of the TCP congestion window size at the Server node. Zoom in on the time period from 102 to 116 seconds into the simulation. Identify the slow start and congestion avoidance phases of the TCP process (i.e., specify the start and end time for each of the phases).

Q18. Duplicate the graph of the TCP congestion window size at the Server node. Zoom in on the time period from 103.8 to 104.4 seconds. Explain how each of the TCP flavors reacts to a single packet loss.

Q19. Duplicate the graph of the TCP congestion window size at the Server node. Zoom in on the time period from 105.8 to 106.4 seconds. Explain how each of the TCP flavors reacts to a second single packet loss.

Q20. Duplicate the graph of the TCP congestion window size at the Server node. Zoom in on the time period from 107.8 to 108.4 seconds. Explain how each of the TCP flavors reacts to multiple packet losses.

Q21. Examine the collected statistics for and draw a graph of the TCP Retransmission Count at the Server node. Do you think that the collected simulation results are consistent with the configured packet discard behavior? Why or why not?

Q22. Examine the collected statistics for and draw a graph of the FTP download response time. Which TCP flavor resulted in the lowest FTP download response time? Why? Which TCP flavor resulted in the largest FTP download response time? Why?

Q23. Examine the collected statistics for and draw a graph of the throughput on the link between Client and Discarder nodes. Do you think that the collected simulation results are consistent with the configured packet discard and observed congestion window behaviors? Why or why not?

Laboratory Assignment #9: IP Addressing and Network Address Translation

Recommended reading for this laboratory assignment: Chapters 1 through 7, 9, and 10.

L9.1 INTRODUCTION

In this laboratory assignment, you will practice IP address allocation with subnetting as well as the configuration of a Network Address Translation (NAT) device. Specifically, you will develop and configure a model of a network with several subnets and a NAT device in the network's access gateway.

Consider the following situation. An Internet Service Provider (ISP) sold a block of IP addresses to a local company called Software R' Us (SRU). The ISP allocated the following range of IP addresses to SRU: 197.13.45.128/27. The SRU network administrators distributed the allocated IP addresses among the company's departments according to the following requirements (i.e., the number of IP addresses needed):

- Application Development (AD): 12 usable IP addresses
- Application Testing (AT): four usable IP addresses
- Administration and Accounting (AA): three usable IP addresses

In this laboratory assignment, you will perform the following tasks:

- Determine the proper IP address allocation for each department within SRU.
- Divide SRU's network into subnets.
- Configure "by hand" (i.e., without using the Auto-Assign option) each of the subnets with the proper IP addresses.
- Specify the network and port address translation rules on the gateway router for one of SRU's subnets.

You will test the final configuration by running a simple simulation and verifying that each of the nodes is able to send and receive data.

L9.2 PRELIMINARY COMPUTATION

In this portion of the laboratory assignment, you will compute the IP address allocation for each department in SRU. Assume that each department is placed within its own subnet. Note that the number of required IP addresses may not correspond to the number of allocated IP addresses. Only usable addresses can be allocated to the nodes. Nonusable IP addresses are addresses that have special meaning such as broadcast address (i.e., host part of IP address is all 1s) or address of this (sub)-network (i.e., host part of IP address is all 0s). To complete this portion of the assignment, you need to answer the following questions:

Questions

Q1. How many IP addresses have been sold to SRU? How many of these addresses are usable (i.e., can be assigned to a node)?

Q2. For each SRU department subnet:
   a. How many IP addresses should be allocated to satisfy the department's address requirement?
   b. What is the network address for this department's subnet?
   c. What is the broadcast address for this department's subnet?
   d. What is the mask value for this department's subnet?
   e. What are the first and the last usable IP addresses in this department's subnet? Please use Classless Inter-Domain Routing (CIDR) notation.
   f. What is the number of unused addresses left in this department's subnet?

L9.3 SIMULATION SETUP

In this portion of the assignment, you will create a simulation model of the SRU network.

1. Create a new project and an empty scenario named `Assignment 09` and `IP_Addressing`, respectively (Section 1.6.2).
2. Create the network topology as shown in Figure L9.1 using node and link models specified in Table L9.1 (Sections 2.3 and 2.4).
3. Verify link connectivity (Section 2.6.2).
4. Configure `The Internet` cloud to discard 0.01% of all arriving packets and to introduce an additional delay of 100 milliseconds to all traffic traveling through it.
5. Set IP addresses and the corresponding subnet masks (Sections 9.2 and 9.3):
   a. Set the IP address and mask of `Internet Server` to `137.34.78.23/25`.
   b. Set the IP addresses and masks on the interfaces of `The Internet` node:
      i. Interface to `Internet Server`: `137.34.78.24/25`.
      ii. Interface to `SRU Gateway`: `146.56.12.1/29`.
   c. Set the IP addresses and masks on the interfaces of `SRU Gateway` node:
      i. Interface to `The Internet`: `146.56.12.2/29`.
      ii. Interface to `AD LAN`: last usable IP address allocated for the AD subnet and the corresponding AD subnet mask value.
      iii. Interface to `AT LAN`: last usable IP address allocated for the AT subnet and the corresponding AT subnet mask value.

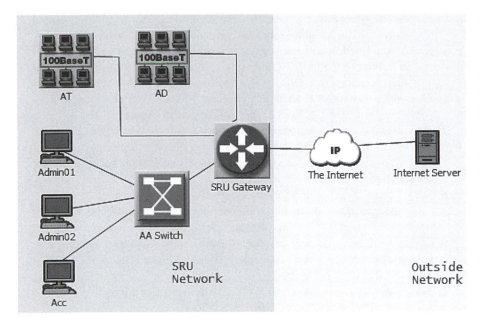

FIGURE L9.1 Network topology for this assignment.

TABLE L9.1
Summary of Network Topology

Object Name	Object Model
AT, AD	*100BaseT_LAN* node object
SRU Gateway	*ethernet4_slip8_gtwy* node object
Admin01, Admin02, Acc	*ethernet_wkstn* node object
AA Switch	*ethernet16_switch* node object
All links in SRU Network	*100BaseT* link object
The Internet	*ip32_cloud* node object
Internet Server	*ppp_server* node object
All Links in Outside Network SRU Gateway <-> The Internet	*PPP_DS3* link object
All IP traffic demands	*ip_traffic_flow* demand object

---

    iv. Interface to AA Switch: last usable IP address allocated for the AA subnet and the corresponding AA subnet mask value.

d. Set the IP address and mask of the AD LAN object to the first usable address in the AD subnet and the corresponding mask value.

    e. Set the IP address and mask of the AT LAN object to the first usable address in the AT subnet and the corresponding mask value.

    f. Set the IP addresses and masks of the `Admin01`, `Admin02`, and `Acc` end nodes to the first, second, and third usable addresses allocated for the AA subnet, respectively, and the corresponding AA subnet mask.

6. Configure `AD` and `AT` LAN objects to have 12 and 4 workstations, respectively (i.e., attribute **LAN…Number of Workstations**).

7. Deploy two IP traffic demand objects from `AD` and `AT` LANs to `Internet Server`. Configure these demand objects to generate traffic at the rate of 1.53 Mbps (i.e., transmission rate of a T1 line) and 200 packets/second, for the duration of 1 hour (Section 6.6.1).

8. Deploy three IP traffic demand objects from `Admin01`, `Admin02`, and `Acc` nodes to `Internet Server`. Configure all these demand objects to generate traffic at the rate of 0.1 Mbps and 200 packets/second, for the duration of 1 hour.

9. Configure all demands to generate only 1.0% of traffic explicitly (Section 6.6.1).

10. Configure your simulation to collect all **Demand** and all link *point-to-point* Statistics (Sections 6.8 and 4.2).

11. Set the global simulation attribute **IP…IP Interface Addressing Mode** to `Auto Addressed/Export` (Section 4.3).

12. Execute your simulation for 1 hour.

Questions

Q3. How many DES Log messages have been generated by the simulation run? If the total number of created DES Log messages exceeds 4, then it is likely that you may have made a configuration error. In that case, examine every DES Log message carefully, identify the error, correct it, and then rerun the simulation.

Q4. Examine the file exported by OPNET with IP address assignment on all active interfaces in the network (i.e., `Assignment 09-IP_Addressing-DES-1-ip_addresses.gdf`). Please attach this file with your report (Section 9.3.7). The IP address assignment exported from the created OPNET simulation must match your configuration.

Q5. Create a graph that illustrates the amount of traffic, in units of bits per second, generated by each demand.

Q6. Create a graph that illustrates the amount of traffic, in units of bits per second, sent from each of the end nodes in the network. Are there any nodes that do not generate any traffic? Why or why not? (Hint: Make sure that you examine all the end nodes in the simulation).

Q7. Examine the amount of traffic in units of bits per second arriving at `Internet Server` node. How does this amount compare to the total amount of traffic generated by all traffic demands in the network? Why?

L9.4   CONFIGURING DYNAMIC NAT

In this portion of the assignment, you will configure a NAT device on `SRU Gateway`. Assume that the AD department increased tenfold and now cannot have each individual machine assigned a separate global IP address. As a result, the network administrators decided to configure NAT for the AD subnet in SRU Gateway. Since the traffic only flows from AD LAN to Internet Server but not back, you will only specify NAT for packets that originate from AD LAN. Specifically, you will perform the following steps to configure NAT on SRU Gateway:

1. Duplicate the current scenario, naming the new one IP_Addressing_ wNAT.
2. Change the IP address on AD LAN to the private address 192.168.0.1 and set the subnet mask to 255.255.255.0.
3. Set the IP address on the interface between SRU Gateway and AD LAN in SRU Gateway to the private address 192.168.0.2 and set the subnet mask to 255.255.255.0.
4. Change the number of workstations in AD LAN to 120.
5. Configure NAT parameters on SRU Gateway:
   a. Specify the global pool of IP addresses available for translation of the nodes in AD subnet. Let us call it AD pool (Section 10.1.3).
      i. Set the interface name to the name of the interface between SRU Gateway and AD LAN.
      ii. Set the range of IP addresses in the pool to the range of addresses that were allocated to AD subnet initially (in Sections L9.1 and L9.2).
      iii. Set the mask of the address pool to the mask of the AD subnet.
      iv. Set the port numbers to be selected automatically.
   b. Define the translation rule, let us call it Local AD to Global AD (Section 10.1.4):
      i. Set the source IP address of the original packet to 192.168.0.1.
      ii. Set the subnet mask of the original packet to 255.255.255.0.
      iii. Set the source IP address of the translated packet to be dynamically selected from the previously defined AD pool.
   c. Finally, specify and configure the interface on which the NAT rules will be deployed (Section 10.1.5):
      i. Set the name of the interface to the name of the interface between AD LAN and SRU Gateway.
      ii. Set the status of this interface to inside.
      iii. Configure the interface to use the Local AD to Global AD translation rule.
6. Execute the simulation for 1 hour.

Questions

Q8. How does the amount of traffic, in bits per second, arriving at Internet Server node compare to the total amount of traffic generated by all IP demands in the network? Why?

Q9. Did the amount of traffic, in bits per second, generated by an IP demand originating from AD LAN change (i.e., since the number of nodes in AD LAN increased) as compared to the results collected in the previous scenario? Why or why not?

L9.5   CONFIGURING PORT ADDRESS TRANSLATION

In this portion of the assignment, you will configure Port Address Translation (PAT) on SRU Gateway. Consider a situation where the AD department decides to have its employees run the FTP application. Let us deploy this application and observe what happens:

1. Duplicate the current scenario, naming the new one IP_Addressing_ wPAT.
2. Add an *Application Config* object and define a High Load FTP application (Sections 5.3.1 and 5.4.3).
3. Add a *Profile Config* object and create a new user profile, which runs the defined FTP application (Section 7.2).
4. Deploy the defined profile so that AD LAN is the source of the FTP application and Internet Server is the destination (Section 7.4).
5. Execute the simulation.

Questions

Q10. Did the simulation terminate properly? Did the DES Log produce any warning or error messages? If so, what kinds of messages were reported by DES Log? Why were these messages generated? What do you think is the underlying cause of the problem?

Since FTP runs over TCP, it requires both ends of the connection to exchange messages. Our original NAT configuration will not work for FTP traffic because SRU Gateway is not configured to handle the responses from Internet Server. Thus, you will need to specify another set of rules for translating incoming traffic from Internet Server:

1. Configure PAT parameters on SRU Gateway.
   a. Update the pool configuration to translate private IP addresses into a single global IP address but with different port numbers:
      i.   Change the IP address in the pool to the first address in AD subnet.
      ii. Set the mask of the address pool to the mask of the AD subnet.
      iii. Set the port number to be in the range from 2000 to 3000.
   b. Add another translation rule; let us call it Global to Local AD (Section 10.1.4):
      i.   Set the source IP address of the original packet to Any or to the IP address of Internet Server. In the latter case, you then may also want to set the mask of the original packet to the mask of Internet Server.
      ii. Set the destination IP address and mask of the *original* packet to the first address in AD subnet and the corresponding mask value of AD subnet.

       iii. Set the destination IP address of the *translated* packet to `192.168.0.1`, the IP address of AD LAN object.

    c. Configure the interface on which the new translation rule will be deployed (Section 10.1.5):

       i. Add another **Interface Information** row.

       ii. Set the name of the interface to the interface between `SRU Gateway` and `The Internet`.

       iii. Set the status of this interface to `outside`.

       iv. Configure the interface to use `Global to Local AD` translation rule.

2. Execute the simulation for 1 hour.

Questions

Q11. Did the simulation terminate properly? Note: you can ignore a few DES Log messages that may appear toward the end of the simulation and which state "The IP routing table on this node does not have a route to the destination X.X.X.X" or any other related messages. It appears to be some internal OPNET bug, which has very little effect on the simulation.

Q12. What was the throughput in units of bits per second on the link between AD LAN and `SRU Gateway`?

Q13. Was there any traffic arriving to AD LAN? Why or why not?

Q14. How does the total amount of traffic, in units of bits per second, arriving at `SRU Gateway` from AD, AT, and AA subnets, compare to the amount of traffic, in units of bits per second, forwarded by `SRU Gateway` to the Internet?

Laboratory Assignment #10: Providing Quality of Service Support

Recommended reading for this laboratory assignment: Chapters 1 through 7, 9, and 10.

L10.1 INTRODUCTION

In this laboratory assignment, you will investigate techniques for deploying various quality of service (QoS) mechanisms in the network. In particular, you will investigate how various levels of service can be provided through such schemes as random early detection (RED), weighted RED (WRED), weighted fair queuing (WFQ), and WFQ with low latency queue (WFQ-LLQ).

L10.2 INTRODUCTION AND SETTING UP BASELINE SCENARIO

In this part of the assignment, you will set up a baseline scenario in which four different traffic classes compete over limited resources in the network with no QoS support.

1. Create a new project and an empty scenario named `Assignment 10` and `FIFO` respectively (Section 1.6.2).
2. Create the network topology as shown in Figure L10.1 using node and link models specified in Table L10.1 (Sections 2.3 and 2.4).
3. Verify link connectivity (Section 2.6.2).
4. Add a *QoS Attribute Config* object.
5. Add and configure all demand objects as follows:
    a. Transmit data at 1.53 Mbps (i.e., T1 line), attribute **Traffic (bits/sec)**.
    b. Transmit data at 200 packets per second, attribute **Traffic (packets/sec)**.
    c. Model all traffic explicitly, set the attribute **Traffic Mix** to `All Explicit`.
6. Specify the type of traffic carried by each demand object by setting the value of the attribute **Traffic Characteristics…Type of Service** (Section 6.6):
    a. `Client(ToS 0)--> Destination` demand carries `Best Effort (0)` traffic.
    b. `Client(ToS 1)--> Destination` demand carries `Background (1)` traffic.
    c. `Client(ToS 2)--> Destination` demand carries `Standard (2)` traffic.

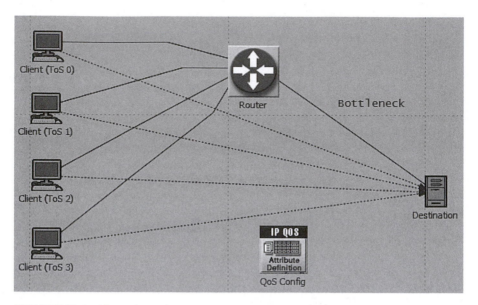

FIGURE L10.1   Network topology.

TABLE L10.1
Summary of Network Topology

Object Name	Object Model
Client (ToS 0) Client (ToS 1) Client (ToS 2) Client (ToS 3)	*ppp_wkstn* node object
Destination	*ppp_server* node object
Router	*ethernet4_slip8_gtwy* node object
All links in the network	*PPP_DS3_int* link object
All demand objects in the network	*ip_traffic_flow* demand object

---

   d. Client (ToS 3) --> Destination demand carries Excellent
       Effort (3) traffic.
7. Set the data rate on the link between Router and Destination to
   5 Mbps.
8. Configure Router to run the FIFO queuing profile (Section 10.4.3):
   a. Determine the number of the Router's interface attached to the link
       between Router and Destination.
   b. Configure the identified interface to use the default FIFO QoS profile.
       Leave the remaining configuration attributes set to their default values.

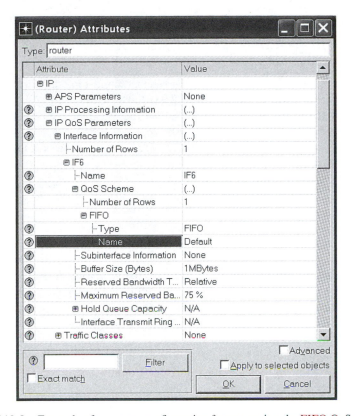

FIGURE L10.2   Example of Router configuration for supporting the FIFO QoS scheme.

Figure L10.2 shows an example of a possible configuration of the Router node, which supports the FIFO QoS scheme.

9. Configure the simulation to collect the following statistics:
   a. All statistics in the Demand Statistics category.
   b. The following Node Statistics from the category **IP interface**:
      i.   **Buffer Usage (bytes)**
      ii. **Queue Delay Variation (sec)**
      iii. **Queuing Delay (sec)**
      iv. **Traffic Dropped (bits/sec)**
      v.   **Traffic Sent (bits/sec)**
      vi. **Traffic Received (bits/sec)**
10. Run the simulation for 10 minutes.

L10.3   FIFO VERSUS RED COMPARISON

In this part of the assignment, you will compare how the FIFO and RED QoS schemes influence the traffic carried through the network.

1. Duplicate the current scenario naming the new one RED (Section 1.7).
2. Create a new FIFO QoS Profile called FIFO w. RED as follows:
    a. Add a new FIFO profile called FIFO w. RED in the QoS Attribute Config node.
    b. Configure the FIFO w. RED profile to support the default RED configuration (i.e., set the attribute **RED Parameters** to RED).
3. Deploy the FIFO w. RED profile on Router's interface attached to the link between Router and Destination as follows:
    a. Expand the attribute **IP...IP QoS Parameters...Interface Information...<interface name>...QoS Scheme...**.
    b. Set the value of the attribute **Type** to FIFO.
    c. Set the value of the attribute **Name** to FIFO w. RED.
4. Run the simulation for 10 minutes.

Questions

Q1. Compare the results collected for each of the scenarios. Examine and display the graph for the FIFO buffer usage on the outgoing interface of the node Router. Which QoS scheme, FIFO or RED, resulted in lower queue occupancy? Why? (Hint: You may need to examine the configuration setting of the RED mechanism.)

Q2. Examine the collected statistic results and display the graph for the FIFO queuing delay on the outgoing interface of the node Router. Which QoS scheme, FIFO or RED, resulted in lower queuing delay? Why?

Q3. Examine the collected statistic results and display the graph for the amount of traffic received, sent, and dropped on the outgoing interface of the node Router. How do these results differ for the FIFO and RED QoS schemes? Why?

Q4. Examine the collected statistic results for the end-to-end delay experienced by the packets of any demand flow. Which QoS scheme, FIFO or RED, resulted in lower end-to-end delay? Why?

L10.4   WEIGHTED RED

In this part of the assignment, you will examine how the WRED QoS schemes influence traffic flow in the network.

1. Duplicate the RED scenario naming the new one WRED.
2. Create a new FIFO QoS Profile called FIFO w. WRED as follows:
    a. Add a new FIFO profile called FIFO w. WRED in the QoS Attribute Config node.
    b. Configure the FIFO w. WRED profile to support the default WRED configuration (i.e., set the attribute **RED Parameters** to WRED).
    c. Configure WRED to discard excess packets of each traffic class with a different probability by setting the value of the attribute **Mark**

Probability Denominator to 1, 5, 10, 20. Recall that you can set this attribute directly to multiple values by separating them with commas.

3. Deploy the FIFO w. WRED profile on Router's interface attached to the link between Router and Destination.

4. Run the simulation for 10 minutes.

Questions

Q5. Examine the collected statistic results and display a graph for the amount of traffic received by the destination node for each of the traffic demand flows. Which traffic class, ToS 0, ToS 1, ToS 2, or ToS 3, delivered the largest amount of traffic? Why?

Q6. Examine the collected statistic results and display the graph for the FIFO buffer usage on the outgoing interface of the node Router. Which QoS scheme, FIFO, RED, or WRED, resulted in the lowest queue occupancy? Why?

Q7. Examine the collected statistic results and display the graph for the FIFO queuing delay on the outgoing interface of the node Router. Which QoS scheme, FIFO, RED, or WRED, resulted in the lowest queuing delay? Why?

Q8. Examine the collected statistics and display the graph for the amount of traffic received, sent, and dropped on the outgoing interface of the node Router. How do these results differ for the FIFO, RED, and WRED QoS schemes? Why?

Q9. Examine the collected statistic results for the end-to-end delay experienced by the packets of any demand flow. Which QoS scheme, FIFO, RED, or WRED, resulted in the lowest end-to-end delay? Why?

Q10. Examine the collected statistic results and display a graph for the packet jitter experienced by the traffic demands of each class. Which traffic class, ToS 0, ToS 1, ToS 2, or ToS 3, experienced the lowest packet jitter? Why? Which QoS scheme, FIFO, RED, or WRED, resulted in the lowest packet jitter for the traffic class ToS 3? Why?

L10.5   WEIGHTED FAIR QUEUING

In this part of the assignment, you will examine how WFQ QoS schemes influence traffic flow in the network.

1. Duplicate the current scenario naming the new one WFQ.
2. Open *QoS Attribute Config* object and create a new WFQ QoS Profile called 4 ToS Classes as follows:
   a. Duplicate the WFQ Profile called ToS Based by right-clicking on the profile's title and, from the menu that appears, selecting the **Duplicate Row** option.
   b. Set the name of the new profile to 4 ToS Classes.
   c. Remove 4 out of the default 8 classes in the new profile by setting the value of the attribute **Number of Rows** to 4.

      d. Examine the configuration for each of the classes and verify that the remaining 4 classes only accept packets marked with ToS values of 0, 1, 2, and 3 respectively.

3. Deploy the 4 ToS Classes profile on the Router's interface attached to the link between Router and Destination.

      a. Expand the attribute **IP...IP QoS Parameters...Interface Information... \<interface name>...QoS Scheme...**.

      b. Set the value of the attribute **Type** to WFQ (Class Based).

      c. Set the value of the attribute **Name** to 4 ToS Classes.

4. Run the simulation for 10 minutes.

Questions

Q11. Examine the collected statistic results and display a graph that illustrates the buffer usage in each of the WFQ subqueues on the outgoing interface of the node Router. Which traffic class, ToS 0, ToS 1, ToS 2, or ToS 3, had the lowest and which had the highest buffer usage? Why?

Q12. Examine the collected statistic results and display a graph that illustrates the queuing delay for each subqueue on the outgoing interface of the node Router. Packets of which traffic class, ToS 0, ToS 1, ToS 2, or ToS 3, experienced the lowest and the highest queuing delay? Why?

Q13. Examine the collected statistic results and display the graphs for the amount of received, sent, and dropped traffic for each subqueue on the outgoing interface of the node Router. Which traffic class, ToS 0, ToS 1, ToS 2, or ToS 3, had the lowest and which had the highest amount of received/sent/dropped traffic? Why?

L10.6 WFQ WITH LOW LATENCY QUEUE

In this part of the assignment, you will examine how configuring one of the WFQ subqueues as a low latency queue influences resource distribution in the WFQ QoS scheme.

1. Duplicate the WFQ scenario naming the new one WFQ_LLQ.

2. Modify the WFQ QoS Profile 4 ToS Classes as follows:

      a. Specify the queue that stores background traffic (i.e., traffic class ToS 1) as a low latency queue by setting the value of its attribute **Queue Category** to Low Latency Queue.

3. Run the simulation for 10 minutes.

Questions

Q14. Answer questions Q11–Q13 for the new scenario where the WFQ queue for the traffic class ToS 1 is designated as a low latency queue. As compared to the results collected for the WFQ scenario, what changes to the resource distribution (e.g., buffer usage, queuing delay, and the amount of traffic received/sent/dropped) among the traffic classes did you observe? Why?

L10.7   CHANGING WFQ CONFIGURATION

In this part of the assignment, you will examine how modifying the default configuration of the WFQ QoS scheme influences resource distribution in WFQ.

1. Duplicate the WFQ scenario naming the new one WFQ_Variations.
2. Modify the WFQ QoS Profile 4 ToS Classes as follows:
   a. Set the weight of the WFQ subqueue that buffers Excellent Effort traffic (i.e., traffic class ToS 3) to 80.
3. Configure the outgoing interface of Router to reserve only 40% of the link bandwidth for the QoS traffic as follows:
   a. Set the value of the attribute **IP...IP QoS Parameters...Interface Information...<interface name>...Maximum Reserved Bandwidth** to 40.
4. Run the simulation for 10 minutes.

Questions

Q15. Answer questions Q11–Q13 for the new scenario where you modify the default configuration of the WFQ QoS scheme. As compared to the results collected for the WFQ scenario, what changes to the resource distribution (e.g., buffer usage, queuing delay, and the amount of traffic received/sent/dropped) among the traffic classes did you observe? Why?

Laboratory Assignment #11: Routing with RIP

Recommended reading for this laboratory assignment: Chapters 1 through 4, 9, and 11.

L11.1 INTRODUCTION

In this laboratory assignment, you will examine the performance of the Routing Information Protocol (RIP) under various configuration conditions. In particular, you will examine how fast routing changes propagate in a network that uses RIP for routing. Furthermore, you will practice creating complex OPNET simulation models, configuring the RIP protocol, failing and recovering links in the network, modifying statistic collection modes, exporting OPNET reports, and others.

L11.2 SETUP OF INITIAL RIP SCENARIO

In this portion of the laboratory assignment, you will create a simulation model of a network for examining RIP performance:

1. Create a new project and an empty scenario named `Assignment 11` and `RIP Normal` respectively (Section 1.6.2).
2. Create the network topology as shown in Figure L11.1 using node and link models specified in Table L11.1 (Sections 2.3 and 2.4).
3. Verify link connectivity (Section 2.6.2).
4. Deploy IPv4 addresses on all active interfaces in the network (i.e., pull-down menu option **Protocols > IP > Addressing > Auto-Assign IPv4 Addresses**; see Section 9.3.4).
5. Configure RIP as a routing protocol on all interfaces in the network (i.e., option **Protocols > IP > Routing > Configure Routing Protocols**; see Section 11.1.1).
6. Set the RIP advertisement mode to `No Filtering` on all router interfaces in the network (Section 11.2.3):
   a. Select all routers in the network.
   b. Expand the attribute **IP Routing Protocols...RIP Parameters... Interface Information** on any of the selected nodes.
   c. Set the value of the attribute **Advertisement Mode** to `No Filtering` on all interfaces in the node.
   d. Select the checkbox *Apply to selected objects* and click the **OK** button to save the configuration changes on all the selected nodes in the network.

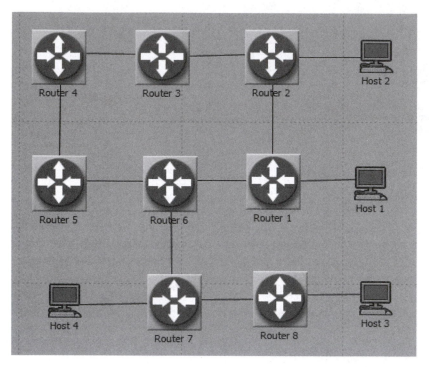

FIGURE L11.1   Network topology for RIP study.

TABLE L11.1
Summary of Network Topology for RIP Study

Object Name	Object Model
Router 1 – Router 8	*ethernet4_slip8_gtwy* node object
Host 1 – Host 4	*ppp_wkstn* node object
All links in the network	*PPP_DS1* link object

---

7. Disable the RIP simulation efficiency mode by setting the value of the global attribute **Simulation Efficiency...RIP Sim Efficiency** to `Disabled` (Section 11.2.5).

8. Configure your simulation to export IP address allocation by setting the global simulation attribute **IP...IP Interface Addressing Mode** to `Auto Addressed/Export` (Section 9.3.7).

9. Configure all the routers to export their routing tables at simulation time 400 seconds and also at the end of the simulation by setting the value of the attribute **Reports...RIP Routing Table** (Section 9.6.3).

10. Configure the simulation to collect the following statistics:
    a.   Global Statistic **Traffic Received (bits/sec)** from category **RIP**. Change the collection mode of this statistic to `all values` (Section 4.2.7).

b. Global Statistic **Convergence Duration** from category **RIP**. Make sure the collection mode for this statistic is also set to `all values` while the draw style is set to `discrete` (Sections 4.2.6 and 4.2.7).

11. Execute the scenario for 15 minutes and verify that it runs without any errors.

L11.3   CONFIGURING ADDITIONAL RIP SCENARIOS

In this portion of the assignment, you will create several additional scenarios for evaluating the performance of RIP under different configurations.

1. Duplicate the scenario `RIP Normal`, naming the new one `RIP with Failure` (Section 1.7). Update the configuration of the `RIP with Failure` scenario as follows:
   a. Add a *Failure/Recovery* object into the workspace (Section 2.7).
   b. Configure your scenario to fail the link between nodes `Router 1` and `Router 6` at simulation time 300 seconds and to recover that link at simulation time 500 seconds (Section 2.7).

2. Duplicate the scenario `RIP with Failure`, naming the new one `RIP Triggered` and update its configuration as follows:
   a. Enable RIP triggered extension mode on all interfaces of all the routers in the network by setting the value of the corresponding attributes named **IP Routing Protocols...RIP Parameters...Interface Information...Triggered Extension** to `Enabled` (Section 11.2.3).

3. Duplicate the scenario `RIP Triggered`, naming the new one `RIP with Split Horizon` and updating its configuration as follows:
   a. Enable split horizon RIP advertising mode on all interfaces of all the routers in the network by setting the value of the corresponding attributes named **IP Routing Protocols...RIP Parameters... InterfaceInformation...Advertisement Mode** to `Split Horizon` (Section 11.2.3).

4. Duplicate the scenario `RIP Triggered`, naming the new one `RIP with Poison Reverse` and updating its configuration as follows:
   a. Enable split horizon with poison reverse RIP advertising mode on all interfaces of all the routers in the network by setting the value of the corresponding attributes named **IP Routing Protocols...RIP Parameters...Interface Information...Advertisement Mode** to `Split Horizon with Poison Reverse` (Section 11.2.3).

5. Execute all scenarios for 15 minutes and examine the collected results.

Questions

Q1. Examine the IP address allocation in the network for the scenario `RIP with Failure` by inspecting the file with the exported IP addresses created by the simulation.

Q2. Examine the simulation results and display graphs for the following statistics:

    a. **RIP...Traffic Received (bits/sec)** plotted in separate panels for the `RIP Normal`, `RIP with Failure`, and `RIP Triggered` scenarios.

    b. **RIP...Convergence Duration** for the above three scenarios plotted in the same panel in stacked statistics mode.

    c. **RIP...Convergence Duration** for the `RIP with Split Horizon` and `RIP with Poison Reverse` scenarios, plotted in the same panel in stacked statistics mode.

Q3. Examine the routing tables at the end of the simulation on all the routers in the `RIP Normal` scenario and answer the following questions:

    a. What route is taken by packets traveling from `Host 2` to `Host 3`? What is the length of this route?

    b. What route is taken by packets traveling from `Host 4` to `Host 1`? What is the length of this route?

Q4. Examine the routing tables at simulation time 400 seconds on all the routers in the `RIP with Failure` Normal scenario and answer the following questions:

    a. What route is taken by packets traveling from `Host 2` to `Host 3`? What is the length of this route?

    b. What route is taken by packets traveling from `Host 4` to `Host 1`? What is the length of this route?

Q5. What are the differences between the amount of traffic received by RIP for the `RIP Normal`, `RIP with Failure`, and `RIP Triggered` scenarios? How do you explain these differences?

Q6. What are the differences between the routing table convergence duration for the `RIP Normal`, `RIP with Failure`, and `RIP Triggered` scenarios? How do you explain these differences?

Q7. What are the differences between the routing table convergence duration for the `RIP with Failure`, `RIP with Split Horizon`, and `RIP with Poison Reverse` scenarios? How do you explain these differences? How do you rate the RIP without Split Horizon, RIP with Split Horizon, and RIP with Poison Reverse in terms of their quickness in reacting to good news and to bad news?

Laboratory Assignment #12: Routing with OSPF

Recommended reading for this laboratory assignment: Chapters 1 through 7, 9, and 11.

L12.1 INTRODUCTION

In this laboratory assignment, you will examine the behavior of the OSPF routing protocol. In particular, you will develop a model of a network with OSPF routing, observe the routes that are used, and examine IP's load balancing feature. In addition, you will divide the network domain into hierarchical areas to examine how such partitioning influences the routes in the network.

L12.2 SETUP OF INITIAL OSPF SCENARIO

In this portion of the laboratory assignment, you will create a simulation model of a network for examining OSPF performance:

1. Create a new project and an empty scenario named `Assignment 12` and `OSPF Flat` respectively (Section 1.6.2).
2. Create the network topology as shown in Figure L12.1 using node and link models specified in Table L12.1 (Sections 2.3 and 2.4).
3. Deploy IPv4 addresses on all active interfaces in the network (i.e., pull-down menu option **Protocols > IP > Addressing > Auto-Assign IPv4 Addresses**; see Section 9.3.4).
4. Configure OSPF as a routing protocol on all interfaces in the network (i.e., option **Protocols > IP > Routing > Configure Routing Protocols**; see Section 11.1.1).
5. Configure costs on the links between the routers as shown in Table L12.2 (i.e., use the option **Protocols > OSPF > Configure Interface Cost...**; see Section 11.3.6).
6. Deploy IP traffic demands (i.e., demand model *ip_traffic_flow*) between the following host node pairs and configure them according to the specification in Table L12.3:
   - `Host A` and `Host C`
   - `Host C` and `Host D`
   - `Host B` and `Host E`
   - `Host B` and `Host F`

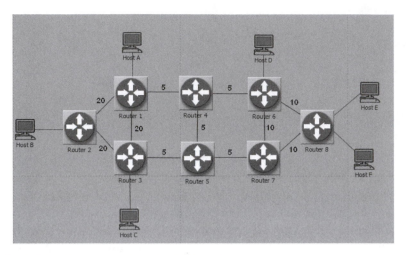

FIGURE L12.1   Network topology for OSPF study.

TABLE L12.1
Summary of Network Topology for OSPF Study

Object Name	Object Model
Router 1 - Router 8	*ethernet4_slip8_gtwy* node object
Host A - Host F	*ethernet_wkstn* node object
Links between routers	*PPP_DS3* link object
Links between hosts and routers	*100BaseT* link objects

TABLE L12.2
Summary of Link Cost Values

Link Name	Link Cost	OSPF Area
Router 1 - Router 3	20	1
Router 1 - Router 2		
Router 2 - Router 3		
Router 1 - Router 4	5	0
Router 4 - Router 5		
Router 4 - Router 3		
Router 4 - Router 6		
Router 5 - Router 7		
Router 6 - Router 7	10	2
Router 6 - Router 8		
Router 7 - Router 8		

TABLE L12.3
Summary of IP Demand Configuration Parameters

Attribute	Value
Traffic (bits/second)	T1_1hour_bps
Traffic (packets/second)	200_pps
All remaining attributes	Default values

---

7. Configure your simulation to export IP address allocation by setting the global simulation attribute **IP…IP Interface Addressing Mode** to Auto Addressed/Export (Section 9.3.7).
8. Configure all the routers to export their routing tables and Link-State Databases at the end of the simulation by setting the values of the attributes **Reports…OSPF Routing Table** and **Reports…OSPF Link State Database** to Export at End of Simulation (Section 9.6.3).
9. Disable the OSPF Simulation Efficiency Mode by setting the value of the global attribute **Simulation Efficiency…OSPF Sim Efficiency** to Disabled (Section 11.3.11).
10. Execute the scenario for 10 minutes.

Questions

Q1. Examine the IP address allocation in the network by inspecting the file with the exported IP addresses created by the simulation.
Q2. Check the Link-State Database on any one router node carefully and make sure that all link costs are what you configured them to be. If they are not, then you should correct them and rerun the scenario. In the Link-State Database, what costs are used for the links between the hosts and the routers?
Q3. In the Link-State Database, identify the entries for Router 3 and Router 6. For these entries, identify the names of the neighboring routers for each Link ID shown in the table.
Q4. Examine the routing tables for the various routers and determine the routes for traffic from Host A to Host C and Host C to Host D. For these routes, list the interface number and router name of each hop, the link cost of each hop, and the total cost for the route.
Q5. Visualize the routes traversed by the packets of the four demands that you had previously configured (i.e., use the option **Protocols > IP > Demands > Display Routes for Configured Demands** or press CRTL+ALT+D). It is generally helpful to view the route of one demand at a time (i.e., toggle the value of the attribute field **Display** in the **Route Report for IP Traffic Flows** window). Also, make sure you turn the route display off before you close the window. Are the routes visualized for demands Host A to Host C and Host C to Host D the same as or different from what you had computed in question Q4?

Q6. List the routes visualized for traffic from Host B to Host E, Host B to Host F, Host E to Host B, and Host F to Host B.

L12.3   OSPF WITH PACKET-BASED LOAD BALANCING OPTION

In this portion of the assignment, you will examine how packet-based load balancing influences traffic routes:

1. Duplicate the scenario OSPF Flat, naming the new one OSPF with packet-based load balancing (Section 1.7).
2. Change the configuration of Router 2 so that it employs the packet-based load balancing option (i.e., change the value of the attribute **IP...IP Routing Parameters...Load Balancing Options** to Packet-Based).
3. Rerun the simulation for 10 minutes and examine the collected results.

Questions

Q7. Visualize the routes taken by the various demands. List the routes that you observe for traffic from Host B to Host E, Host B to Host F, Host E to Host B, and Host F to Host B. Explain what you think is significant about these routes. How are these routes different from the corresponding routes in the scenario OSPF Flat?

Q8. Based on your understanding of the load balancing options, discuss whether or not your observations of these routes are consistent with how load balancing works.

L12.4   OSPF WITH HIERARCHICAL STRUCTURE OPTION

In this scenario, you will examine how the OSPF performance is influenced by partitioning the network into areas:

1. Duplicate the scenario OSPF Flat, naming the new one OSPF Hierarchical.
2. Divide the network into areas as shown in Table L12.2 (i.e., use the option **Protocols > OSPF > Configure Areas**, see Section 11.3.8):
   - Area 1 will consist of the three links on the left with costs 20 each along with the links to Host A, Host B, and Host C.
   - Area 2 will consist of the three links on the right with costs 10 each along with the links to Host D, Host E, and Host F.
   - The backbone area 0 will be composed of the links in the middle with costs 5 each.
3. Verify your area configuration by visualizing the hierarchical structure of the network (i.e., use the option **View > Visualize Protocol Configuration > OSPF Area Configuration...**).
4. Rerun the simulation for 10 minutes and examine the collected results.

Question

Q9. Observe the routes used for all the configured demands. How are the routes in this scenario different from those used in the OSPF Flat scenario? Can you explain the reasons for the differences?

L12.5   OSPF WITH AREA BORDER ROUTERS

In this scenario, you will examine how the OSPF performance is influenced by the presence of Area Border Routers (ABRs):

1. Duplicate the scenario OSPF Hierarchical, naming the new one OSPF with ABR.
2. Configure Router 3 to hide the subnet addresses for Host A, Host B, and Host C and prevent them from being advertised outside Area 0, Area 1, etc. (Section 11.3.9).
3. Rerun the simulation for 10 minutes and examine the collected results.

Questions

Q10. Observe the routes used for all the configured demands. How are the routes in this scenario different from those used in the OSPF Hierarchical scenario? Can you explain the reasons for the differences?
Q11. Which router is the Area Border Router for Area 1?
Q12. Are there any routes that cross over from Area 0 to Area 1 at Router 3? Are there any routes that cross over from Area 1 to Area 0 at Router 3? If so, explain why that happens.
Q13. Are there any routes that are not symmetric? If so, why?

Finally,

1. Duplicate the scenario OSPF with ABR, naming the new one OSPF with ABR2.
2. Configure Router 6 to hide the subnet addresses for Host D, Host E, and Host F and prevent them from being advertised outside Area 2 (Section 11.3.9).
3. Rerun the simulation for 10 minutes and examine the collected results.

Questions

Q14. Observe the routes used for all the configured demands. How are the routes in this scenario different from those used in the OSPF Hierarchical and OSPF with ABR scenarios? Can you explain the reasons for the differences?
Q15. Which router is the Area Border Router for Area 2?

Laboratory Assignment #13: Ethernet

Recommended reading for this laboratory assignment: Chapters 1 through 4 and 12.

L13.1 INTRODUCTION

In this laboratory assignment, you will investigate Ethernet performance in the bus and star network topologies. In addition, this lab will illustrate how to use OPNET's **Rapid Configuration** feature as well as how to collect scalar statistics and set up parametric studies.

L13.2 BUS TOPOLOGY

First, set up and configure a bus network that you will use to evaluate Ethernet performance.

1. Create a new project and an empty scenario named `Assignment 13` and `Ethernet_Bus` respectively (Section 1.6.2). When creating a new scenario using the **Startup Wizard**, use `Office` network scale and include the `ethcoax` model family.
2. Create the network topology as shown in Figure L13.1 using OPNET's **Rapid Configuration** feature available through the **Topology > Rapid Configuration...** menu option (Section 2.5).
   - Configure the node and link models in the bus topology as follows:
     - **Node model**: `ethcoax_station`
     - **Number**: 20
     - **Link model**: `eth_coax`
     - **Tap model**: `eth_tap`
   - Leave the remaining attributes set to their default values.
3. Verify link connectivity (Section 2.6.2).
4. Modify the attributes of the bus model as follows:
   - Change the model name to `eth_coax_adv`.
   - Set **data rate** to 500,000 bps.
   - Set **delay** to 0.05 seconds. Recall that **delay** is an advanced attribute that requires you to select the *Advanced* checkbox before you can view and modify its value.
5. Configure all the nodes in the network to generate traffic according to the following specification. Recall that you can use the **Select Similar Nodes** option and the *Apply to selected objects* checkbox to configure all the nodes in the network simultaneously.

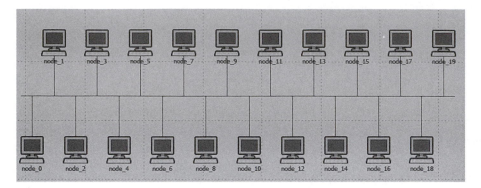

FIGURE L13.1   Ethernet bus topology.

- Configure the **Traffic Generation Parameters** compound attribute as follows:
    - Use the constant distribution when setting attribute values.
    - Set the attribute **ON State Time (seconds)** to 100 seconds.
    - Set the attribute **OFF State Time (seconds)** to 0 seconds.
    - Promote the attribute **Packet Generation Arguments…Interarrival Time (seconds)**.
    - Leave all the remaining attributes set to their default values.
6. Configure the simulation to collect the following statistics:
    - Global Statistics **Traffic Received (bits/sec)** in the category **Traffic Sink**.
        - Configure this statistic to be also collected as a scalar data using sample mean value by selecting the *Generate scalar data* checkbox and using the `sample mean` function. See Figure L13.2 for the desired configuration.
    - Global Statistics **Traffic Sent (bits/sec)** in the category **Traffic Source.**
        - Configure this statistic to be also collected as a scalar data using sample mean value.
    - The following Node Statistics in the category **Ethcoax**:
        - **Collision Count**
        - **Transmission attempts**
    - Node Statistics **Traffic Sent (bits/sec)** in the category **Traffic Source**.
    - Node Statistics **Traffic Received (bits/sec)** in the category **Traffic Sink**.
7. Set the value of the promoted attribute **Interarrival Time (seconds)** for all nodes in the network. Recall that you need to use the wildcard option to perform this task (Section 3.5).
    - Replace the node number in any of the promoted attributes with the wildcard character so that you can specify the value of this attribute in all nodes in the network, simultaneously.
    - Set the promoted attribute with the wildcard character to the following values:
        - `exponential (0.25)`
        - `exponential (0.1)`

         −   `exponential (0.05)`
         −   `exponential (0.025)`
         −   `exponential (0.02)`
         −   `exponential (0.01)`
         −   `exponential (0.005)`

8. Run the simulation for 20 seconds.

Questions

To simplify interpretation of the collected results, for each question, you may want to display the statistic results in overlaid form using the average function (Section 4.4).

Q1. Examine and display the global statistic for the amount of traffic sent in the network. Which value of packet interarrival time resulted in the highest and lowest amounts of traffic sent? Why? Recall that OPNET numbers the DES runs according to the order of the promoted attribute values. Thus, if you have specified the values of the promoted attribute **Interarrival Time (seconds)** in the same order as described in step 7, then the DES-1 run will correspond to the packet interarrival time set to `exponential (0.25)`, DES-2 to `exponential (0.1)`, and so on.

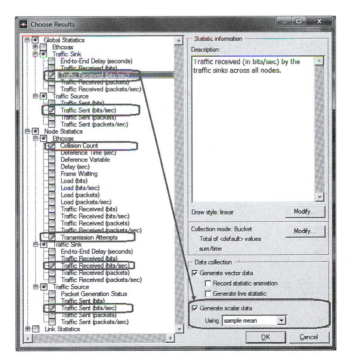

FIGURE L13.2   Configuring the simulation to collect scalar statistics.

Q2. For any node in the network, examine and display the Node Statistic **Collision Count** in the category **Ethcoax**. Which simulation run resulted in the highest and in the second highest number of collisions? Why? Do you think that these results agree with the data reported for the amount of traffic received in the network? Why or why not?

Q3. For any node in the network, examine and display the Node Statistic **Transmission Attempts** in the category **Ethcoax**. Which simulation run resulted in the highest and in the second highest number of transmission attempts? Why? Do you think that these results agree with the data reported for the amount of traffic received in the network? Why or why not?

Q4. Create a graph that displays the amount of data sent versus the amount of data received. To create such a graph, perform the following steps:

- Click on the *DES Parametric Studies* tab, as shown in Figure L13.3.
- Expand the **Global Statistics...Traffic Sink...Traffic Received (bits/sec)** statistic results. Click on the sample mean category and click the button **Set As Y-Series**.
- Expand the **Global Statistics...Traffic Source...Traffic Sent (bits/sec)** statistic results. Click on the sample mean category and click the button **Set As X-Series**.
- Click the **Show** button to display the graph.

The resulting graph should look similar to that shown in Figure L13.3. Explain the Ethernet behavior as depicted in the graph.

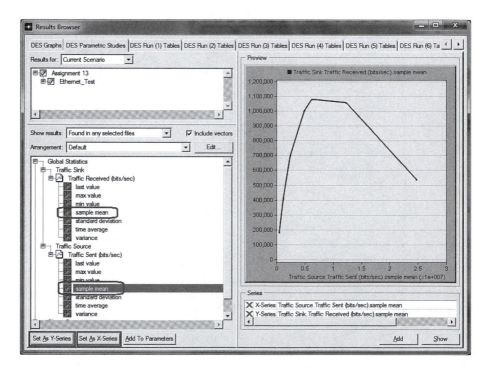

FIGURE L13.3 Analyzing the collected results using parametric studies.

L13.3   STAR TOPOLOGY

In this part of the assignment, you will examine how Ethernet performs in the star topology with the promiscuous mode enabled and disabled. First, you will create a simulation scenario that models the Ethernet network with promiscuous mode disabled.

1. Create a new empty scenario named `Ethernet_Star`. When creating a new scenario using the **Startup Wizard**, use `Office` network scale and the `ethernet_advanced` model family.
2. Create the network topology as shown in Figure L13.4 using OPNET's **Rapid Configuration** feature according to the specifications listed in the following (Section 2.5):
   - **Center node model**: `ethernet32_hub_adv`
   - **Periphery node model**: `ethernet_station_adv`
   - **Number**: 20
   - **Link model**: `100BaseT_adv`
   - Leave the remaining attributes set to their default values.

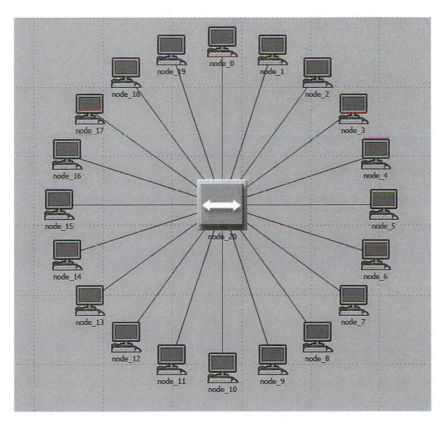

FIGURE L13.4   Ethernet star topology.

3. Verify link connectivity (Section 2.6.2).
4. Configure *all* the nodes in the network to generate traffic according to the following specification:
   - Configure **Traffic Generation Parameters** as follows:
     - Set the attribute **ON State Time (seconds)** to constant(100).
     - Set the attribute **OFF State Time (seconds)** to constant(0).
     - Set the attribute **Packet Generation Arguments...Interarrival Time (seconds)** to exponential(0.01).
     - Leave all the remaining attributes set to their default values.
     - By default, promiscuous mode is disabled in the Ethernet nodes, so no changes are needed for configuring this feature.
5. Configure the simulation to collect the following statistics:
   - Global statistics **Traffic Received (bits/sec)** in the category **Traffic Sink**.
   - Global statistics **Traffic Sent (bits/sec)** in the category **Traffic Source**.
   - Global statistics **Delay (sec)** in the category **Ethernet**
   - The following node statistics in the category **Ethernet:**
     - **Collision Count**
     - **Delay (sec)**
     - **Load (bits/sec)**
     - **Traffic Received (bits/sec)**
     - **Transmission Attempts**
   - Link *point-to-point* **throughput (bits/sec)** statistics in both directions.

Now, you will create another scenario that models an identically configured Ethernet network but with promiscuous mode enabled in all nodes in the network:

6. Duplicate the current scenario, naming the new scenario Ethernet_ Star_Promiscuous.
7. Enable promiscuous mode on all nodes in the network by setting the attribute **Ethernet...Ethernet Parameters...Promiscuous Mode** to Enabled. Recall that this attribute is only available in advanced node models.
8. Execute both simulations for 20 seconds. You may want to use the **Manage Scenarios...** option (Section 4.3.5).

Questions

Q5. Examine the simulation results collected for the above two scenarios and display graphs that compare the **Load (bits/sec)**, **Collision Count**, and **Transmission Attempts** statistic results for any node in the network. Also display graphs that compare the **throughput (bits/sec)** statistic on any link in the network. Do you observe any differences in these statistic results reported by the Ethernet_Star and Ethernet_Star_ Promiscous scenarios? Why or why not?

Q6. Examine the simulation results collected for the above two scenarios and display a graph that compares the **Traffic Received (bits/sec)** statistic for any node in the network. Do you observe any differences between the results collected in these two scenarios? Why or why not?

Q7. Examine the Global Statistics collected for these two scenarios. Explain the observed behavior.

Q8. Examine the simulation results collected for the above two scenarios and display a graph that compares the **Delay (sec)** statistic for any node in the network. Do you observe any differences between the results collected in these two scenarios? Why or why not?

L13.4   HUB VERSUS SWITCH

In this part of the assignment, you will examine how Ethernet performance is influenced by the type of the central node in a star network topology.

1. Duplicate the `Ethernet_Star` scenario, naming the new scenario `Ethernet_Star_vs_Switch`.
2. Change the attribute **model** of the center node in the network to `ethernet128_switch_adv`. Recall that the attribute **model** is an advanced attribute and you need to have the checkbox *Advanced* selected for it to become visible.
3. Run the simulation for this new scenario and examine the collected statistics.

Questions

Q9. What do you expect will happen in the network in terms of throughput and delay when you replace the central hub in the start topology with a switch? Why?

Q10. Compare the Global Statistic **Traffic Sent (bits/sec)** in the category **Traffic Source** collected in scenarios `Ethernet_Star` and `Ethernet_Star_vs_Switch`. Did you observe any differences in the amount of traffic sent through the network in these scenarios? Why or why not?

Q11. Compare the Global Statistic **Traffic Received (bits/sec)** in the category **Traffic Sink** collected in scenarios `Ethernet_Star` and `Ethernet_Star_vs_Switch`. Did you observe any differences in the amount of traffic received by the destination nodes in the network in these scenarios? Why or why not?

Q12. Compare the traffic delay (i.e., Global Statistic **Delay (sec)** in the category **Ethernet**) in scenarios `Ethernet_Star` and `Ethernet_Star_vs_Switch`. Which scenario resulted in lower delay and why?

Q13. Examine the throughput on a link between any peripheral node and the central node in both directions. In which direction, if any, was the examined link throughput statistically different in the compared scenarios? Which scenario, if any, resulted in lower link throughput and why?

Laboratory Assignment #14: Wireless Communication

Recommended reading for this laboratory assignment: Chapters 1 through 4 and 12.

L14.1 INTRODUCTION

In this laboratory assignment, you will investigate the performance of wireless communication networks. In particular, in the first part of this laboratory assignment, you will conduct an experimental study to determine the maximum communication range between two mobile nodes configured with the default WLAN settings. Next, you will examine how transmission power influences the communication range. Finally, in the last part of this assignment, you will compare the performance of the AODV and DSR MANET routing protocols.

Please note that you need to have a license for the wireless module to complete this assignment. Contact OPNET Support to obtain additional module licenses for education or scholarly research.

L14.2 DETERMINING COMMUNICATION RANGE

First, you will set up and configure a simple two-node network topology where the source node continually sends traffic to the destination, while the destination node moves away from the source. At some point during the simulation, the destination node will move too far away from the source for any traffic to go through. Based on the time at which communication stops and the speed with which the destination node moves away, you will determine the maximum communication distance between these two nodes.

1. Create a new project and an empty scenario named `Assignment 14` and `Communication Range`, respectively (Section 1.6.2). When creating a new scenario with the **Startup Wizard**, use `Campus` network scale and include `wireless_lan`, `wireless_lan_adv`, and `MANET` model families.
2. Create the network topology as shown in Figure L14.1:
   - Place two `wlan_station_adv` mobile nodes in the workspace and name them `source` and `destination`.
   - Specify the locations of the nodes so that they are 0.1 km apart on the *x*-axis (e.g., 12.7 and 12.8), while the *y* coordinate values are the same (e.g., 2.8). You can specify the node *x* and *y* coordinates through advanced attributes **x position** and **y position** (i.e., you must select the *Advanced* checkbox first).

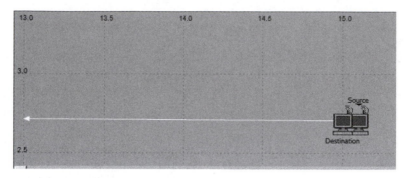

FIGURE L14.1   Simple two-node topology.

3. Define a new node trajectory as follows (Section 12.8.1):
   • Name the new trajectory `Distance` and configure it to be defined rela-
     tive to the node's position (i.e., select the checkbox titled *Coordinates
     are relative to node's position*).
   • Set the length of the trajectory to be about 2000 meters on a straight
     line, which is parallel to the line between the two mobile nodes.
   • Set the node's speed to 2 meters per second.
4. Configure the `destination` node:
   • Set the MAC address of the node (attribute **Wireless LAN…Wireless
     LAN MAC Address**) to `123456`.
   • Set the node's **trajectory** to `Distance`.
5. Configure the `source` node:
   • Set **Destination Address** to `123456`.
   • Configure **Traffic Generation Parameters** as follows:
     – Set the attribute **Start Time (seconds)** to `constant(5)`.
     – Set the attribute **ON State Time (seconds)** to `constant(2000)`.
     – Set the attribute **OFF State Time (seconds)** to `constant(0)`.
     – Leave all the remaining attributes set to their default values.
6. Configure the simulation to collect all Node Statistics in the category
   Wireless LAN.
7. Run the simulation for 18 minutes.

Questions

Q1. Examine and display the statistics for the amount of traffic received by the
    destination node and the amount of traffic sent by the source node. At which
    point of time during the simulation did the destination node stop receiv-
    ing any data? Based on this value, compute the maximum communication
    distance between the two nodes. Recall that we configured the destination
    node to travel at the speed of 2 meters per second and that initially the
    nodes were placed 100 meters apart.
Q2. As you have observed in the graph displayed for Q1, after the nodes move
    too far apart for any data to successfully arrive at the destination, the

sending rate of the source node increased. Write down your hypothesis as to why this occurred. Examine and display the following statistics collected at the source node: **Backoff Slots (slots)**, **Retransmission Attempts (packet)**, **Control Traffic Rcvd (bits/sec)**, and **Queue Size (packets)**. Also examine and display in a single overlaid graph **Data Traffic Sent (bits/sec)** and **Load (bits/sec)** statistics. Based on the results presented by these statistics, explain what has occurred. Do these results support your initial hypothesis? Why or why not?

L14.3   COMMUNICATION RANGE VERSUS TRANSMISSION POWER

In this part of the assignment, you will examine how transmission power influences communication range.

1. Duplicate the current scenario, naming the new one `Communication Range vs Power`.
2. Promote the **Transmit Power (W)** attribute in the source node.
3. Set the promoted attribute to the following values: `0.001`, `0.002`, `0.005`, and `0.01`.
4. Execute the simulation for 18 minutes.

Question

Q3. Examine the simulation results and determine the communication range for each value of transmit power. Draw a graph that shows communication range versus transmit power.

L14.4   MANET COMMUNICATION: DSR VERSUS AODV

In this part of the assignment, you will compare the performance of the AODV and DSR routing protocols.

1. Create a new scenario named `AODV` using `Campus` network scale and including the `MANET` model family.
2. Use the Wireless Deployment Wizard (Section 12.9) to create the MANET network topology shown in Figure L14.2. When deploying the wireless network, use default settings unless specified otherwise:
   • Set the wireless technology to `WLAN (Ad-hoc)` as shown in Figure L14.3.
   • Set the size of the deployment area to 9,000,000 m² and use the 4 × 4 grid node placement, as shown in Figure L14.4.
   • Use the default settings for node mobility, which are shown in Figure L14.5 (no configuration change should be needed).

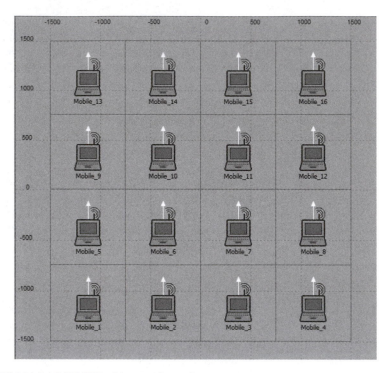

FIGURE L14.2   MANET grid network topology.

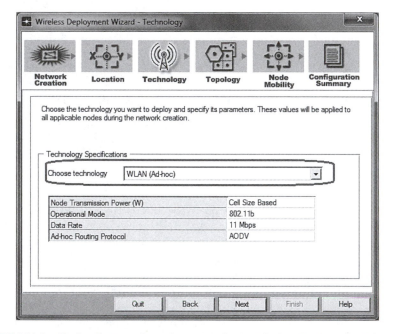

FIGURE L14.3   **Technology** configuration screen in the Wireless Deployment Wizard.

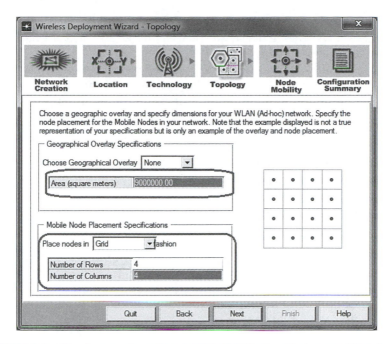

FIGURE L14.4   Topology configuration screen in the Wireless Deployment Wizard.

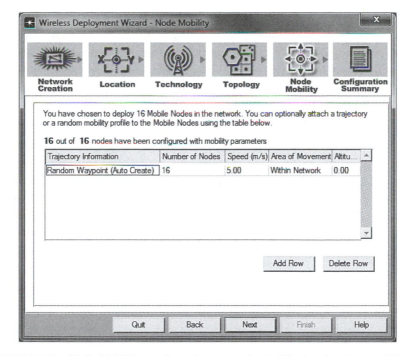

FIGURE L14.5   Node Mobility configuration screen in the Wireless Deployment Wizard.

3. Configure all nodes in the network to generate traffic:
   - Select all nodes in the network and edit the attributes of any node that was selected.
   - Enable the *Apply to selected objects* checkbox.
   - Set the attribute **MANET Traffic Generation Parameters...Number of Rows** to 1, which enables the default traffic generation pattern (i.e., starting from time 100 seconds, communicate with a randomly selected node, generating packets of size 1024 bytes every 1 second).
   - Click the **OK** button to apply these changes to all nodes in the network.
4. Configure the simulation to collect all global **AODV**, **MANET**, and **Wireless LAN** statistics (AODV is the default routing protocol in a MANET).
5. Duplicate the current scenario, naming the new one DSR.
6. Configure all the nodes in the DSR scenario to run the DSR routing protocol by setting the attribute **AD-HOC Routing Parameters...AD-HOC Routing Protocol** to DSR.
7. Configure the DSR scenario to collect global **DSR** statistics instead of **AODV**.
8. Execute both AODV and DSR scenarios for 200 seconds.

Questions

Q4. Provide a brief description of the AODV and DSR MANET routing protocols, highlighting their differences and similarities.

Q5. Compare and display such Global Statistics as **Number of Hops per Route**, **Route Discovery Time**, and **Routing Traffic Received (bits/sec)** for the AODV and DSR scenarios. Are the reported results consistent with the expected behavior for AODV and DSR in the MANET environment where all the nodes move around and communicate with any other randomly selected node in the network? Why or why not?

Q6. Compare and display such global MANET statistics as **Delay (sec)**, **Traffic Received (bits/sec)**, and **Traffic Sent (bits/sec)** for the AODV and DSR scenarios. Are the reported results consistent with the expected behavior for the AODV and DSR routing protocols? Why or why not?

Q7. Compare and display such global Wireless LAN statistics as **Data Dropped (Retry Threshold Exceeded) (bits/sec)**, **Delay (sec)**, **Load (bits/sec)**, **Retransmission Attempts (packets)**, and **Throughput (bits/sec)** for the AODV and DSR scenarios. Are the reported results consistent with the expected behavior for the AODV and DSR routing protocols? Why or why not?

Q8. Do you think that the collected results would illustrate a slightly different AODV and DSR behavior if you ran the simulations again but with a different seed value. Why or why not? Rerun the AODV and DSR scenarios with five different seed values (you can specify multiple seed values through the **Configure/Run DES** dialog box). Use the collected results to support or refute your hypothesis about the influence of the seed value on collected results.

Q9. Compare AODV and DSR behaviors in the MANET network with only two communicating nodes:
- Duplicate the AODV and DSR scenarios naming the new ones as AODV_ one_comm_pair and DSR_one_comm_pair respectively.
- Modify each new scenario as follows:
  - Disable traffic generation on all nodes in the network.
  - Set the IP address of node Mobile_16 to 192.0.1.1.
  - Configure node Mobile_1 to send traffic to Mobile_16 by setting the attribute **Destination IP address** to 192.0.1.1.
- Run the new scenarios and answer questions Q5 through Q8 again for the scenarios with one communicating node only.

Q10. Compare the AODV and DSR behaviors in the MANET network with the nodes remaining static throughout the simulation:
- Duplicate the AODV and DSR scenarios naming the new ones as AODV_ static and DSR_static respectively.
- Modify both new scenarios so all nodes in the network have their trajectories set to NONE (i.e., the nodes will not move around).
- Run the new scenarios and answer questions Q5 through Q8 again for the scenarios where the nodes do not move during the simulation.

Index